Cognitive Radio Networks: Technologies, Challenges and Applications

Cognitive Radio Networks: Technologies, Challenges and Applications

Guest Editors

Jia Liu
Kechen Zheng
Xiaoying Liu
Zhao Li
Yang Xu

MDPI

Basel • Beijing • Wuhan • Barcelona • Belgrade • Novi Sad • Cluj • Manchester

Guest Editors

Jia Liu
Center for Strategic Cyber Resilience Research and Development
National Institute of Informatics
Tokyo
Japan

Kechen Zheng
School of Computer Science and Technology
Zhejiang University of Technology
Hangzhou
China

Xiaoying Liu
School of Computer Science and Technology
Zhejiang University of Technology
Hangzhou
China

Zhao Li
School of Cyber Engineering
Xidian University
Xi'an
China

Yang Xu
School of Computer Science and Technology
Xidian University
Xi'an
China

Editorial Office
MDPI AG
Grosspeteranlage 5
4052 Basel, Switzerland

This is a reprint of the Special Issue, published open access by the journal *Sensors* (ISSN 1424-8220), freely accessible at: https://www.mdpi.com/journal/sensors/special_issues/LMGJ4IH614.

For citation purposes, cite each article independently as indicated on the article page online and as indicated below:

Lastname, A.A.; Lastname, B.B. Article Title. *Journal Name* **Year**, *Volume Number*, Page Range.

ISBN 978-3-7258-7551-1 (Hbk)
ISBN 978-3-7258-7552-8 (PDF)
https://doi.org/10.3390/books978-3-7258-7552-8

Contents

About the Editors

Jia Liu

Jia Liu is an Associate Professor with the Center for Strategic Cyber Resilience Research and Development, the National Institute of Informatics, Japan. He has long been engaged in research and education in the field of wireless communications. His current research interests include beyond 5G and 6G technologies and communication networks, space-air-ground integrated networks, AI native network architecture and protocol design, large language models for network automation and optimization, distributed intelligence and edge learning over wireless networks, intelligent multimedia transmission and QoS assurance, trustworthy and resilient networked systems, and network economics. He has published high-quality research papers in leading journals and conferences, and received the IEEE Sapporo Section Encouragement Award in 2016 and 2020.

Kechen Zheng

Kechen Zheng is an Associate Professor with the School of Computer Science and Technology, Zhejiang University of Technology, China. He has expertise in a multitude of research methods in the field of wireless communications. His research interests include wireless sensor networks, symbiotic radio, energy harvesting, wireless communication, and edge computing. He received the Best Paper Award at the International Conference on Networking and Network Applications in 2021, and was included in the "World's Top 2% Scientists List" in 2025.

Xiaoying Liu

Xiaoying Liu is an Associate Professor with the School of Computer Science and Technology, Zhejiang University of Technology, China. She has long been engaged in research in the field of wireless communications and computer networks. Her research interests include Internet of Things, energy harvesting, wireless communication, mobile edge computing, age of information, and cognitive radio networks. She has published about 50 high-quality research papers in IEEE journals and conferences and received the Best Paper Award at the International Conference on Networking and Network Applications in 2021.

Zhao Li

Zhao Li is an associate professor in the School of Cyber Engineering, Xidian University. He has published over 70 academic papers in prestigious international journals and conferences in the field of wireless communications, including *IEEE TWC*, *IEEE TMC*, *IEEE TIFS*, *IEEE Internet of Things Journal*, *Computer Networks*, *Sensors*, IEEE INFOCOM, IEEE GLOBECOM, and IEEE ICC. He also holds more than 50 patents in China. His research interests include wireless communication, 5G communication systems, interference management, IoT, and physical layer security. Dr. Li has served as Program Chair, Workshop Chair, and TPC Member in several IEEE conferences, including IEEE EUC, IEEE WF-IoT, IEEE WCNC, and IEEE ICPADS. He is also a peer reviewer for numerous IEEE Transactions journals and other IEEE publications. As an independent PI, Dr. Li has led several projects of the research institute in China. He has also actively participated in NSF projects in both Shaanxi Province and China.

Yang Xu

Yang Xu is an associate professor in the School of Computer Science and Technology, Xidian University, China. She has long been engaged in research and teaching in the areas of mobile communications, information security, and space-air-ground integrated networks. Her current research interests include intelligent routing protocol design, AI-driven network security, and the integration of large language models and machine learning techniques for intelligent network management and optimization, etc. She has published a number of influential papers in leading international journals and conferences. In addition, she has led and participated in several competitive research projects related to wireless networking, information security, and intelligent network systems. Her research aims to advance secure, intelligent, and efficient communication technologies for next-generation networked systems.

Preface

With the rapid growth of wireless communication services, mobile Internet applications, and the Internet of Things, the demand for wireless spectrum resources has increased dramatically. Traditional static spectrum allocation policies often lead to inefficient utilization of spectrum, as many licensed frequency bands remain underused in time and space while other bands experience severe congestion. Cognitive radio technology has emerged as a promising paradigm to address this challenge by enabling dynamic spectrum access, intelligent spectrum management, and adaptive communication mechanisms. Through spectrum sensing, learning, and decision making, cognitive radio networks can identify spectrum opportunities and utilize them efficiently while minimizing interference to licensed users.

In recent years, the development of cognitive radio networks has been closely connected with advances in machine learning, artificial intelligence, and intelligent signal processing. These technologies have enhanced the capability of wireless systems to sense complex spectrum environments, predict spectrum availability, and optimize spectrum access strategies. At the same time, the integration of cognitive radio networks with emerging networking paradigms and application scenarios has further expanded the scope of research in this field. Topics such as cooperative spectrum sensing, adaptive spectrum sharing, energy-efficient cognitive communication, and intelligent spectrum management have attracted increasing attention from both academia and industry.

Against this background, this Reprint brings together all peer-reviewed articles published in the Special Issue "Cognitive Radio Networks: Technologies, Challenges and Applications". This Reprint aims to present recent advances in the analysis, design, and implementation of cognitive radio networks, highlighting both methodological developments and practical solutions. The collected contributions cover a wide range of research topics, including spectrum sensing, spectrum sharing, spectrum learning and prediction, machine learning techniques for cognitive radio networks, energy harvesting cognitive communication systems, and signal processing methods based on high-order cumulants.

The articles in this Reprint investigate a variety of important problems in cognitive radio networks, such as cooperative spectrum sensing, interference detection, adaptive spectrum allocation, deep learning-based signal identification, privacy-preserving spectrum data processing, and intelligent resource management. In addition to algorithmic and analytical studies, several contributions also explore practical system implementations and application scenarios, including wireless relay communications, civil aviation spectrum monitoring, cognitive Internet of Things environments, and intelligent transportation-related sensing and detection systems.

By presenting diverse research perspectives and technological approaches, this Reprint provides a comprehensive overview of recent developments and challenges in cognitive radio networks. It is intended for researchers, engineers, and practitioners working in wireless communications, signal processing, and intelligent networking, as well as graduate students interested in spectrum-efficient communication technologies. We hope that this Reprint will serve as a useful reference for understanding current research trends and for promoting further advances in cognitive radio networks.

Jia Liu, Kechen Zheng, Xiaoying Liu, Zhao Li, and Yang Xu
Guest Editors

sensors

MDPI

Editorial

Cognitive Radio Networks: Technologies, Challenges and Applications

Yang Xu [1], Kechen Zheng [2,*], Xiaoying Liu [2], Zhao Li [3] and Jia Liu [4]

1 School of Computer Science and Technology, Xidian University, Xi'an 710071, China; yxu@xidian.edu.cn
2 School of Computer Science and Technology, Zhejiang University of Technology, Hangzhou 310023, China; xiaoyingliu@zjut.edu.cn
3 School of Cyber Engineering, Xidian University, Xi'an 710071, China; zli@xidian.edu.cn
4 Center for Strategic Cyber Resilience Research and Development, National Institute of Informatics, Tokyo 101-8430, Japan; jliu@nii.ac.jp
* Correspondence: kechenzheng@zjut.edu.cn

1. Introduction and Current Trends in the Field

In recent years, Cognitive Radio Networks (CRNs) have emerged as a transformative solution to address the growing demand for wireless spectrum resources and the inefficiencies of traditional static spectrum allocation policies. With the proliferation of wireless devices and the rapid expansion of Internet of Things (IoT) ecosystems, the scarcity of available spectrum resources has become a critical bottleneck. Cognitive Radio, empowered by dynamic spectrum management and intelligent decision-making capabilities, offers a promising way of maximizing spectrum utilization while minimizing interference [1–3]. This Special Issue brings together cutting-edge research and innovative solutions aimed at advancing CRN technologies, addressing existing challenges, and exploring diverse applications across industries.

One of the most prominent trends in CRNs is the integration of Machine Learning (ML) and Artificial Intelligence (AI) techniques to enhance spectrum sensing, access, and management. AI-driven models are being employed to predict spectrum availability, optimize resource allocation, and detect unauthorized spectrum usage with improved accuracy [4,5]. Additionally, deep learning approaches are being increasingly utilized for anomaly detection, cognitive decision-making, and dynamic adaptation to rapidly changing wireless environments [6].

Another significant trend is the integration of CRNs with emerging wireless paradigms such as Software-Defined Networking (SDN) and Network Function Virtualization (NFV) [7]. These technologies enable centralized control and programmable network architectures, facilitating seamless coordination and dynamic resource allocation in large-scale CRN applications. Furthermore, the use of blockchain technology is gaining traction in CRNs to ensure secure and transparent spectrum trading and management [8].

CRNs are also witnessing significant progress in addressing coexistence challenges with Primary Users (PUs) and other Secondary Users (SUs). Techniques such as cooperative spectrum sensing and interference management are being refined to minimize Age of Information (AoI) performance and optimize spectrum-sharing mechanisms [9]. Energy efficiency remains a crucial focus area, with researchers exploring low-power spectrum sensing and access techniques to extend the operational lifespan of battery-constrained cognitive devices.

Finally, the application domains of CRNs are expanding beyond traditional telecommunications into sectors such as public safety, disaster recovery, smart cities, and industrial IoT [10,11]. These applications require robust, reliable, and highly adaptive communication networks, and CRNs are increasingly viewed as a key enabler in these scenarios. As we move forward, the convergence of CRNs with other advanced technologies, such as edge computing and terahertz communication, will further unlock their potential to address the ever-growing demands of future wireless ecosystems [12].

In summary, the field of CRNs continues to evolve with the integration of advanced technologies, the exploration of new applications, and the growing focus on security and privacy. We believe the articles in this Special Issue will serve as a valuable resource for researchers, practitioners, and policymakers working toward the advancement of wireless communication systems.

2. Scope of Special Issue and Contributions

This Special Issue on "Cognitive Radio Networks: Technologies, Challenges and Applications" aims to collate and disseminate state-of-the-art research advances in the analysis, design, optimization, implementation, and standardization of Cognitive Radio Networks. As a result of the open call for papers, papers related to ML, neural networks, and energy harvesting have been accepted after a rigorous peer-review process and assessed for their technical merit and relevance. The accepted papers cover applied issues in the following fields:

- Spectrum sensing, spectrum sharing, and spectrum learning and prediction;
- Machine learning techniques for Cognitive Radio Networks;
- Energy-harvesting Cognitive Radio Networks;
- Challenges and issues in designing Cognitive Radio Networks and communications;
- High-order cumulants in Cognitive Radio Networks.

These contributions are outlined in more detail below.

In contribution 1, Cervantes-Junco et al. proposed a decision-making algorithm with geographic mobility (DMAGM), which considerably reduces latency and computational complexity, to improve the communication performance of CRNs. They further proposed an improved version of the DMAGM, i.e., a feedback decision-making algorithm with geographic mobility (FDMAGM), where the feedback of the output is incorporated into the algorithm so that it can continually readapt based on the received feedback. In practical scenarios, the DMAGM improves the latency performance, and the FDMAGM improves accuracy and stability.

In contribution 2, Luo et al. employed a spectrum-based encoder to transform unstructured data into the latent space and presented a partial privacy-preserving framework that utilizes the designed interpretable model to restrict access for undesired task-independent attacks while preserving the utility of target tasks. Evaluations on multiple standard datasets show that the framework achieves competitive tradeoffs between privacy and utility.

In contribution 3, Dgani et al. employed a feature-based method that utilizes high-order cumulants (HOCs) as key features for automatic modulation classification (AMC) and explored the best classification feature for each modulation in CRNs. The proposed method performs well with different signal-to-noise ratios (SNRs) and channel conditions and offers a cost-effective and high-quality solution for AMC in various civilian and military applications.

In contribution 4, Rugini et al. analyzed the performance of centralized cooperative spectrum sensing, where SUs collaboratively detect the presence or absence of a primary signal based on compressed measurements, and obtained simplified closed-form expressions that calculate the number of sensors, number of samples, compression ratio, and SNR as a function of the detection probability and false alarm probability. The obtained results have both high accuracy and low complexity and are useful for applications with few sensors and low sample sizes.

In contribution 5, Kim et al. proposed a multitask learning-based deep signal identification network that addresses the deep signal identification problem in advanced spectrum sensing systems for CRNs. Based on performance investigations in terms of task correlation, and model size and computational efficiency evaluations, the comprehensive results emphasize the potential and practical utility of the proposed network for solving deep signal identification problems.

In contribution 6, Zhang et al. proposed a one-timescale iteration-adjustable unlicensed spectrum allocation algorithm where the step size and timescale parameter can be jointly adjusted, and they further proposed a two-timescale iteration-adjustable joint frequency selection and frequency allocation algorithm depending on slow-changing statistical channel state information. Moreover, the advantages of the proposed algorithms were validated in terms of the derived convergence conditions of the two-timescale algorithm and the derived upper bound of the corresponding convergence error.

In contribution 7, Wang et al. analyzed the fourth-order cyclic cumulant (FOCC) of signals with different modulation modes for a narrow-band wireless communication system and proposed an adjacent-frequency interference detection algorithm according to the FOCC of the received signal. The performance of the proposed algorithm in terms of correct rate and computational complexity was analyzed and compared with traditional methods. In a rail transit wireless communication scenario, the proposed algorithm overcomes the Doppler effect and works well at medium SNRs.

In contribution 8, Yu et al. proposed an environment-aware robust algorithm that employs an occasional small packet and designs a rate-adaptive scheme for the IEEE 802.11 standard. The oscillation detection mechanism in the proposed algorithm effectively suppresses abnormal rate fluctuations and further improves system throughput. Compared with similar methods, the proposed algorithm improves the throughput performance of wireless networks in mobile environments.

Contribution 9 is a review article that focuses on Cognitive Radio technology and its applications in the field of civil aviation. Zheng et al. examined the current state of Cognitive Radio technology, including ongoing research and development efforts, regulatory issues, and potential challenges to widespread adoption, and they explored potential applications of Cognitive Radio technology in civil aviation by enabling the intelligent monitoring and management of radio signals.

In contribution 10, based on the energy-harvesting cognitive Internet of Things model, Jiang et al. studied the free-riding problem of SUs and established a penalty mechanism to stimulate SUs to sense the spectrum normally during the sensing process. To improve spectrum utilization, they proposed a two-layer game-based cooperative spectrum sensing and access method that dynamically coordinates the strategies of SUs, and this significantly improved the overall throughput compared with traditional algorithms.

In contribution 11, Molina-Tenorio et al. considered a multilayer perception in a centralized CRN to monitor a multiband spectrum in real time and implemented it in a real wireless communication environment. The results show that the proposed CRN accurately locates PUs with the characteristic bandwidth, carrier frequency and power, avoiding the hidden terminal problem.

In contribution 12, Zhang et al. proposed a buffer-aided relay selection scheme based on deep Q-learning that uses neural networks to fit the Q-function, and they verified the reliability and security performances of the proposed scheme in terms of the connection outage probability and secrecy outage probability. Monte Carlo simulation results demonstrate that the proposed scheme achieves reliable and secure communications compared with similar schemes when applied in two-hop wireless relay networks.

In contribution 13, to solve the problem of vehicles not being detected in dense scenes, Wang et al. developed a new vehicle detection model by adding a normalization-based attention module to the classical YOLOv5s model, and they proposed a real-time small-target vehicle-tracking method that embeds the feature extraction process into the prediction head for training. Adding the spatial attention mechanism and channel attention mechanism module into the proposed model improves the detection accuracy for vehicles in real-time, enhances efficiency, and promotes in-depth research in the fields of vehicle detection and tracking.

3. Conclusions

As Guest Editors, we are very pleased with the final outcome of this Special Issue (SI) and look forward to seeing fellow researchers and members of the scientific community enjoy reading the articles within. Furthermore, we would like to extend our special thanks to the managing team of the Sensors journal for their continuous effort and support throughout all stages of editing this Special Issue, including the initial preparation and planning, as well as the submission and review processes for all candidate manuscripts. Finally, we are honored to have received outstanding research papers from the contributing authors and deeply appreciate the reviewers for their assistance, timely feedback, and valuable suggestions.

Conflicts of Interest: The authors declare no conflicts of interest.

List of Contributions:

1. Cervantes-Junco, G.B.; Rodriguez-Colina, E.; Palacios-Luengas, L.; Pascoe-Chalke, M.; Lara-Velázquez, P.; Marcelín-Jiménez, R. Decision-making algorithm with geographic mobility for cognitive radio. *Sensors* **2024**, *24*, 1540. https://doi.org/10.3390/s24051540.
2. Luo, Q.; Li, H. Respecting partial privacy of unstructured data via spectrum-based encoder. *Sensors* **2024**, *24*, 1015. https://doi.org/10.3390/s24031015.
3. Dgani, B.; Cohen, I. Efficient cumulant-based automatic modulation classification using machine learning. *Sensors* **2024**, *24*, 701. https://doi.org/10.3390/s24020701.
4. Rugini, L.; Banelli, P. Performance analysis of centralized cooperative schemes for compressed sensing. *Sensors* **2024**, *24*, 661. https://doi.org/10.3390/s24020661.
5. Kim, H.; Kim, Y.-J.; Kim, W.-T. Multitask learning-based deep signal identification for advanced spectrum sensing. *Sensors* **2023**, *23*, 9806. https://doi.org/10.3390/s23249806.
6. Zhang, Y.; Wu, W.; He, W.; Zhao, N. Algorithm design and convergence analysis for coexistence of cognitive radio networks in unlicensed spectrum. *Sensors* **2023**, *23*, 9705. https://doi.org/10.3390/s23249705.
7. Wang, P.; Yao, J.; Pu, Y.; Zhang, S.; Wen, L. Study of interference detection of rail transit wireless communication system based on fourth-order cyclic cumulant. *Sensors* **2023**, *23*, 8291. https://doi.org/10.3390/s23198291.
8. Yu, W.; Wang, L.; Song, J.; He, L.; Wang, Y. Environment-aware rate adaptation based on occasional request and robust adjustment in 802.11 networks. *Sensors* **2023**, *23*, 7889. https://doi.org/10.3390/s23187889.
9. Zheng, R.; Li, X.; Chen, Y. An overview of cognitive radio technology and its applications in civil aviation. *Sensors* **2023**, *23*, 6125. https://doi.org/10.3390/s23136125.

10. Jiang, K.; Ma, C.; Lin, R.; Wang, J.; Jiang, W.; Hou, H. Free-rider games for cooperative spectrum sensing and access in CIoT networks. *Sensors* **2023**, *23*, 5828. https://doi.org/10.3390/s23135828.
11. Molina-Tenorio, Y.; Prieto-Guerrero, A.; Aguilar-Gonzalez, R.; Lopez-Benitez, M. Cooperative multiband spectrum sensing using radio environment maps and neural networks. *Sensors* **2023**, *23*, 5209. https://doi.org/10.3390/s23115209.
12. Zhang, C.; Liao, X.; Wu, Z.; Qiu, G.; Chen, Z.; Yu, Z. Deep Q-learning-based buffer-aided relay selection for reliable and secure communications in two-hop wireless relay networks. *Sensors* **2023**, *23*, 4822. https://doi.org/10.3390/s23104822.
13. Wang, J.; Dong, Y.; Zhao, S.; Zhang, Z. A high-precision vehicle detection and tracking method based on the attention mechanism. *Sensors* **2023**, *23*, 724. https://doi.org/10.3390/s23020724.

References

1. Song, Z.; She, Y.; Yang, J.; Peng, J.; Gao, Y.; Tafazolli, R. Nonuniform Sampling Pattern Design for Compressed Spectrum Sensing in Mobile Cognitive Radio Networks. *IEEE Trans. Mobile. Comput.* **2024**, *23*, 8680–8693. [CrossRef]
2. Ge, J.; Liang, Y.-C.; Wang, S.; Sun, C. RIS-Assisted Cooperative Spectrum Sensing for Cognitive Radio Networks. *IEEE Trans. Wireless Commun.* **2024**, *23*, 12547–12562. [CrossRef]
3. Bhattacharjee, S.; Acharya, T.; Bhattacharya, U. Cognitive radio based spectrum sharing models for multicasting in 5G cellular networks: A survey. *Comput. Netw.* **2022**, *208*, 108870. [CrossRef]
4. Kaur, A.; Kumar, K. Imperfect CSI Based Intelligent Dynamic Spectrum Management Using Cooperative Reinforcement Learning Framework in Cognitive Radio Networks. *IEEE Trans. Mobile. Comput.* **2022**, *21*, 1672–1683. [CrossRef]
5. Chen, L.; Lu, Z.; Zhou, P.; Xu, J. Learning Optimal Sniffer Channel Assignment for Small Cell Cognitive Radio Networks. In Proceedings of 2020 IEEE International Conference on Computer Communications (INFOCOM), Toronto, ON, Canada, 6–9 July 2020; pp. 656–665.
6. Chen, Z.; Zhang, Z.; Xiao, Z.; Yang, Z.; Jin, R. Deep Learning-Based Multi-User Positioning in Wireless FDMA Cellular Networks. *IEEE J. Sel. Areas Commun.* **2023**, *41*, 3848–3862. [CrossRef]
7. Cicioğlu, M.; Cicioğlu, S.; Çalhan, A. SDN-enabled cognitive radio network architecture. *IET Commun.* **2020**, *14*, 3153–3160. [CrossRef]
8. Luong, N.C.; Anh, T.T.; Xiong, Z.; Niyato, D.; Kim, D.I. Joint Time Scheduling and Transaction Fee Selection in Blockchain-Based RF-Powered Backscatter Cognitive Radio Network. *Comput. Netw.* **2022**, *214*, 109135. [CrossRef]
9. Liu, X; Li, X; Zheng, K; Liu, J. AoI minimization of ambient backscatter-assisted EH-CRN with cooperative spectrum sensing. *Comput. Netw.* **2024**, *245*, 110389. [CrossRef]
10. Wen, Y.; Liu, L.; Li, J.; Li, Y.; Wang, K.; Yu, S.; Guizani, M. Covert Communications Aided by Cooperative Jamming in Overlay Cognitive Radio Networks. *IEEE Trans. Mobile Comput.* **2024**, *23*, 12878–12891. [CrossRef]
11. Shahid, S.M.; Kwon, S. Distributed Robust Channel Allocation for Clustered Cognitive Radio-Based IoT Networks Using Graph Theory. *Comput. Netw.* **2022**, *218*, 109406. [CrossRef]
12. Wang, D.; Zhou, F.; Lin, W.; Ding, Z.; Al-Dhahir, N. Cooperative Hybrid Nonorthogonal Multiple Access-Based Mobile-Edge Computing in Cognitive Radio Networks. *IEEE Trans. Cogn. Commun. Netw.* **2022**, *8*, 1104–1117. [CrossRef]

sensors

MDPI

Article

Performance Analysis of Centralized Cooperative Schemes for Compressed Sensing

Luca Rugini * and Paolo Banelli

Department of Engineering, University of Perugia, 06125 Perugia, Italy; paolo.banelli@unipg.it
* Correspondence: luca.rugini@unipg.it

Abstract: This paper presents a performance analysis of centralized spectrum sensing based on compressed measurements. We assume cooperative sensing, where unlicensed users individually perform compressed sensing and send their results to a fusion center, which makes the final decision about the presence or absence of a licensed user signal. Several cooperation schemes are considered, such as and-rule, or-rule, majority voting, soft equal-gain combining (EGC). The proposed analysis provides simplified closed-form expressions that calculate the required number of sensors, the required number of samples, the required compression ratio, and the required signal-to-noise ratio (SNR) as a function of the probability of detection and the probability of the false alarm of the fusion center and of the sensors. The resulting expressions are derived by exploiting some accurate approximations of the test statistics of the fusion center and of the sensors, equipped with energy detectors. The obtained results are useful, especially for a low number of sensors and low sample sizes, where conventional closed-form expressions based on the central limit theorem (CLT) fail to provide accurate approximations. The proposed analysis also allows the self-computation of the performance of each sensor and of the fusion center with reduced complexity.

Keywords: compressed sensing; cooperative spectrum sensing; energy detector; number of samples; number of sensors; probability of detection; probability of false alarm; required SNR

1. Introduction

Cooperative spectrum sensing (CSS) [1–3] is a key component of a cognitive radio (CR) network and aims to discover whether some resources are available for signal transmission or not in a given area. Using CSS, unlicensed secondary users (SUs) collaboratively detect the presence or absence of a signal transmitted by a licensed primary user (PU) to understand whether a given channel is already occupied by a PU or free and hence available for SU signal transmission.

One crucial factor of CSS is the amount of data used for sensing purposes. In general, both false alarm (FA) and detection performances improve by increasing the number of SUs and the time spent for sensing at the expense of increased computational processing. Hence, two fundamental parameters are the number of SU sensors and the number of samples used by an SU. Compressed sensing [4,5] is often employed to reduce the number of samples to the reduce complexity. The need for power consumption reduction also requires sensors that are in sleeping mode (i.e., inactive) for a certain fraction of time, to save energy.

A second major factor of CSS is the sensing algorithm of the SUs. The choice of this algorithm mostly depends on the available information about the structure of the PU signal [6–13]. Matched filtering and feature extraction presuppose that the SUs have some knowledge about the PU signal properties. On the other hand, an energy detector (ED) simply monitors the received energy in a specific time–frequency resource, and therefore is more versatile and suitable for those users without knowledge of the PU signal properties. The popularity of the ED is widespread [13–27], so that the ED is often also used as a baseline for comparison purposes. Consequently, this paper focuses on SU sensors equipped with EDs.

The third important factor of CSS is the type of cooperation. Indeed, the sensing results acquired by the SUs may be combined either by a centralized fusion center (FC) or using a distributed approach. Herein, we focus on a centralized approach wherein an FC collects the reports of the SUs. In addition, the fusion operation can be performed in several different ways [28–52]: this paper mainly focuses on *hard reporting (HR)*, where the SUs report their hard decisions to the FC. Anyway, we also consider a *soft reporting (SR)* case, where the SUs report their soft decisions to the FC, which aggregates these values using equal-gain combining (EGC).

Although there exist several performance analyses of CSS with EDs [15,16,18,19,23–27], most of the existing approaches either involve complicated multidimensional functions or have low accuracy. For instance, the exact calculation of the receiver operating characteristic (ROC) of the ED requires the inversion of a chi-squared distribution function [16,19,23–25], but this computation can be energy-demanding for low-cost sensors. On the other hand, a popular alternative approach is the Gaussian approximation (GA) for the test statistic of the ED [15,18,26,27]; however, the accuracy of the obtained ROC is reduced, especially when the number of samples is low. Similarly, when the FC applies a *majority-voting (MV) rule*, an exact analysis of the required number of SUs involves complicated functions like the incomplete beta function [53] and its inverse, whereas GA approaches based on the central limit theorem (CLT) may be inaccurate, especially when the number of sensors is low.

This paper aims at closing the gap between the accurate-but-complicated performance analyses and simple-but-inaccurate performance analyses of CSS with EDs. Specifically, this paper proposes simple low-complexity accurate approximations in such a way that the performance assessment can be also self-performed by low-cost sensors and by a low-energy FC. The proposed approximations are obtained by modifying and adapting the methods of probability distribution approximation derived several decades ago [54–63]. The main contributions of this paper can be summarized as follows.

- For the HR case, we propose closed-form expressions for the number of active sensors required by an FC to achieve a target performance, assuming SUs with a given performance. Several fusion rules are considered: *and-rule*, *or-rule*, and *MV-rule*. In contrast to existing work, the proposed expressions have both accuracy and low complexity.
- For the HR case, we propose closed-form expressions for the number of compressed samples required by an ED to achieve a target performance, assuming a given PU signal-to-noise ratio (SNR). Again, the proposed expressions have both accuracy and low complexity, thereby enabling low-energy self-computation at the SU side. Previous work in [64–66] only includes a limited subset of expressions, mainly for conventional non-cooperative sensing, whereas this paper derives a complete performance analysis valid for cooperative compressed sensing.
- For the SR case, we propose closed-form expressions for the aggregate number of samples required by an FC to achieve a target performance. Also, in this case, the proposed expressions combine accuracy with low complexity.

The remainder of this paper is structured as follows. Section 2 outlines the state of the art, while Section 3 introduces the system model. Section 4 develops the approximated performance analysis. Numerical results are presented in Section 5 to validate the proposed approximations, and some concluding remarks are drawn in Section 6.

2. State of the Art

In this section, we focus on the literature about CSS performance analysis in terms of the required number of users and required number of samples. For what concerns the number of sensors to warranty (at the FC) a predefined probability of detection and a predefined probability of FA, let us assume HR with *MV-rule* and sensors with the same statistics. In this case, the probabilities of detection and of FA are expressed as a sum [6,15] whose result is expressed by the incomplete beta function [53]. The most common approach for inverting this function is to apply general-purpose software routines, such as in [29,30,33,34,41,45,67–69], without an explicit closed-form solution and without complexity analysis. Therefore, the complexity of the FC performance self-estimation may potentially be large, especially when accurate self-estimation is required. A low-complexity approximation applies the CLT, but several other approaches are possible, with greater accuracy and tolerable complexity. For the *MV-rule*, Section 4.1 of this paper will present six low-complexity approximations, whose accuracy will be compared in Section 5.

For what concerns the number of samples required to obtain (at the SU) the desired probabilities of detection and FA, a summary of relevant results is given in [15], which considers two models for the PU signal received by the SUs, namely the deterministic and random models. In the literature, the deterministic model is more popular [17,20,21], and sometimes includes random fading [19,23,25,26] and diversity [9,24]. On the other side, the random model for the PU signal [6] is less common. However, the deterministic model is only appropriate when the SUs are perfectly synchronous and do not apply compressed sensing. When the SU sensors are partially asynchronous or apply compressed sensing, the EDs of the different SUs collect samples that are different in time; therefore, a random PU signal model seems more reasonable. In any case, the results in [15] show that the exact value for the required sample size is significantly different from the CLT-approximated sample size obtained using a Gaussian approach for both deterministic and random models. For the deterministic model, some improved approximations are proposed in [17,20,21], but for the random model, there exist less results [64–66]. Section 4.2 of this paper will present three low-complexity approaches, whose accuracy will be compared in Section 5.

3. System Model

First, in Section 3.1, we briefly describe the CSS system model without using mathematical equations, in order to give a quick general overview. Second, in Section 3.2, we present a detailed statistical signal model based on mathematical equations, which are instrumental for the performance derivation of the considered CSS scheme in Section 4.

3.1. Overview of Centralized Cooperative Sensing

We consider a CR network with multiple sensors acting as SUs and a centralized FC. We assume that the SUs and the FC are located in a specific area. The aim is to sense the presence or absence of a PU signal in a given frequency band. It is assumed that the PU is located outside the area where the SUs and the FC are located. The centralized CSS is performed in three phases. In the first phase, the SUs perform compressed spectrum sensing using an ED. The sensing results of the SUs may be either binary hard decisions or soft decisions. In the second phase, the SUs, or a fraction of them, report their sensing results to the FC, which collects the received results. In the case of sensing results with binary hard decisions, we refer to this second phase as HR. In the case of sensing results with soft decisions, we refer to this second phase as SR. In the third phase, the FC makes a final decision based on the reports received from the SUs. In the case of HR, the FC collects the received binary decisions of the SUs and performs a final decision using a given rule, such as the *and-rule*, the *or-rule*, or the *MV-rule*. In the case of SR, the FC collects the received nonbinary results and performs a soft EGC to obtain the test statistics for the final decision.

3.2. Statistical Signal Model

We consider a CR network with K SUs equipped with EDs. As in [14], we assume that the PU transmitter is located far away from the area where the K SUs are located. We assume that all the SUs have the same activity rate a, with $0 \leq a \leq 1$; hence, $1 - a$ represents the fraction of time of a generic SU staying in sleeping mode. Therefore, the number of active sensors S can be expressed as

$$S = aK. \tag{1}$$

We assume that the activity rate a is chosen in such a way that S is always integer. The S active sensors perform compressed sensing by observing the received signal in a given bandwidth, while the remaining $K - S$ sensors stay in sleeping mode. The signal vector $\mathbf{x}_i = [x_{1,i}, \ldots, x_{N,i}]^T$ received by the ith active sensor can be expressed by

$$\mathbf{x}_i = \alpha \mathbf{s}_i + \mathbf{w}_i, \qquad i = 1, \ldots, S, \tag{2}$$

where $\mathbf{s}_i = [s_{1,i}, \ldots, s_{N,i}]^T$ is the signal vector sent by the PU received by the ith sensor and $\mathbf{w}_i = [w_{1,i}, \ldots, w_{N,i}]^T$ represents the noise gathered by the ith sensor. The vector dimension N in (2) is the sample size, assumed fixed for all the SUs. In (2), α represents the hypothesis on the PU signal with $\alpha = 0$ when the PU signal is absent (hypothesis H_0) and $\alpha = 1$ when the PU signal is present (Hypothesis H_1). Both signal and noise are assumed to be circularly symmetric complex Gaussian and white, with zero mean and with covariance $E\{\mathbf{s}_i\mathbf{s}_i^H\} = \sigma_s^2\mathbf{I}_N$ and $E\{\mathbf{w}_i\mathbf{w}_i^H\} = \sigma_w^2\mathbf{I}_N$. Since the distance from the PU is similar for all the sensors, σ_s^2 is assumed to be equal for all the sensors. In addition, the sensors are of the same type and located in the same area; hence, σ_w^2 is assumed to be equal for all the SUs. Originating from different devices, the signal and noise perceived by the ith sensor are assumed to be uncorrelated, i.e., $E\{\mathbf{s}_i\mathbf{n}_i^H\} = \mathbf{0}_{N\times N}$. We also assume that the signal and noise received by different sensors are uncorrelated, i.e., $E\{\mathbf{s}_i\mathbf{n}_l^H\} = \mathbf{0}_{N\times N}$, $E\{\mathbf{s}_i\mathbf{s}_l^H\} = \mathbf{0}_{N\times N}$, $E\{\mathbf{n}_i\mathbf{n}_l^H\} = \mathbf{0}_{N\times N}$, for $i \neq l$. These assumptions are reasonable, because different noise terms are generated by different devices, and the signal terms received by the different SUs are different copies of the same PU signal that is passed through different channels, each one with a different delay and a different phase shift.

The ith sensor performs compressed sensing using an $M \times N$ compression matrix $\mathbf{\Phi}_i$, where $M < N$, as expressed by

$$\mathbf{y}_i = \mathbf{\Phi}_i\mathbf{x}_i = \alpha\mathbf{\Phi}_i\mathbf{s}_i + \mathbf{\Phi}_i\mathbf{w}_i, \tag{3}$$

where $\mathbf{y}_i = [y_{1,i}, \ldots, y_{M,i}]^T$ is the compressed vector of the ith active sensor. We define the compression ratio $c = M/N$, with $0 < c < 1$, such that its inverse $1/c$ represents the undersampling factor from the full sample size N to the compressed sample size $M = cN$. We assume that the compression matrices have rank$(\mathbf{\Phi}_i) = M$ and satisfy $\mathbf{\Phi}_i\mathbf{\Phi}_i^H = c^{-1}\mathbf{I}_M$. A class of matrices that satisfy this condition is the set of the multi-coset samplers [5]. In this case, $\mathbf{\Phi}_i\mathbf{\Phi}_l^H = \mathbf{0}_{M\times M}$ when the multi-coset time delay of the user i is different from the user l.

The ED of the ith active sensor calculates the received signal energy E_i, expressed by

$$E_i = ||\mathbf{y}_i||^2 = \mathbf{y}_i^H\mathbf{y}_i = \sum_{m=1}^{M} |y_{m,i}|^2. \tag{4}$$

In the case of HR, the ith active sensor compares E_i with a threshold η and produces the binary decision D_i expressed by

$$D_i = 1 \quad \text{if } E_i \geq \eta, \tag{5}$$

$$D_i = 0 \quad \text{otherwise}. \tag{6}$$

Equation (5) is valid when the ith ED decides that a PU signal is present, whereas (6) is valid when the ith ED decides that a PU signal is absent. We assume that the active sensors use the same threshold η. The choice of this threshold will be discussed in Section 4.2, where useful expressions are derived for the threshold self-computation. After the hard decisions, the S active sensors report their binary decisions to the FC. Then, the FC collects the S binary decisions and counts them according to

$$T = \sum_{i=1}^{S} D_i. \tag{7}$$

Therefore, T in (7) represents the number of EDs that detected the presence of a PU signal. The FC makes the final decision on the presence of a PU signal by comparing T with a threshold τ, as expressed by

$$\hat{\alpha} = 1 \quad \text{if } T \geq \tau, \tag{8}$$

$$\hat{\alpha} = 0 \quad \text{otherwise.} \tag{9}$$

Equation (8) is valid when the FC decides that a PU signal is present, whereas (9) is valid when the FC decides that a PU signal is absent. Depending on the choice of the FC threshold τ, in Section 4.1, we will consider three cases:

- *And-rule*, where $\tau = S$;
- *Or-rule*, where $\tau = 1$;
- *MV-rule*, where $\tau = (S+1)/2$, with S being an odd integer.

In the case of SR, we consider the unquantized case where the ith active sensor directly sends its computed energy E_i to the FC. Then, the FC collects all the energy results using equal-gain combining (EGC), leading to

$$\tilde{T} = \sum_{i=1}^{S} E_i = \sum_{i=1}^{S} \sum_{m=1}^{M} |y_{m,i}|^2. \tag{10}$$

Successively, the FC compares $\tilde{T}$ with a threshold $\tilde{\tau}$ to detect the presence or absence of a PU signal, as in (8) and (9). The choice of $\tilde{\tau}$ will be discussed in Section 4.3. Note that, in this case, the FC acts like a single ED that collects the cumulative energy of the S active sensors that act in parallel: this is equivalent to a single ED that uses the SM compressed samples of all the S active sensors.

4. Performance Analysis

We first consider the HR case by analyzing the FC performance in Section 4.1 and the SU performance in Section 4.2. Successively, we consider the SR case in Section 4.3.

4.1. HR Case, FC Performance

In case of HR, the FC receives S binary random variables D_i from the S active sensors. We assume that these variables are correctly received by the FC. As explained in Section 3.2, the active sensors are EDs with the same threshold and the same noise power; hence, the S sensors have the same probability of FA, denoted by p_{FA}. Therefore, each D_i is a Bernoulli random variable with $\Pr\{D_i = 1|H_0\} = p_{\text{FA}}$. The noise samples of different devices are independent, and consequently, under the hypothesis H_0, the random variable T in (7) has a binomial distribution given by

$$\Pr\{T = i|H_0\} = \binom{S}{i} p_{\text{FA}}^i (1 - p_{\text{FA}})^{S-i}. \tag{11}$$

The probability of FA at the FC, denoted by π_{FA}, can be calculated as

$$\pi_{\text{FA}} = \Pr\{\hat{\alpha} = 1|H_0\} = \sum_{i=\tau}^{S} \binom{S}{i} p_{\text{FA}}^i (1 - p_{\text{FA}})^{S-i}. \tag{12}$$

We now consider the hypothesis H_1. The S sensors are located in the same area, and therefore, the SNR γ, defined as $\gamma = \sigma_s^2/\sigma_w^2$, is the same for all the S sensors, which have the same probability of detection, denoted by p_D. This means that each D_i is a Bernoulli random variable with $\Pr\{D_i = 1|H_1\} = p_D$. Since the S sensors receive the PU signal from different channels, the PU signal samples received from different SUs are assumed to be uncorrelated, i.e., $E\{\mathbf{s}_i\mathbf{s}_l^H\} = \mathbf{0}_{N\times N}$ for $i \neq l$. Having modeled the signal as Gaussian, the vectors $\mathbf{s}_i$ and $\mathbf{s}_l$ are also independent. Consequently, the random variable T under the hypothesis H_1 in (7) also has a binomial distribution given by

$$\Pr\{T = i|H_1\} = \binom{S}{i} p_D^i (1-p_D)^{S-i}. \tag{13}$$

The probability of detection at the FC, denoted by π_D, can be calculated as

$$\pi_D = \Pr\{\hat{\alpha} = 1|H_1\} = \sum_{i=\tau}^{S} \binom{S}{i} p_D^i (1-p_D)^{S-i}. \tag{14}$$

Our aim is to invert (12) and (14) to find the required number of active sensors S that warranties a desired performance π_{FA} and π_D at the FC, for a fixed performance p_{FA} and p_D of the SU sensors.

4.1.1. And-Rule

In the case of the *and-rule*, the FC threshold is $\tau = S$, and hence (12) and (14) become, respectively,

$$\pi_{FA} = \Pr\{\hat{\alpha} = 1|H_0\} = \Pr\{T = S|H_0\} = p_{FA}^S, \tag{15}$$

$$\pi_D = \Pr\{\hat{\alpha} = 1|H_1\} = \Pr\{T = S|H_1\} = p_D^S. \tag{16}$$

When $S > 1$, both probabilities at the FC are lower than the corresponding probabilities at the SU. Therefore, the *and-rule* is appropriate for an FC that aims to reduce the probability of FA, although this also reduces the probability of detection, with respect to the single SU. Assuming that the FC requires a probability of FA π_{FA} not larger than a target probability $\bar{\pi}_{FA}$, and a probability of detection π_D not smaller than a target probability $\bar{\pi}_D$, we obtain that the number of active sensors S must satisfy

$$\frac{\ln \bar{\pi}_{FA}}{\ln p_{FA}} = S_{\min} \leq S \leq S_{\max} = \frac{\ln \bar{\pi}_D}{\ln p_D}, \tag{17}$$

where $S_{\min}$ and $S_{\max}$ denote the minimum and the maximum number of active sensors to satisfy the FC performance. Note that (17) implies that $p_D \geq \bar{\pi}_D$. If the probabilities in (17) are chosen incorrectly, it may be impossible to satisfy both inequalities in (17).

Bearing in mind (1), we can use (17) to find the requirement on the number of total sensors K (both active and sleeping) for a fixed activity rate a, as expressed by

$$\frac{S_{\min}}{a} = K_{\min} \leq K \leq K_{\max} = \frac{S_{\max}}{a}, \tag{18}$$

where $S_{\min}$ and $S_{\max}$ are defined in (17). When low-cost sensors want to save energy by staying in sleeping mode more often, their activity rate a reduces and consequently a larger number of total sensors K is required due to the increase in both limits in (18). Alternatively, if the number of total sensors K is fixed, we can find the required activity rate as

$$\frac{S_{\min}}{K} = a_{\min} \leq a \leq a_{\max} = \frac{S_{\max}}{K}. \tag{19}$$

Therefore, if the number of total sensors K reduces, the activity rate a should increase to warranty the FC performance; in other words, each sensor must stay awake for a longer time.

If we want both S and K to be fixed (and hence, also the activity rate a fixed), to warranty the FC target performance, from (15) and (16), the probability performance of each SU must satisfy

$$p_{\rm FA} \leq \sqrt[aK]{\bar{\pi}_{\rm FA}}, \tag{20}$$

$$p_{\rm D} \geq \sqrt[aK]{\bar{\pi}_{\rm D}}. \tag{21}$$

4.1.2. Or-Rule

In the case of the *or-rule*, the FC threshold is $\tau = 1$, and hence (12) and (14) become, respectively

$$\pi_{\rm FA} = \Pr\{\hat{\alpha} = 1|H_0\} = 1 - \Pr\{T = 0|H_0\} = 1 - (1 - p_{\rm FA})^S, \tag{22}$$

$$\pi_{\rm D} = \Pr\{\hat{\alpha} = 1|H_1\} = 1 - \Pr\{T = 0|H_1\} = 1 - (1 - p_{\rm D})^S. \tag{23}$$

When $S > 1$, both probabilities at the FC are greater than the corresponding probabilities at the SU. Therefore, the *or-rule* is appropriate for an FC that aims to increase the probability of detection, although this also increases the probability of FA, with respect to the single SU. Assuming that the FC requires a probability of FA $\pi_{\rm FA}$ that is no larger than a target probability $\bar{\pi}_{\rm FA}$, and a probability of detection $\pi_{\rm D}$ not smaller than a target probability $\bar{\pi}_{\rm D}$, we obtain that the number of active sensors S must satisfy

$$\frac{\ln(1 - \bar{\pi}_{\rm D})}{\ln(1 - p_{\rm D})} = S_{\min} \leq S \leq S_{\max} = \frac{\ln(1 - \bar{\pi}_{\rm FA})}{\ln(1 - p_{\rm FA})}. \tag{24}$$

The main difference of (24) with (17) is the exchanged role of the probabilities of FA, which now determine the maximum number of active sensors $S_{\max}$, and the probabilities of detection, which now determine the minimum number of active sensors $S_{\min}$. This exchanged role can be clarified by defining the corresponding probabilities of missed detection (MD), $p_{\rm MD} = 1 - p_{\rm D}$ and $\pi_{\rm MD} = 1 - \pi_{\rm D}$, which in this *or-rule* case, play a similar role played by the probabilities of FA when using the *and-rule*.

Again, we can use (24) to find the required number of total sensors K for a fixed activity rate a, and the required activity rate when K is fixed: in both cases, the obtained equations are identical to the corresponding Equations (18) and (19), but with the new definitions of $S_{\min}$ and $S_{\max}$ provided by (24). Also, in this *or-rule* case, when low-cost sensors save energy by frequently staying in sleeping mode, their activity rate a is low and consequently a larger number of total sensors K is required; vice versa, if the number of total sensors K reduces, the activity rate a should increase, leading to more power consumption of each SU sensor.

When both S and K are fixed, to warranty the FC target performance, from (22) and (23), the probability performance of each SU must satisfy

$$p_{\rm FA} \leq 1 - \sqrt[aK]{1 - \bar{\pi}_{\rm FA}}, \tag{25}$$

$$p_{\rm D} \geq 1 - \sqrt[aK]{1 - \bar{\pi}_{\rm D}}. \tag{26}$$

4.1.3. MV-Rule

The *MV-rule* uses a threshold $\tau = (S + 1)/2$ with S odd, and this complicates the inversion of the cumulative binomial functions (12) and (14). To perform the inversion, we make use of the suitable approximations of the binomial distribution, most of which have existed for decades. According to the CLT, a first approach is to use a GA obtained by matching the first two moments, which lead to [54]

$$\sum_{i=\frac{S+1}{2}}^{S} \binom{S}{i} p^i (1-p)^{S-i} \approx Q\left(\sqrt{\frac{(1-2p)^2}{4p(1-p)}S}\right), \qquad p < \frac{1}{2}, \tag{27}$$

$$\sum_{i=\frac{S+1}{2}}^{S} \binom{S}{i} p^i (1-p)^{S-i} \approx 1 - Q\left(\sqrt{\frac{(2p-1)^2}{4p(1-p)}S}\right), \qquad p \geq \frac{1}{2}, \tag{28}$$

where

$$Q(z) = \frac{1}{\sqrt{2\pi}} \int_z^{+\infty} e^{-t^2/2} dt \tag{29}$$

is the right tail of the standard Gaussian probability density function, also known as Gaussian Q-function. Assuming that the FC requires a probability of FA π_{FA} not larger than a target probability $\bar{\pi}_{\text{FA}}$, and a probability of detection π_{D} that is no smaller than a target probability $\bar{\pi}_{\text{D}}$ from (12), (14) and (27), we obtain that the number of active sensors S must satisfy

$$S \geq S_{\min,\text{FA}} = \frac{4p_{\text{FA}}(1-p_{\text{FA}})}{(1-2p_{\text{FA}})^2} \left[Q^{-1}(\bar{\pi}_{\text{FA}})\right]^2, \tag{30}$$

$$S \geq S_{\min,\text{D}} = \frac{4p_{\text{D}}(1-p_{\text{D}})}{(2p_{\text{D}}-1)^2} \left[Q^{-1}(1-\bar{\pi}_{\text{D}})\right]^2. \tag{31}$$

Therefore, differently from the *and-rule* and the *or-rule*, there is no limit on the maximum number of active sensors. Note that the GA produces the simple results (30) and (31) whose calculation only requires the inverse of the Q-function. It is also worth noting that, when $p_{\text{FA}} = 1 - p_{\text{D}}$ and $\bar{\pi}_{\text{FA}} = 1 - \bar{\pi}_{\text{D}}$, we have $S_{\min,\text{FA}} = S_{\min,\text{D}}$. Bounds on the number of total sensors K and on the activity rate a can be found by incorporating (1) into (30) and (31), similarly to the derivation of (18) and (19) from (17). Equations (30) and (31) can be used to obtain the maximum probability of FA at the SU and the minimum probability of detection at the SU, for a given number of active SUs S, as expressed by

$$p_{\text{FA}} \leq \frac{1}{2}\left\{1 - \frac{Q^{-1}(\bar{\pi}_{\text{FA}})}{\sqrt{S + [Q^{-1}(\bar{\pi}_{\text{FA}})]^2}}\right\}, \tag{32}$$

$$p_{\text{D}} \geq \frac{1}{2}\left\{1 + \frac{Q^{-1}(1-\bar{\pi}_{\text{D}})}{\sqrt{S + [Q^{-1}(1-\bar{\pi}_{\text{D}})]^2}}\right\}. \tag{33}$$

A second possibility for the *MV-rule* is to use an arcsine approximation. According to [55], the most accurate arcsine approximation of a binomial distribution is the one that includes a Borges refinement, as expressed by

$$\sum_{i=\frac{S+1}{2}}^{S} \binom{S}{i} p^i (1-p)^{S-i} \approx Q\left(\left[\frac{\pi}{2} - 2\arcsin(\sqrt{p})\right]\sqrt{S+\frac{1}{3}}\right), \qquad p < \frac{1}{2}, \tag{34}$$

$$\sum_{i=\frac{S+1}{2}}^{S} \binom{S}{i} p^i (1-p)^{S-i} \approx 1 - Q\left(\left[\frac{\pi}{2} - 2\arcsin(\sqrt{1-p})\right]\sqrt{S+\frac{1}{3}}\right), \qquad p \geq \frac{1}{2}, \tag{35}$$

which, by (12) and (14), leads to

$$S \geq S_{\min,\text{FA}} = \left[\frac{2Q^{-1}(\bar{\pi}_{\text{FA}})}{\pi - 4\arcsin(\sqrt{p_{\text{FA}}})}\right]^2 - \frac{1}{3}, \tag{36}$$

$$S \geq S_{\min,\text{D}} = \left[\frac{2Q^{-1}(1-\bar{\pi}_{\text{D}})}{\pi - 4\arcsin(\sqrt{1-p_{\text{D}}})}\right]^2 - \frac{1}{3}, \tag{37}$$

where we have assumed that the probabilities of FA are lower than one half and the probabilities of detection are greater than one half. As for the GA, Equations (36) and (37) permit the calculations of the maximum probability of FA at the SU and the minimum probability of detection at the SU, for a given number of active SUs S, as expressed by

$$p_{\mathrm{FA}} \leq \frac{1}{2}\left\{1-\sin\left[\frac{Q^{-1}(\bar{\pi}_{\mathrm{FA}})}{\sqrt{S+\frac{1}{3}}}\right]\right\}, \tag{38}$$

$$p_{\mathrm{D}} \geq \frac{1}{2}\left\{1+\sin\left[\frac{Q^{-1}(1-\bar{\pi}_{\mathrm{D}})}{\sqrt{S+\frac{1}{3}}}\right]\right\}. \tag{39}$$

A third invertible approximation of the cumulative binomial function was derived by Camp and Paulson [55], as expressed by

$$\sum_{i=\frac{S+1}{2}}^{S}\binom{S}{i}p^i(1-p)^{S-i} \approx Q\left(R(p)\frac{9-\frac{2}{S+1}}{3\sqrt{\frac{2}{S+1}}}\right), \qquad p<\frac{1}{2}, \tag{40}$$

$$\sum_{i=\frac{S+1}{2}}^{S}\binom{S}{i}p^i(1-p)^{S-i} \approx 1-Q\left(R(1-p)\frac{9-\frac{2}{S+1}}{3\sqrt{\frac{2}{S+1}}}\right), \qquad p\geq\frac{1}{2}, \tag{41}$$

where $R(p)$ is the ratio defined as

$$R(p) = \frac{1-(\frac{p}{1-p})^{1/3}}{\sqrt{1+(\frac{p}{1-p})^{2/3}}}. \tag{42}$$

We now use (12) with (14) and $p_{\mathrm{FA}} < 1/2$ and $p_{\mathrm{D}} > 1/2$, jointly with (40) and (41). Assuming a probability of FA $\pi_{\mathrm{FA}} < 1/2$ at the FC that is no greater than the target probability $\bar{\pi}_{\mathrm{FA}}$ and a probability of detection $\pi_{\mathrm{D}} > 1/2$ at the FC that is no lower than the target probability $\bar{\pi}_{\mathrm{D}}$, we obtain

$$\frac{S+\frac{7}{9}}{\sqrt{S+1}} \geq \frac{\sqrt{2}Q^{-1}(\bar{\pi}_{\mathrm{FA}})}{3R(p_{\mathrm{FA}})}, \tag{43}$$

$$\frac{S+\frac{7}{9}}{\sqrt{S+1}} \geq \frac{\sqrt{2}Q^{-1}(1-\bar{\pi}_{\mathrm{D}})}{3R(1-p_{\mathrm{D}})}, \tag{44}$$

which produces the results

$$S \geq S_{\mathrm{min,FA}} = U^2(p_{\mathrm{FA}},\bar{\pi}_{\mathrm{FA}}) + U(p_{\mathrm{FA}},\bar{\pi}_{\mathrm{FA}})\sqrt{U^2(p_{\mathrm{FA}},\bar{\pi}_{\mathrm{FA}})+\frac{4}{9}} - \frac{7}{9}, \tag{45}$$

$$S \geq S_{\mathrm{min,D}} = U^2(1-p_{\mathrm{D}},1-\bar{\pi}_{\mathrm{D}}) + U(1-p_{\mathrm{D}},1-\bar{\pi}_{\mathrm{D}})\sqrt{U^2(1-p_{\mathrm{D}},1-\bar{\pi}_{\mathrm{D}})+\frac{4}{9}} - \frac{7}{9}, \tag{46}$$

where

$$U(p,\bar{\pi}) = \frac{Q^{-1}(\bar{\pi})}{3R(p)}, \tag{47}$$

where $Q^{-1}(z)$ is the inverse of (29) and $R(p)$ is defined in (42).

The three methods discussed so far (Gaussian, arcsine, and Camp–Paulson approximations) directly approximate the binomial distribution. Another set of approximations can be obtained by exploiting the relation between the binomial distribution and the incomplete beta function, as expressed in Equation 26.5.24 of [53]:

$$\sum_{i=\tau}^{S} \binom{S}{i} p^i (1-p)^{S-i} = I_p(\tau, S-\tau+1), \tag{48}$$

where $I_p(z_1, z_2)$ is the (normalized) incomplete beta function defined as [53]

$$I_p(z_1, z_2) = \frac{1}{B(z_1, z_2)} \int_0^p t^{z_1-1}(1-t)^{z_2-1} dt, \tag{49}$$

where $B(z_1, z_2)$ is the beta function [53]. The *MV-rule* uses a threshold $\tau = (S+1)/2$ with S odd: this leads to $I_p(\frac{S+1}{2}, \frac{S+1}{2})$ in (48), for which several invertible approximations exist [53,56–59]. The method of Fisher [56,57] uses the approximation

$$\sum_{i=\frac{S+1}{2}}^{S} \binom{S}{i} p^i (1-p)^{S-i} \approx Q\Big(L(p)\sqrt{S}\Big), \qquad p < \frac{1}{2}, \tag{50}$$

$$\sum_{i=\frac{S+1}{2}}^{S} \binom{S}{i} p^i (1-p)^{S-i} \approx 1 - Q\Big(L(1-p)\sqrt{S}\Big), \qquad p \geq \frac{1}{2}, \tag{51}$$

where $L(p)$ is defined as

$$L(p) = \frac{1}{2} \ln \frac{1-p}{p}. \tag{52}$$

Using (12) and (14) with (50) and (51), and assuming $p_{\text{FA}} < 1/2$, $p_{\text{D}} > 1/2$, $\pi_{\text{FA}} \leq \bar{\pi}_{\text{FA}} < 1/2$ and $\pi_{\text{D}} \geq \bar{\pi}_{\text{D}} > 1/2$, we obtain

$$S \geq S_{\text{min,FA}} = \left[\frac{Q^{-1}(\bar{\pi}_{\text{FA}})}{L(p_{\text{FA}})}\right]^2, \tag{53}$$

$$S \geq S_{\text{min,D}} = \left[\frac{Q^{-1}(1-\bar{\pi}_{\text{D}})}{L(1-p_{\text{D}})}\right]^2. \tag{54}$$

If we assume a given number of active sensors S, the maximum probability of FA required at the SU and the minimum probability of detection required at the SU are given by

$$p_{\text{FA}} \leq \frac{1}{1+\exp\{2Q^{-1}(\bar{\pi}_{\text{FA}})/\sqrt{S}\}}, \tag{55}$$

$$p_{\text{D}} \geq \frac{\exp\{2Q^{-1}(1-\bar{\pi}_{\text{D}})/\sqrt{S}\}}{1+\exp\{2Q^{-1}(1-\bar{\pi}_{\text{D}})/\sqrt{S}\}}, \tag{56}$$

where $\bar{\pi}_{\text{FA}}$ and $\bar{\pi}_{\text{D}}$ are the FC target probability of FA and the FC target probability of detection, respectively.

On the other hand, the method of Cochran [56,57] refines the method of Fisher, such that the approximation of the binomial sum becomes

$$\sum_{i=\frac{S+1}{2}}^{S} \binom{S}{i} p^i (1-p)^{S-i} \approx Q\left(\frac{L(p)}{\sqrt{1+\frac{1}{6}L^2(p)}}\sqrt{S+\frac{1}{2}}\right), \qquad p < \frac{1}{2}, \tag{57}$$

$$\sum_{i=\frac{S+1}{2}}^{S} \binom{S}{i} p^i (1-p)^{S-i} \approx 1 - Q\left(\frac{L(1-p)}{\sqrt{1+\frac{1}{6}L^2(1-p)}}\sqrt{S+\frac{1}{2}}\right), \qquad p \geq \frac{1}{2}, \tag{58}$$

where $L(p)$ is defined in (52). Note that, from (52), $L(1-p) = -L(p)$, hence $L^2(1-p) = L^2(p)$. Using the same assumptions $p_{\text{FA}} < 1/2$, $p_{\text{D}} > 1/2$, $\pi_{\text{FA}} \leq \bar{\pi}_{\text{FA}} < 1/2$

and $\pi_D \geq \bar{\pi}_D > 1/2$, (57) and (58) jointly with (12) and (14) lead to the following bounds for the minimum number of active sensors:

$$S \geq S_{\text{min,FA}} = \left[\frac{Q^{-1}(\bar{\pi}_{\text{FA}})}{L(p_{\text{FA}})}\right]^2 + \frac{[Q^{-1}(\bar{\pi}_{\text{FA}})]^2}{6} - \frac{1}{2}, \tag{59}$$

$$S \geq S_{\text{min,D}} = \left[\frac{Q^{-1}(1-\bar{\pi}_{\text{D}})}{L(1-p_{\text{D}})}\right]^2 + \frac{[Q^{-1}(1-\bar{\pi}_{\text{D}})]^2}{6} - \frac{1}{2}. \tag{60}$$

In this case, the required probability of FA and the required probability of detection at the SU, assuming a fixed number of active sensors S, are

$$p_{\text{FA}} \leq \frac{1}{1+\exp\left\{2Q^{-1}(\bar{\pi}_{\text{FA}})/\sqrt{S-\frac{1}{6}\{[Q^{-1}(\bar{\pi}_{\text{FA}})]^2-3\}}\right\}}, \tag{61}$$

$$p_{\text{D}} \geq \frac{\exp\left\{2Q^{-1}(1-\bar{\pi}_{\text{D}})/\sqrt{S-\frac{1}{6}\{[Q^{-1}(1-\bar{\pi}_{\text{D}})]^2-3\}}\right\}}{1+\exp\left\{2Q^{-1}(1-\bar{\pi}_{\text{D}})/\sqrt{S-\frac{1}{6}\{[Q^{-1}(1-\bar{\pi}_{\text{D}})]^2-3\}}\right\}}, \tag{62}$$

where $\bar{\pi}_{\text{FA}}$ and $\bar{\pi}_{\text{D}}$ are again the target probabilities (of FA and of detection, respectively) required at the FC.

Another approximation is the one proposed by Carter [58,59] which, with our notation, can be expressed as

$$\sum_{i=\frac{S+1}{2}}^{S}\binom{S}{i}p^i(1-p)^{S-i} \approx Q\left(\sqrt{3\left[\sqrt{\left(S-\frac{1}{2}\right)^2+\frac{2}{3}L^2(p)S^2}-S+\frac{1}{2}\right]}\right), \quad p<\frac{1}{2}, \tag{63}$$

$$\sum_{i=\frac{S+1}{2}}^{S}\binom{S}{i}p^i(1-p)^{S-i} \approx 1-Q\left(\sqrt{3\left[\sqrt{\left(S-\frac{1}{2}\right)^2+\frac{2}{3}L^2(p)S^2}-S+\frac{1}{2}\right]}\right), \quad p\geq\frac{1}{2}, \tag{64}$$

where $L(p)$ is defined in (52). In (64), we used $L^2(p) = L^2(1-p)$. Using again $p_{\text{FA}} < 1/2$, $p_{\text{D}} > 1/2$, $\pi_{\text{FA}} \leq \bar{\pi}_{\text{FA}} < 1/2$ and $\pi_{\text{D}} \geq \bar{\pi}_{\text{D}} > 1/2$, we can join (12) and (14) with (63) and (64) to obtain the minimum number of required sensors expressed by

$$S \geq S_{\text{min,FA}} = \frac{1}{2}\left[\frac{Q^{-1}(\bar{\pi}_{\text{FA}})}{L(p_{\text{FA}})}\right]^2\left\{1+\sqrt{1+2L^2(p_{\text{FA}})\left\{\frac{1}{3}-\frac{1}{[Q^{-1}(\bar{\pi}_{\text{FA}})]^2}\right\}}\right\}, \tag{65}$$

$$S \geq S_{\text{min,D}} = \frac{1}{2}\left[\frac{Q^{-1}(1-\bar{\pi}_{\text{D}})}{L(1-p_{\text{D}})}\right]^2\left\{1+\sqrt{1+2L^2(p_{\text{D}})\left\{\frac{1}{3}-\frac{1}{[Q^{-1}(1-\bar{\pi}_{\text{D}})]^2}\right\}}\right\}, \tag{66}$$

where $\bar{\pi}_{\text{FA}}$ and $\bar{\pi}_{\text{D}}$ are the target probabilities required at the FC. When the number of active sensors S is fixed, the required performance of the SU sensors can be expressed by

$$p_{\text{FA}} \leq \frac{1}{1+\exp\left\{2Q^{-1}(\bar{\pi}_{\text{FA}})\sqrt{S+\frac{1}{6}\{[Q^{-1}(\bar{\pi}_{\text{FA}})]^2-3\}/S}\right\}}, \tag{67}$$

$$p_{\text{D}} \geq \frac{\exp\left\{2Q^{-1}(1-\bar{\pi}_{\text{D}})\sqrt{S+\frac{1}{6}\{[Q^{-1}(1-\bar{\pi}_{\text{D}})]^2-3\}/S}\right\}}{1+\exp\left\{2Q^{-1}(1-\bar{\pi}_{\text{D}})\sqrt{S+\frac{1}{6}\{[Q^{-1}(1-\bar{\pi}_{\text{D}})]^2-3\}/S}\right\}}. \tag{68}$$

Summarizing, for the *MV-rule*, we presented six approximations, and each one gives a couple of lower bounds on the minimum number of required active sensors S for a target performance at the FC, expressed by (30) and (31), (36) and (37), (45) and (46), (53) and (54), (59) and (60), as well as (65) and (66). The two bounds in each couple become equal $(S_{\text{min,FA}} = S_{\text{min,D}})$ when $p_{\text{D}} = 1 - p_{\text{FA}}$ and $\bar{\pi}_{\text{D}} = 1 - \bar{\pi}_{\text{FA}}$. Similar bounds on the minimum number of required total sensors and on the activity rate can be obtained using (1). The accuracy of the above six approximations is compared in Section 5, using numerical examples.

4.2. HR Case, SU Performance

Section 4.1 has detailed the performance of the FC that receive HR information from the SU, and proposed analytical expressions that linked the probability of FA (and of detection) of the FC to the number of sensors and to the probability of FA (and of detection) of the active sensors. In contrast, this Section 4.2 focuses on the performance of the SU sensors with HR and provides analytical expressions that link the probability of FA (and of detection) of the SU sensor to the number of sensors and to the SNR.

Assuming that the ith SU is equipped with an ED, under the hypothesis H_0, the energy E_i in (4) is a chi-squared random variable, with a number of degrees of freedom equal to $2M = 2cN$, where N is the total number of samples and c is the compression ratio. Therefore, according to (5), the probability of FA of the SU sensor can be expressed as

$$p_{\text{FA}} = \Pr\{D_i = 1 | H_0\} = 1 - F_{2M}\left(\frac{2\eta}{\sigma_{\text{w}}^2}\right), \tag{69}$$

where

$$F_{2M}(z) = \frac{1}{\Gamma(M)} \int_0^{z/2} t^{M-1} e^{-t} dt, \tag{70}$$

where $\Gamma(M)$ denotes the gamma function [53]. According to (69), the ED threshold is expressed by

$$\eta = \frac{\sigma_{\text{w}}^2}{2} F_{2M}^{-1}(1 - p_{\text{FA}}), \tag{71}$$

where $F_{2M}^{-1}(z)$ is the inverse function of $F_{2M}(z)$ with respect to the argument under parentheses.

Under the hypothesis H_1, the energy E_i in (4) is a chi-squared random variable with the same number of degrees of freedom than under H_0, but with increased power due to the presence of a PU signal. Therefore, according to (5), the probability of detection of the SU sensor can be expressed as

$$p_{\text{D}} = \Pr\{D_i = 1 | H_1\} = 1 - F_{2M}\left(\frac{2\eta}{\sigma_{\text{s}}^2 + \sigma_{\text{w}}^2}\right). \tag{72}$$

By inserting (71) into (72), we obtain the receiver operating characteristic (ROC) of the SU sensor as

$$p_{\text{D}} = 1 - F_{2M}\left(\frac{F_{2M}^{-1}(1 - p_{\text{FA}})}{1 + \gamma}\right), \tag{73}$$

where $\gamma = \sigma_{\text{s}}^2 / \sigma_{\text{w}}^2$ is the sensing SNR.

Our aim is to invert (73) to find the required number of samples N that warranties a desired performance p_{FA} and p_{D} at the SU for a fixed SNR. In principle, we could use an iterative algorithm to find the minimum compressed sample size M that warranties the SU performance, and then calculate the required sample size as $N = M/c$. However, this iterative algorithm would require the multiple evaluation of integral functions like $F_{2M}(z)$ and $F_{2M}^{-1}(z)$; hence, this approach is not appropriate for low-complexity SU sensors that want to self-calculate the required sample size. Therefore, whenever possible, we look for invertible low-complexity accurate approximations of the ROC (73). When iterative algorithms are unavoidable, we resort to the low-complexity approximations of $F_{2M}(z)$ and $F_{2M}^{-1}(z)$. Specifically, we consider the three following approaches.

- *Standard GA*, where a chi-squared random variable is approximated by a Gaussian random variable using the CLT (see [15] and references therein).
- *Power-of-Gaussian approximation (PGA)*, where a chi-squared random variable is approximated by the power of a Gaussian random variable [60,61,64], with a suitable power exponent.
- *Polynomial approximation (PA)*, where a polynomial of the rth root of a chi-squared random variable is approximated by a Gaussian random variable [62,63], using suitable degree and coefficients.

A comparison of these three approaches for non-cooperative non-compressed sensing was introduced in the conference paper [65]. Herein, we extend the earlier preliminary results of [65,66] and derive a complete performance analysis valid for all three approaches when applied to compressed sensing.

4.2.1. Standard GA

The standard GA can be obtained by approximating (70) as [15]

$$F_{2M}(z) \approx 1 - Q\left(\left[\frac{z}{2M} - 1\right]\sqrt{M}\right), \tag{74}$$

which leads to the approximated inverse function

$$F_{2M}^{-1}(z) \approx 2M\left[\frac{Q^{-1}(1-z)}{\sqrt{M}} + 1\right]. \tag{75}$$

Equations (75) and (71) allow the ED to self-compute the threshold as

$$\eta \approx M\sigma_{\mathrm{W}}^2\left[\frac{Q^{-1}(p_{\mathrm{FA}})}{\sqrt{M}} + 1\right]. \tag{76}$$

Note that the approximated expression (76) is much simpler than the exact expression (71), because the inverse Q-function is a one-dimensional function that is easier to compute than $F_{2M}^{-1}(1 - p_{\mathrm{FA}})$. Equation (74) can be combined with (73) to obtain the performance of the SU sensor: when the number of samples N and the SNR γ are fixed, (73) and (74) lead to the GA ROC

$$p_{\mathrm{D}} \approx Q\left(\frac{Q^{-1}(p_{\mathrm{FA}}) - \gamma\sqrt{cN}}{1+\gamma}\right). \tag{77}$$

When the SU probabilities of FA p_{FA} and detection p_{D} are fixed, from (77), we obtain the required number of samples N as a function of the SNR γ and of the compression ratio c, as expressed by

$$N \geq \frac{1}{c}\left[\frac{Q^{-1}(p_{\mathrm{FA}}) - (1+\gamma)Q^{-1}(p_{\mathrm{D}})}{\gamma}\right]^2. \tag{78}$$

Moreover, when the SU probabilities of FA p_{FA} and detection p_{D} are fixed, from (77) we obtain the required SNR γ as a function of the sample size N and of the compression ratio c, as expressed by

$$\gamma \geq \frac{Q^{-1}(p_{\mathrm{FA}}) - Q^{-1}(p_{\mathrm{D}})}{Q^{-1}(p_{\mathrm{D}}) + \sqrt{cN}}. \tag{79}$$

4.2.2. PGA

In general, GA approaches can only be accurate in the middle of the Gaussian shape, or when the number of samples is large. To improve the accuracy near the tails of a probability density function (pdf), PGA approaches can be used, especially when the number of samples is low. In a nutshell, PGA approaches replace a chi-squared random variable with the rth power of a Gaussian random variable, with r integer. These PGA approaches are also called root transformations, because a Gaussian random variable approximates the rth root of a chi-squared random variable. When $r = 1$, PGA reduces to GA. The improved accuracy of PGA approaches with $r > 1$ has been proven in [61] and in [62]. Specifically, [61] has shown that the Kullback–Leibler divergence between the pdf of the rth root of a chi-squared random variable and the Gaussian pdf is minimized for $3 \leq r \leq 4$; in addition, [62] has compared the cumulants of the two pdfs, showing that the best match is obtained when $2 \leq r \leq 4$. Therefore, it is expected that a performance analysis based on a PGA approach will be more accurate than a performance analysis based on the GA.

The PGA can be obtained by approximating (70) as [60]

$$F_{2M}(z) \approx 1 - Q\left(\left[\sqrt[r]{\frac{z}{2M}} - \frac{2Mr^2 - r + 1}{2Mr^2}\right] r\sqrt{M}\right), \tag{80}$$

which can be used in place of (74) to improve the accuracy. In this case, the inverse function is approximated as

$$F_{2M}^{-1}(z) \approx 2M\left[\frac{Q^{-1}(1-z)}{r\sqrt{M}} + \frac{2Mr^2 - r + 1}{2Mr^2}\right]^r. \tag{81}$$

Therefore, combining (81) with (71), the ED can easily self-compute its threshold as

$$\eta \approx M\sigma_{\mathrm{w}}^2\left[\frac{Q^{-1}(p_{\mathrm{FA}})}{r\sqrt{M}} + \frac{2Mr^2 - r + 1}{2Mr^2}\right]^r, \tag{82}$$

which coincides with (76) when $r = 1$. Equation (80) can be combined with (73) to obtain the performance of the SU sensor: when the number of samples N and the SNR γ are fixed, after some calculus, (73) and (80) lead to the PGA ROC as

$$p_{\mathrm{D}} \approx Q\left(\frac{Q^{-1}(p_{\mathrm{FA}}) - (\sqrt[r]{1+\gamma} - 1)\left(r\sqrt{cN} - \frac{r-1}{2r\sqrt{cN}}\right)}{\sqrt[r]{1+\gamma}}\right). \tag{83}$$

When $r = 1$, (83) reduces to (77). When the SU probabilities of FA p_{FA} and detection p_{D} are fixed, from (83), we obtain the required number of samples N as a function of the SNR γ and of the compression ratio c, as expressed by

$$N \geq \frac{1}{c}\left(b_r + \frac{r-1}{2r^2} + \sqrt{b_r^2 + \frac{r-1}{r^2} b_r}\right), \tag{84}$$

where

$$b_r = \frac{1}{2}\left[\frac{Q^{-1}(p_{\mathrm{FA}}) - \sqrt[r]{1+\gamma}\, Q^{-1}(p_{\mathrm{D}})}{r(\sqrt[r]{1+\gamma} - 1)}\right]^2. \tag{85}$$

In addition, when the SU probabilities of FA p_{FA} and detection p_{D} are fixed, from (83), we obtain the required SNR γ as a function of the sample size N and of the compression ratio c as expressed by

$$\gamma \geq \left[\frac{Q^{-1}(p_{\mathrm{FA}}) + r\sqrt{cN} - \frac{r-1}{2r\sqrt{cN}}}{Q^{-1}(p_{\mathrm{D}}) + r\sqrt{cN} - \frac{r-1}{2r\sqrt{cN}}}\right]^r - 1. \tag{86}$$

When $r = 1$, (84) reduces to (78), while (86) boils down to (79).

4.2.3. PA

PA approaches try to increase the accuracy of PGA [62,63] by constructing a polynomial expression whose variable is the rth root of a chi-squared random variable; then, this polynomial is approximated by a Gaussian random variable. Therefore, a PA approach is also called a linear combination of power (or root) transformations.

A first PA, originally developed by Goria [62], is obtained by first defining the fourth root transformation

$$x = \sqrt[4]{\frac{z}{2M}}, \tag{87}$$

where z is a chi-squared random variable with $2M$ degrees of freedom, and then choosing the polynomial expression

$$g = x^2 + 4x. \tag{88}$$

Successively, g in (88) is approximated as a Gaussian random variable with a mean $5 - \frac{1}{2M}$ and variance $\frac{9}{4M}$. Consequently, the chi-squared distribution (70) is approximated as

$$F_{2M}(z) \approx 1 - Q\left(\left[\sqrt{\frac{z}{2M}} + 4\sqrt[4]{\frac{z}{2M}} - \left(5 - \frac{1}{2M}\right)\right]\frac{2}{3}\sqrt{M}\right), \tag{89}$$

which can be inverted with respect to z as

$$F_{2M}^{-1}(z) \approx 2Mv^4(1 - z, M), \tag{90}$$

where

$$v(z, M) = \sqrt{\frac{3}{2\sqrt{M}}Q^{-1}(z) + 9 - \frac{1}{2M}} - 2. \tag{91}$$

Hence, using (90) together with (71), the threshold can be computed by the SU as

$$\eta \approx M\sigma_{\mathrm{w}}^2 v^4(p_{\mathrm{FA}}, M). \tag{92}$$

Note that (92) is much easier to compute than (71). From (73), the approximated ROC is given by

$$p_{\mathrm{D}} \approx Q\left(\frac{2\sqrt{cN}}{3}\left[\frac{v^2(p_{\mathrm{FA}}, cN)}{\sqrt{1+\gamma}} + \frac{4v(p_{\mathrm{FA}}, cN)}{\sqrt[4]{1+\gamma}} - 5 + \frac{1}{2cN}\right]\right). \tag{93}$$

When the SU probabilities of FA p_{FA} and detection p_{D} are fixed, from (93), we obtain the required SNR γ as a function of the sample size N and of the compression ratio c, as expressed by

$$\gamma \geq \left[\frac{v(p_{\mathrm{FA}}, cN)}{v(p_{\mathrm{D}}, cN)}\right]^4 - 1, \tag{94}$$

where $v(p_{\mathrm{FA}}, cN)$ and $v(p_{\mathrm{D}}, cN)$ are calculated using (91).

When p_{FA}, p_{D}, and the SNR γ are fixed, the approximated ROC (93) can also be used to obtain the required number of samples N as a function of the compression ratio c. A closed-form expression for the required sample size N would be complicated; therefore, we resort to an iterative algorithm. Since $N = M/c$, the iterative algorithm looks for the minimum compressed sample size M that satisfies a given p_{FA} and a given p_{D}. The algorithm starts with a tentative value $M = 1$, and calculates two possible thresholds η_{FA} and η_{D}, as expressed by

$$\eta_{\mathrm{FA}} = M\sigma_{\mathrm{w}}^2 v^4(p_{\mathrm{FA}}, M), \tag{95}$$

$$\eta_{\mathrm{D}} = M\sigma_{\mathrm{w}}^2 (1+\gamma) v^4(p_{\mathrm{D}}, M), \tag{96}$$

where the function $v(z, M)$ is defined in (91). Equation (95) represents the minimum value of the ED threshold to warranty a probability of FA p_{FA} using M samples, while (96) is the maximum value of the ED threshold to warranty a probability of detection p_{D} using M samples. If $\eta_{\mathrm{FA}} \leq \eta_{\mathrm{D}}$, then any threshold η in the interval $\eta_{\mathrm{FA}} \leq \eta \leq \eta_{\mathrm{D}}$ satisfies both probability requirements; hence, the tentative M is a sufficient number for the required compressed sample size. Instead, if $\eta_{\mathrm{FA}} > \eta_{\mathrm{D}}$, then it is impossible to find a unique threshold η that satisfies both probability requirements, and hence the tentative M is insufficient. Therefore, when $\eta_{\mathrm{FA}} > \eta_{\mathrm{D}}$, the iterative algorithm doubles the tentative value of M, and performs a new computation of (95) and (96). The two updated thresholds are compared again, and, if $\eta_{\mathrm{FA}} > \eta_{\mathrm{D}}$ again, then the algorithm doubles again the tentative value of M. After a finite number of doublings, the algorithm finds a value of $M = M_{\max}$ such that $\eta_{\mathrm{FA}} \leq \eta_{\mathrm{D}}$, and this value can be used to set an upper bound $N_{\max} = M_{\max}/c$ on the required sample size. At this point, the minimum value of M that satisfies both probability requirements surely lies in the interval $M_{\max}/2 < M \leq M_{\max}$, so the iterative algorithm can use an interval-halving method to refine the search. Consequently, the algorithm sets a new tentative value $M = 3M_{\max}/4$, which is in the midway of the interval $M_{\max}/2 < M \leq M_{\max}$, and updates the two thresholds (95) and (96) again. If $\eta_{\mathrm{FA}} > \eta_{\mathrm{D}}$, then the new search interval reduces to $3M_{\max}/4 < M \leq M_{\max}$, whereas, if $\eta_{\mathrm{FA}} \leq \eta_{\mathrm{D}}$, then the new search interval reduces to $M_{\max}/2 < M \leq 3M_{\max}/4$. Then, the algorithm continues again by selecting a tentative M in the midway of the resulting interval, and by updating the two thresholds. The iterative algorithm ends when the size of the search interval becomes equal to one, say $M_{\mathrm{final}} - 1 < M \leq M_{\mathrm{final}}$, yielding the compressed sample size $M = M_{\mathrm{final}}$ as the final solution. At the end of the iterative algorithm, the required sample size is set to $N = M_{\mathrm{final}}/c$. It is noteworthy that the iterative algorithm has logarithmic complexity with respect to M_{final}: the number of iterations is upper bounded by $2\lceil \log_2(M_{\mathrm{final}})\rceil$, where $\lceil x\rceil$ is the integer ceiling function, because the number of doublings is upper bounded by $\lceil \log_2(M_{\mathrm{final}})\rceil$, while the number of halvings is upper bounded by $\lceil \log_2(M_{\mathrm{final}})\rceil - 1$.

A second PA, originally developed by Canal [63], is obtained by first defining the sixth-root transformation

$$x = \sqrt[6]{\frac{z}{2M}} \tag{97}$$

and then choosing the polynomial

$$g = \frac{x^3}{3} - \frac{x^2}{2} + x. \tag{98}$$

Successively, g in (98) is approximated as a Gaussian random variable with the mean $\frac{5}{6} - \frac{1}{18M}$ and variance $\frac{1}{36M}$. Consequently, the chi-squared distribution (70) is approximated as

$$F_{2M}(z) \approx 1 - Q\left(\left[\frac{1}{3}\sqrt{\frac{z}{2M}} - \frac{1}{2}\sqrt[3]{\frac{z}{2M}} + \sqrt[6]{\frac{z}{2M}} - \left(\frac{5}{6} - \frac{1}{18M}\right)\right]6\sqrt{M}\right), \tag{99}$$

which can be inverted, using Cardano's method for cubic equations, as

$$F_{2M}^{-1}(z) \approx 2Mq^6(1-z, M), \tag{100}$$

where

$$q(z,M) = \frac{1+\sqrt[3]{\sqrt{\bar{q}^2(z,M)+27}+\bar{q}(z,M)}-\sqrt[3]{\sqrt{\bar{q}^2(z,M)+27}-\bar{q}(z,M)}}{2}, \tag{101}$$

$$\bar{q}(z,M) = \frac{2}{\sqrt{M}}Q^{-1}(z) + 5 - \frac{2}{3M}. \tag{102}$$

Consequently, using (71) and (100), in this second case, the ED can obtain the threshold as

$$\eta \approx M\sigma_{\mathrm{w}}^2 q^6(p_{\mathrm{FA}}, M). \tag{103}$$

From (73), the approximated ROC is given by

$$p_{\rm D} \approx Q\left(6\sqrt{cN}\left[\frac{q^3(p_{\rm FA}, cN)}{3\sqrt{1+\gamma}} - \frac{q^2(p_{\rm FA}, cN)}{2\sqrt[3]{1+\gamma}} + \frac{q(p_{\rm FA}, cN)}{\sqrt[6]{1+\gamma}} - \frac{5}{6} + \frac{1}{18cN}\right]\right). \quad (104)$$

When the SU probabilities of FA $p_{\rm FA}$ and detection $p_{\rm D}$ are fixed, (104) gives the required SNR γ as a function of the sample size N and of the compression ratio c, as expressed by

$$\gamma \geq \left[\frac{q(p_{\rm FA}, cN)}{q(p_{\rm D}, cN)}\right]^6 - 1, \quad (105)$$

where $q(p_{\rm FA}, cN)$ and $q(p_{\rm D}, cN)$ are calculated using (101) and (102). When $p_{\rm FA}$, $p_{\rm D}$, and the SNR γ are fixed, the approximated ROC (104) can be used to obtain the required number of samples N as a function of the compression ratio c, using the same iterative algorithm detailed above, provided that the two thresholds $\eta_{\rm FA}$ and $\eta_{\rm D}$ are updated using

$$\eta_{\rm FA} = M\sigma_{\rm w}^2 q^6(p_{\rm FA}, M), \quad (106)$$

$$\eta_{\rm D} = M\sigma_{\rm w}^2(1+\gamma)q^6(p_{\rm D}, M), \quad (107)$$

instead of (95) and (96).

4.3. SR Case, FC Performance

This section considers the SR case, where the SUs do not make any hard decision and transmit soft information to the FC, specifically the energy values collected by the EDs. According to (10), the FC combines the received energy values using EGC, therefore the FC acts like a single ED with MS samples, where S is the number of active SUs and M is the compressed sample size of each ED.

Under the hypothesis H_0, the variable $\tilde{T}$ in (10) is a chi-squared random variable with $2MS = 2acKN$ degrees of freedom, where a is the activity factor, c is the compression ratio, K is the total number of users, and N is the total number of samples of each ED. Therefore, the probability of FA of the FC can be expressed as

$$\pi_{\rm FA} = \Pr\{\tilde{T} \geq \tilde{\tau} | H_0\} = 1 - F_{2acKN}\left(\frac{2\tilde{\tau}}{\sigma_{\rm w}^2}\right), \quad (108)$$

hence, the threshold $\tilde{\tau}$ of the FC can be expressed by

$$\tilde{\tau} = \frac{\sigma_{\rm w}^2}{2} F_{2acKN}^{-1}(1 - \pi_{\rm FA}). \quad (109)$$

Under the hypothesis H_1, the variable $\tilde{T}$ in (10) is again a chi-squared random variable with $2MS = 2acKN$ degrees of freedom; therefore, the probability of detection of the FC can be expressed as

$$\pi_{\rm D} = \Pr\{\tilde{T} \geq \tilde{\tau} | H_1\} = 1 - F_{2acKN}\left(\frac{2\tilde{\tau}}{\sigma_{\rm s}^2 + \sigma_{\rm w}^2}\right), \quad (110)$$

hence, the ROC is expressed by

$$\pi_{\rm D} = 1 - F_{2acKN}\left(\frac{F_{2acKN}^{-1}(1 - \pi_{\rm FA})}{1+\gamma}\right), \quad (111)$$

where $\gamma = \sigma_{\rm s}^2/\sigma_{\rm w}^2$ is the sensing SNR of each ED.

Also, for the performance of the FC with SR, we can use the same approximations used in Section 4.2, i.e., GA, PGA, and PA. Below, we summarize the PGA approach, which includes the GA as a special case when $r = 1$. The analysis could be extended to the PA case with minor modifications, and therefore, for the sake of brevity, we omit the details for the PA case.

The PGA is obtained as

$$F_{2acKN}(z) \approx 1 - Q\left(\left[\sqrt[r]{\frac{z}{2acKN}} - \frac{2acKNr^2 - r + 1}{2acKNr^2}\right] r\sqrt{acKN}\right), \tag{112}$$

while the FC threshold can be computed as

$$\tilde{\tau} \approx acKN\sigma_{\mathrm{w}}^2 \left[\frac{Q^{-1}(\pi_{\mathrm{FA}})}{r\sqrt{acKN}} + \frac{2acKNr^2 - r + 1}{2acKNr^2}\right]^r. \tag{113}$$

Equation (112) permits the expression of the ROC as

$$\pi_{\mathrm{D}} \approx Q\left(\frac{Q^{-1}(\pi_{\mathrm{FA}}) - (\sqrt[r]{1+\gamma} - 1)\left(r\sqrt{acKN} - \frac{r-1}{2r\sqrt{acKN}}\right)}{\sqrt[r]{1+\gamma}}\right). \tag{114}$$

In this SR case, the product between the total number of sensors K and the total number of samples N of each ED is bounded by

$$KN \geq \frac{1}{ac}\left(\tilde{b}_r + \frac{r-1}{2r^2} + \sqrt{\tilde{b}_r^2 + \frac{r-1}{r^2}\tilde{b}_r}\right), \tag{115}$$

where

$$\tilde{b}_r = \frac{1}{2}\left[\frac{Q^{-1}(\pi_{\mathrm{FA}}) - \sqrt[r]{1+\gamma}\,Q^{-1}(\pi_{\mathrm{D}})}{r(\sqrt[r]{1+\gamma} - 1)}\right]^2. \tag{116}$$

Equation (114) also gives the required SNR γ as

$$\gamma \geq \left[\frac{Q^{-1}(\pi_{\mathrm{FA}}) + r\sqrt{acKN} - \frac{r-1}{2r\sqrt{acKN}}}{Q^{-1}(\pi_{\mathrm{D}}) + r\sqrt{acKN} - \frac{r-1}{2r\sqrt{acKN}}}\right]^r - 1. \tag{117}$$

In all Equations (112)–(117), the value of r can be selected as desired.

5. Numerical Results: Validation and Discussion

We use numerical results to validate the performance analysis of Section 4. Specifically, we want to compare the accuracy of the different approximations used in our proposed performance analysis.

We start with the FC performance in the HR case. For the *and-rule* and *or-rule* cases, the performance analysis is exact, so there is no approximation to validate. Hence, we focus on the *MV-rule*, where approximations are present. Figure 1 shows the minimum number of required active sensors $S_{\mathrm{min,FA}}$ as a function of the SU probability of FA p_{FA} to obtain a probability of FA at the FC equal to $\pi_{\mathrm{FA}} = 10^{-7}$. Since the *MV-rule* requires an odd integer number of sensors, $S_{\mathrm{min,FA}}$ has been rounded accordingly, using the ceiling function. Figure 1 reveals that the Carter approximation (65) provides the best estimate of the exact number of required active sensors. Also, the approximations (53) and (59), derived from Fisher and Cochran studies, respectively, produce good results in agreement with the exact results. On the contrary, the GA (30) underestimates the minimum number of required active sensors, while the arcsine approximation (36) overestimates the number of active sensors. Therefore, the approximations based on the incomplete beta distribution (Fisher, Cochran, and Carter) are more accurate than the approximations based on the binomial distribution (GA, arcsine, and Camp–Poulson). We verified that this behavior is a general result that is also valid for other values of π_{FA}, and also valid for $S_{\mathrm{min,D}}$, whose results (not shown in this paper for the sake of brevity) have the same trends as those of $S_{\mathrm{min,FA}}$ shown in this paper.

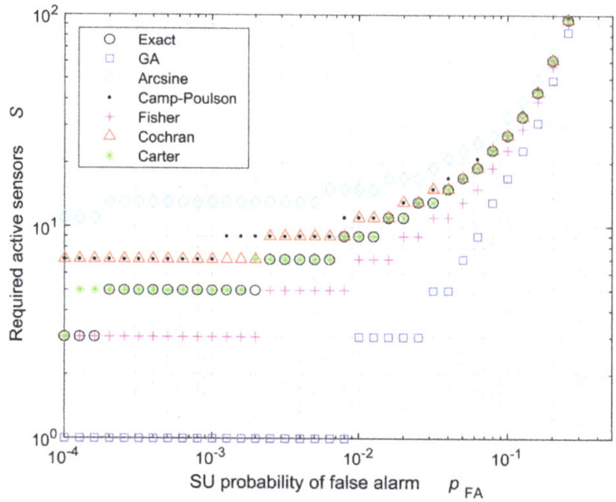

Figure 1. Required number of active sensors to obtain a probability of FA equal to $\pi_{\mathrm{FA}} = 10^{-7}$ at the FC in the HR case. All the results are rounded towards the next odd integer.

To better appreciate the accuracy of the different approximations, Figure 2 displays the same approximated performance of Figure 1, but without applying the rounding operation to the next odd integer, leading to a non-integer number of required active sensors $S_{\mathrm{min,FA}}$. In this case, since the binomial distribution provides integer results, the exact number of required active sensors has been obtained by replacing the binomial distribution with the corresponding incomplete beta distribution, which provides non-integer results. Figure 2 confirms that the Carter approximation provides the best estimate of the required number of active sensors, and that the accuracy of the Carter approximation is greater than those of the other approximations (including Fisher and Cochran ones). Note that the GA provides accurate results only when the exact number of required active sensors is large, as predicted by the CLT. Consequently, the use of a GA performance analysis should be avoided when the number of active sensors is low.

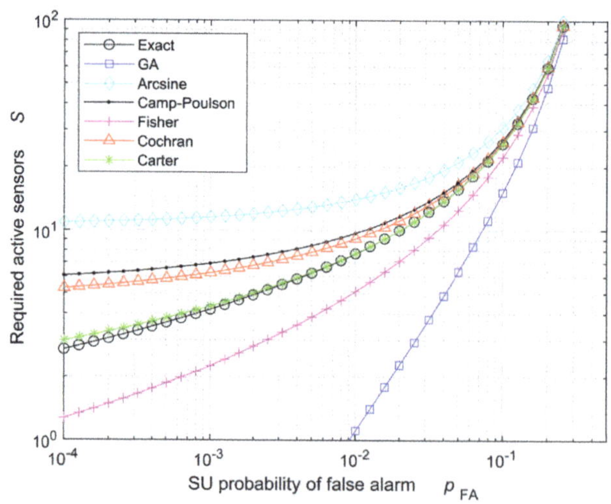

Figure 2. Required number of active sensors to obtain a probability of FA equal to $\pi_{\mathrm{FA}} = 10^{-7}$ at the FC, in the HR case. All the results are non-integer.

Figure 3 and Table 1 exhibit the minimum number of total sensors $K_{\text{min,FA}}$ required as a function of the SU probability of FA p_{FA}, when the activity rate is $a = 1/20$ and the probability of FA at the FC is equal to $\pi_{\text{FA}} = 10^{-7}$. The results for the active number of sensors are rounded towards the next odd integer; therefore, all the results in Figure 3 and Table 1 are multiples of $1/a = 20$. The results in Figure 3 are similar to those in Figure 1, but now the number of required sensors is larger, because most of the sensors are in sleeping mode: when $a = 1/20$, only 5% of the total sensors are active.

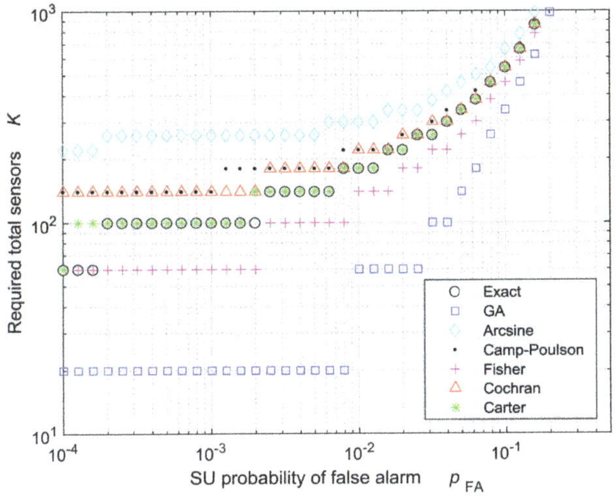

Figure 3. Required number of total sensors to obtain a probability of FA equal to $\pi_{\text{FA}} = 10^{-7}$ at the FC, when the activity rate is $a = 1/20$, in the HR case. All the results are rounded such that the number of active sensors is an odd integer.

Table 1. Required number of total sensors in Figure 3.

	$p_{\text{FA}} = 10^{-4}$	$p_{\text{FA}} = 10^{-3}$	$p_{\text{FA}} = 10^{-2}$	$p_{\text{FA}} = 10^{-1}$
Exact	60	100	180	540
GA	20	20	60	340
Arcsine	220	260	300	660
Camp–Poulson	140	140	220	540
Fisher	60	60	140	460
Cochran	140	140	220	540
Carter	60	100	180	540

We now consider the SU performance in the HR case. Specifically, we want to assess the accuracy of GA, PGA, and PA approaches when the SU ED self-estimates its probability of detection and the number of total samples N. Figure 4 illustrates the ROC approximations obtained with the different approaches, when the total number of samples is $N = 60$, the compression ratio is $c = 1/10$, and the sensing SNR is $\gamma = 8$ dB. Figure 4 points out that both the PA approaches give an accurate estimation of the SU probability of detection p_{D}. The PGA approach yields a good approximation in the cubic case ($r = 3$), while the other powers ($r = 2$ and $r = 4$) provide less accurate results. The GA is the worst approximation, since the number of compressed samples is too low ($M = cN = 6$) for invoking the CLT. Similar results would be obtained with different values of the number of samples and of the SNR.

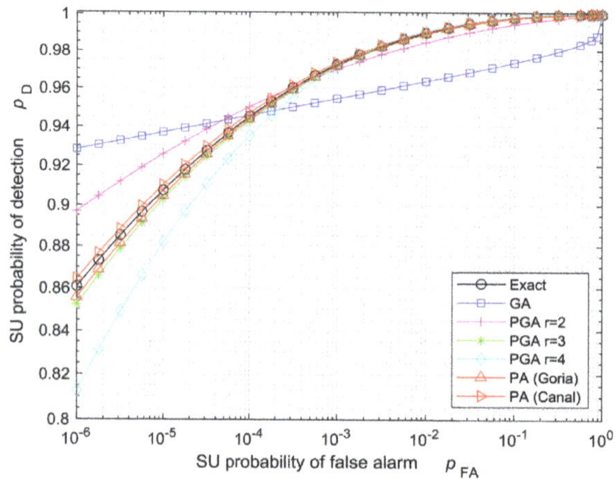

Figure 4. ROC of the SU ED when the number of total samples is $N = 60$, the compression ratio is $c = 1/10$, and the sensing SNR is $\gamma = 8$ dB.

Figure 5 highlights the results of Figure 4 by showing the relative error (RE) ϵ_{rel} on the ROC, defined by

$$\epsilon_{\text{rel}} = \frac{|p_{\text{D}}^{(\text{approx})} - p_{\text{D}}^{(\text{exact})}|}{p_{\text{D}}^{(\text{exact})}}, \tag{118}$$

where $p_{\text{D}}^{(\text{exact})}$ is the exact probability of detection, and $p_{\text{D}}^{(\text{approx})}$ is its approximation. Figure 5 gives an idea about the trend of the PA approaches, whose RE reduces for the increasing probability of FA. The same behavior happens for the PGA with $r = 3$, which has a RE below 10^{-2} like the PA approaches. On the contrary, the GA and the PGA with $r = 2$ and $r = 4$ have a non-monotonic RE, which can exceed 10^{-2} in this scenario. Note that, for some specific values of p_{FA}, the RE shown in Figure 5 can be quite low also for GA and PGA with $r = 2$ and $r = 4$: this behavior only happens for the selected values of p_{FA}, identified by the intersection between the approximated ROC and the exact ROC in Figure 4. Therefore, occasionally, GA and PGA with $r = 2$ and $r = 4$ can produce accurate results, but this accuracy is not generally valid for all the p_{FA} values.

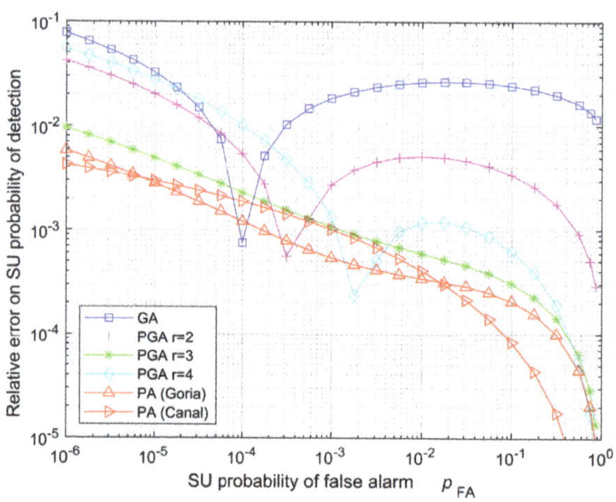

Figure 5. RE on the ROC of the SU ED when the number of total samples is $N = 60$, the compression ratio is $c = 1/10$, and the sensing SNR is $\gamma = 8$ dB.

Figure 6 and Table 2 present the results concerning the approximated number of total samples N self-estimated by the SU ED when the probability of FA is $p_{\mathrm{FA}} = 10^{-4}$, the probability of detection is $p_{\mathrm{D}} = 1 - 10^{-3}$, and the compression ratio is $c = 1/10$. The results for the compressed number of samples are rounded towards the next integer; therefore, all the results in Figure 6 and Table 2 are multiples of $1/c = 10$. In this case, the PA approaches and the PGA approaches with $r = 3$ and $r = 4$ yield accurate estimates of the number of samples N. It is worth noting that, in these PA cases, the required number of samples are obtained using the iterative algorithm described in Section 4.2.3, while the PGA results are derived using (84). In addition, note that, at a large SNR γ, the GA largely overestimates the required number of samples, thereby increasing the sensing time more than necessary. This would consequently reduce the capacity of transmission of the SU due to the unnecessary additional time spent for sensing.

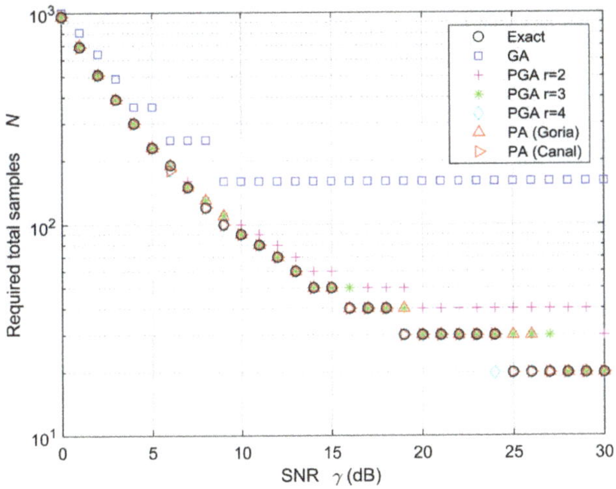

Figure 6. Required number of total samples to obtain a probability of FA equal to $p_{\mathrm{FA}} = 10^{-4}$ and a probability of detection equal to $p_{\mathrm{D}} = 1 - 10^{-3}$ at the SU, when the compression ratio is $c = 1/10$, in the HR case.

Table 2. Required number of total samples in Figure 6.

	SNR = 0 dB	SNR = 10 dB	SNR = 20 dB	SNR = 30 dB
Exact	960	90	30	20
GA	1000	160	160	160
PGA, $r = 2$	960	100	40	30
PGA, $r = 3$	960	90	30	20
PGA, $r = 4$	960	90	30	20
PA (Goria)	960	90	30	20
PA (Canal)	960	90	30	20

We now conclude this section by focusing on the SR case. In this case, only the FC performance is of interest, because the SUs do not make any hard decision. Figure 7 underlines the required product KN as a function of the sensing SNR γ, where K is the total number of sensors, and N is the total number of samples of each SU. An activity rate $a = 1/4$ and a compression ratio $c = 1/3$ are assumed. In this case, the exact result is rounded using the ceiling function, such that the number S of active sensors times the number of compressed samples M is integer, while the approximated values are non-integer. Figure 7 emphasizes that both PGA approaches with $r = 3$ and $r = 4$ provide accurate results, with minor differences due to the rounding operation. To simplify the readability of Figure 7, the PA results are not shown: in this SR case, an iterative algorithm would produce PA results very close to the PGA results with $r = 3$, similarly to the HR case of Figure 6. Again, the GA approach largely overestimates the product KN, leading to an incorrect estimation with either too many required sensors, or too many required samples per sensor, or both. For instance, when $\gamma = 15$ dB, the product KN estimated by the GA is more than doubled with respect to the exact value, as shown in Figure 7.

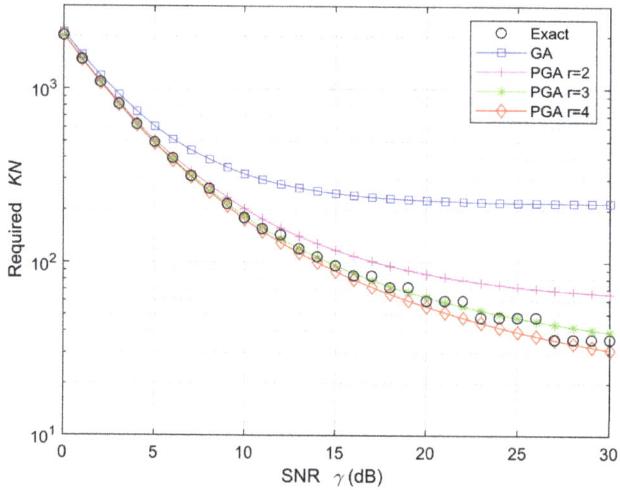

Figure 7. Required product between the number of total sensors K and the number of total samples N to obtain a probability of FA equal to $\pi_{\text{FA}} = 10^{-6}$ and a probability of detection equal to $\pi_{\text{D}} = 1 - 10^{-5}$ at the FC, when the activity rate is $a = 1/4$ and the compression ratio is $c = 1/3$, in the SR case. The exact result is rounded such that the number of active sensors times the number of compressed samples is integer, while the approximated values are non-integer.

6. Conclusions

This paper has proposed useful analytical approximations that characterize the performance of centralized CSS. Specifically, this paper has derived accurate low-complexity closed-form expressions that calculate the required number of sensors and the required number of samples as a function of the probability of detection and the probability of FA of the FC and of the sensors. In contrast with most of the existing literature, this paper does not leverage on the CLT, therefore the proposed analysis is accurate even when the number of active sensors is low and when the sample size is reduced. Therefore, each sensor and the FC can self-compute its own performance with reduced complexity. Future work may extend the proposed approach to sensors equipped with multiple antennas that sense multiple frequency bands.

Author Contributions: Conceptualization, L.R. and P.B.; methodology, L.R.; software, L.R.; validation, L.R.; formal analysis, L.R.; investigation, L.R.; resources, L.R. and P.B.; data curation, L.R.; writing—original draft preparation, L.R.; writing—review and editing, L.R. and P.B.; visualization, L.R.; supervision, P.B.; project administration, L.R. and P.B.; funding acquisition, L.R. and P.B. All authors have read and agreed to the published version of the manuscript.

Funding: This work was supported by University of Perugia with the Fondo di Ricerca di Base 2022 project "Energy-Efficient Networking, Signal Processing and Communications (EFESO)" and by Ministero dell'Istruzione, dell'Università e della Ricerca (MIUR) with the PRIN 2017 project "Liquid Edge Computing Based on Distributed Machine Learning and Millimeter-Wave Radio Access (LIQUID EDGE)".

Institutional Review Board Statement: Not applicable.

Informed Consent Statement: Not applicable.

Data Availability Statement: The data that support the findings of this study are available from the corresponding author, L.R., upon reasonable request.

Acknowledgments: The first author thanks G. Leus and E. Lagunas for their collaborations on earlier conference papers on subjects similar to that of the present paper.

Conflicts of Interest: The authors declare no conflicts of interest.

Abbreviations

The following abbreviations are used in this manuscript:

CLT	Central limit theorem
CR	Cognitive radio
CSS	Cooperative spectrum sensing
ED	Energy detector
EGC	Equal-gain combining
FA	False alarm
FC	Fusion center
GA	Gaussian approximation
HR	Hard reporting
MD	Missed detection
MV	Majority voting
PA	Polynomial approximation
pdf	Probability density function
PGA	Power-of-Gaussian approximation
PU	Primary user
RE	Relative error
ROC	Receiver operating characteristic
SNR	Signal-to-noise ratio
SR	Soft reporting
SU	Secondary user

References

1. Akyildiz, I.F.; Lo, B.F.; Balakrishnan, R. Cooperative Spectrum Sensing in Cognitive Radio Networks: A Survey. *Phys. Commun.* **2011**, *4*, 40–62. [CrossRef]
2. Joshi, G.; Nam, S.; Kim, S. Cognitive Radio Wireless Sensor Networks: Applications, Challenges and Research Trends. *Sensors* **2013**, *13*, 11196–11228. [CrossRef] [PubMed]
3. Cichoń, K.; Kliks, A.; Bogucka, H. Energy-Efficient Cooperative Spectrum Sensing: A Survey. *IEEE Commun. Surv. Tuts.* **2016**, *18*, 1861–1886. [CrossRef]
4. Davenport, M.A.; Boufounos, P.T.; Wakin, M.B.; Baraniuk, R.G. Signal Processing with Compressive Measurements. *IEEE J. Sel. Topics Signal Process.* **2010**, *4*, 445–460. [CrossRef]
5. Song, Z.; Yang, J.; Zhang, H.; Gao, Y. Approaching Sub-Nyquist Boundary: Optimized Compressed Spectrum Sensing Based on Multicoset Sampler for Multiband Signal. *IEEE Trans. Signal Process.* **2022**, 70, 4225–4238. [CrossRef]
6. Yucek, T.; Arslan, H. A Survey of Spectrum Sensing Algorithms for Cognitive Radio Applications. *IEEE Commun. Surv. Tuts.* **2009**, *11*, 116–130. [CrossRef]
7. Arjoune, Y.; Kaabouch, N. A Comprehensive Survey on Spectrum Sensing in Cognitive Radio Networks: Recent Advances, New Challenges, and Future Research Directions. *Sensors* **2019**, *19*, 126. [CrossRef]
8. Nasser, A.; Al Haj Hassan, H.; Abou Chaaya, J.; Mansour, A.; Yao, K. Spectrum Sensing for Cognitive Radio: Recent Advances and Future Challenge. *Sensors* **2021**, *21*, 2408. [CrossRef]
9. Lorincz, J.; Ramljak, I.; Begušić, D. A Survey on the Energy Detection of OFDM Signals with Dynamic Threshold Adaptation: Open Issues and Future Challenges. *Sensors* **2021**, *21*, 3080. [CrossRef]
10. Men, S.; Chargé, P.; Fu, Z. Dynamic Robust Spectrum Sensing Based on Goodness-of-Fit Test Using Bilateral Hypotheses. *Drones* **2023**, *7*, 18. [CrossRef]
11. Guimarães, D. Modified Gini Index Detector for Cooperative Spectrum Sensing over Line-of-Sight Channels. *Sensors* **2023**, *23*, 5403. [CrossRef] [PubMed]

12. Fernando, X.; Lăzăroiu, G. Spectrum Sensing, Clustering Algorithms, and Energy-Harvesting Technology for Cognitive-Radio-Based Internet-of-Things Networks. *Sensors* **2023**, *23*, 7792. [CrossRef] [PubMed]
13. Axell, E.; Leus, G.; Larsson, E.G.; Poor, H.V. Spectrum Sensing for Cognitive Radio: State-of-the-art and Recent Advances. *IEEE Signal Process. Mag.* **2012**, *29*, 101–116. [CrossRef]
14. Zhang, W.; Mallik, R.K.; Letaief, K.B. Optimization of Cooperative Spectrum Sensing with Energy Detection in Cognitive Radio Networks. *IEEE Trans. Wireless Commun.* **2009**, *8*, 5761–5766. [CrossRef]
15. Umar, R.; Sheikh, A.U.; Deriche, M. Unveiling the Hidden Assumptions of Energy Detector Based Spectrum Sensing for Cognitive Radios. *IEEE Commun. Surv. Tuts.* **2013**, *16*, 713–728. [CrossRef]
16. Atapattu, S.; Tellambura, C.; Jiang, H. Energy Detection Based Cooperative Spectrum Sensing in Cognitive Radio Networks. *IEEE Trans. Wireless Commun.* **2011**, *10*, 1232–1241. [CrossRef]
17. Banjade, V.R.S.; Tellambura, C.; Jiang, H. Approximations for Performance of Energy Detector and p-norm Detector. *IEEE Commun. Lett.* **2015**, *19*, 1678–1681. [CrossRef]
18. Sobron, I.; Diniz, P.; Martins, W.; Velez, M. Energy Detection Technique for Adaptive Spectrum Sensing. *IEEE Trans. Commun.* **2015**, *63*, 617–627. [CrossRef]
19. Chandrasekaran, G.; Kalyani, S. Performance Analysis of Cooperative Spectrum Sensing Over κ-μ Shadowed Fading. *IEEE Wireless Commun. Lett.* **2015**, *4*, 553–556. [CrossRef]
20. Lesnikov, V.; Naumovich, T.; Chastikov, A.; Dubovcev, D. Approximation of the Central Chi-Squared Distribution for On-Line Computation of the Threshold for Energy Detector. In Proceedings of the 2016 IEEE East–West Design and Test Symposium (EWDTS), Yerevan, Armenia, 14–17 October 2016. [CrossRef]
21. Lesnikov, V.; Naumovich, T.; Chastikov, A. Computation of the Energy Detector Threshold for Various Approximations of Noise Power Distribution. In Proceedings of the 2017 IEEE East–West Design and Test Symposium (EWDTS), Novi Sad, Serbia, 29 September–2 October 2017. [CrossRef]
22. Sobron, I.; Eizmendi, I.; Martins, W.; Diniz, P.; Ordiales, J.; Velez, M. Implementation Issues of Adaptive Energy Detection in Heterogeneous Wireless Networks. *Sensors* **2017**, *17*, 932. [CrossRef]
23. Li, J.; Li, B.; Liu, M. Performance Analysis of Cooperative Spectrum Sensing over Large and Small Scale Fading Channels. *AEÜ Intl. J. Electron. Commun.* **2017**, *78*, 90–97. [CrossRef]
24. Bhatt, M.; Soni, S.K. A Unified Performance Analysis of Energy Detector over α-η-μ/Lognormal and α-κ-μ/Lognormal Composite Fading Channels with Diversity and Cooperative Spectrum Sensing. *AEÜ Intl. J. Electron. Commun.* **2018**, *94*, 367–376. [CrossRef]
25. Cao, K.; Qian, P.; An, J.; Wang, L. Accurate and Practical Energy Detection over α-μ Fading Channels. *Sensors* **2020**, *20*, 754. [CrossRef] [PubMed]
26. Lopez-Benitez, M.; Toma, O.H.; Patel, D.K.; Umebayashi, K. Sample Size Analysis of Energy Detection under Fading Channels. In Proceedings of the 2020 IEEE Wireless Communications and Networking Conference Workshops (WCNCW), Seoul, Republic of Korea, 25–28 May 2020. [CrossRef]
27. Lorincz, J.; Ramljak, I.; Begušić, D. Algorithm for Evaluating Energy Detection Spectrum Sensing Performance of Cognitive Radio MIMO-OFDM Systems. *Sensors* **2021**, *21*, 6881. [CrossRef] [PubMed]
28. Maleki, S.; Leus, G. Censored Truncated Sequential Spectrum Sensing for Cognitive Radio Networks. *IEEE J. Sel. Areas Commun.* **2013**, *31*, 364–378. [CrossRef]
29. Maleki, S.; Chepuri, S.P.; Leus, G. Optimization of Hard Fusion Based Spectrum Sensing for Energy-Constrained Cognitive Radio Networks. *Phys. Commun.* **2013**, *9*, 193–198. [CrossRef]
30. Shakir, M.Z.; Rao, A.; Alouini, M.-S. Generalized Mean Detector for Collaborative Spectrum Sensing. *IEEE Trans. Commun.* **2013**, *61*, 1242–1253. [CrossRef]
31. Hamza, D.; Aissa, S.; Aniba, G. Equal Gain Combining for Cooperative Spectrum Sensing in Cognitive Radio Networks. *IEEE Trans. Wireless Commun.* **2014**, *13*, 4334–4345. [CrossRef]
32. Gupta, K.; Merchant, S.N.; Desai, U.B. A Novel Multistage Decision Fusion for Cognitive Sensor Networks Using AND and OR Rules. *Digital Signal Process.* **2015**, *42*, 27–34. [CrossRef]
33. Alaa, A.M.; Nasr, O.A. Globally Optimal Cooperation in Dense Cognitive Radio Networks. *Wireless Pers. Commun.* **2015**, *84*, 885–899. [CrossRef]
34. Duan, M.; Zeng, Z.; Guo, C.; Liu, F. User Selection for Cooperative Spectrum Sensing in Mobile Cognitive Radios. In Proceedings of the 2015 IEEE/CIC International Conference on Communications in China (ICCC), Shenzhen, China, 2–4 November 2015. [CrossRef]
35. Do, T.-N.; An, B. A Soft-Hard Combination-Based Cooperative Spectrum Sensing Scheme for Cognitive Radio Networks. *Sensors* **2015**, *15*, 4388–4407. [CrossRef]
36. Pradhan, P.M.; Panda, G. Information Combining Schemes for Cooperative Spectrum Sensing: A Survey and Comparative Performance Analysis. *Wireless Pers. Commun.* **2017**, *94*, 685–711. [CrossRef]
37. Men, S.; Chargé, P.; Pillement, S. Cooperative Spectrum Sensing with Small Sample Size in Cognitive Wireless Sensor Networks. *Wireless Pers. Commun.* **2017**, *96*, 1871–1885. [CrossRef]

38. Aquino, G.; Guimarães, D.; Mendes, L.; Pimenta, T. Combined Pre-Distortion and Censoring for Bandwidth-Efficient and Energy-Efficient Fusion of Spectrum Sensing Information. *Sensors* **2017**, *17*, 654. [CrossRef] [PubMed]
39. Fu, Y.; Yang, F.; He, Z. A Quantization-Based Multibit Data Fusion Scheme for Cooperative Spectrum Sensing in Cognitive Radio Networks. *Sensors* **2018**, *18*, 473. [CrossRef]
40. Qian, X.; Hao, L.; Ni, D.; Tran, Q. Hard Fusion Based Spectrum Sensing over Mobile Fading Channels in Cognitive Vehicular Networks. *Sensors* **2018**, *18*, 475. [CrossRef]
41. Tong, J.; Jin, M.; Guo, Q.; Li, Y. Cooperative Spectrum Sensing: A Blind and Soft Fusion Detector. *IEEE Trans. Wireless Commun.* **2018**, *17*, 2726–2737. [CrossRef]
42. Costa, L.; Guimarães, D.; De Souza, R.; Bomfin, R. Cooperative Spectrum Sensing with Coded and Uncoded Decision Fusion under Correlated Shadowed Fading Report Channels. *Sensors* **2019**, *19*, 51. [CrossRef]
43. Liu, S.; Wang, K.; Liu, K.; Chen, W. Noncoherent Decision Fusion over Fading Hybrid MACs in Wireless Sensor Networks. *Sensors* **2019**, *19*, 120. [CrossRef]
44. Mi, Y.; Lu, G.; Li, Y.; Bao, Z. A Novel Semi-Soft Decision Scheme for Cooperative Spectrum Sensing in Cognitive Radio Networks. *Sensors* **2019**, *19*, 2522. [CrossRef]
45. Luo, X.; Zhao, W.; Li, H.; Jin, M.; Cui, G. Fusion Test Statistics Based Mixture Detector for Spectrum Sensing. *IEEE Trans. Veh. Technol.* **2022**, *71*, 3315–3319. [CrossRef]
46. Wang, Y.; Ma, S.; Chen, Q.; Zhuang, J.; Jiang, D. A Geodesic Projection-Based Data Fusion Scheme for Cooperative Spectrum Sensing. *Digital Signal Process.* **2023**, *137*, 104006. [CrossRef]
47. Zhuang, J.; Wang, Y.; Peng, S.; Zhang, S.; Liu, Y. Siegel Distance-Based Fusion Strategy and Differential Evolution Algorithm for Cooperative Spectrum Sensing. *Digital Signal Process.* **2023**, *142*, 104215. [CrossRef]
48. Liu, X.; Jia, M.; Gu, X.; Tan, X. Optimal Periodic Cooperative Spectrum Sensing Based on Weight Fusion in Cognitive Radio Networks. *Sensors* **2013**, *13*, 5251–5272. [CrossRef] [PubMed]
49. Do, T.-N.; An, B. Cooperative Spectrum Sensing Schemes with the Interference Constraint in Cognitive Radio Networks. *Sensors* **2014**, *14*, 8037–8056. [CrossRef]
50. Guimarães, D.; Aquino, G. Resource-Efficient Fusion over Fading and Non-Fading Reporting Channels for Cooperative Spectrum Sensing. *Sensors* **2015**, *15*, 1861–1884. [CrossRef] [PubMed]
51. Guimarães, D.; Aquino, G.; Cattaneo, M. Resource-Efficient Fusion with Pre-Compensated Transmissions for Cooperative Spectrum Sensing. *Sensors* **2015**, *15*, 10891–10908. [CrossRef] [PubMed]
52. Banavathu, N.R.; Khan, M.Z.A. On Throughput Maximization of Cooperative Spectrum Sensing Using the M-out-of-K Rule. In Proceedings of the IEEE 89th Vehicular Technology Conference (VTC2019-Spring), Kuala Lumpur, Malaysia, 28 April–1 May 2019. [CrossRef]
53. Abramowitz, M.; Stegun, I.A. *Handbook of Mathematical Functions with Formulas, Graphs, and Mathematical Tables*; Dover: New York, NY, USA, 1972.
54. Raff, M.S. On Approximating the Point Binomial. *J. Am. Stat. Assoc.* **1956**, *51*, 293–303. [CrossRef]
55. Gebhardt, F. Some Numerical Comparisons of Several Approximations to the Binomial Distribution. *J. Am. Stat. Assoc.* **1969**, *64*, 1638–1646. [CrossRef]
56. Cochran, W.G. Note on an Approximate Formula for the Significance Levels of z. *Ann. Math. Stat.* **1940**, *11*, 93–95. [CrossRef]
57. Thompson, C.M.; Pearson, E.S.; Comrie, L.J.; Hartley, H.O. Tables of Percentage Points of the Incomplete Beta-Function. *Biometrika* **1941**, *32*, 151–181. [CrossRef]
58. Carter, A.H. Approximation to Percentage Points of the z-Distribution. *Biometrika* **1947**, *34*, 352–358. [CrossRef] [PubMed]
59. Aroian, L.A. On the Levels of Significance of the Incomplete Beta Function and the F-Distributions. *Biometrika* **1950**, *37*, 219–223. [CrossRef] [PubMed]
60. Wilson, E.B.; Hilferty, M.M. The Distribution of Chi-Square. *Proc. Natl. Acad. Sci. USA* **1931**, *17*, 684–688. [CrossRef] [PubMed]
61. Hawkins, D.M.; Wixley, R.A.J. A Note on the Transformation of Chi-Squared Variables to Normality. *Amer. Stat.* **1986**, *40*, 296–298. [CrossRef]
62. Goria, M.N. On the Fourth Root Transformation of Chi-Square. *Austral. J. Stat.* **1992**, *34*, 55–64. [CrossRef]
63. Canal, L. A Normal Approximation for the Chi-Square Distribution. *Computat. Stat. Data Analysis* **2005**, *48*, 803–808. [CrossRef]
64. Rugini, L.; Banelli, P.; Leus, G. Small Sample Size Performance of the Energy Detector. *IEEE Commun. Lett.* **2013**, *17*, 1814–1817. [CrossRef]
65. Rugini, L.; Banelli, P.; Leus, G. Spectrum Sensing Using Energy Detectors with Performance Computation Capabilities. In Proceedings of the 24th European Signal Processing Conference (EUSIPCO), Budapest, Hungary, 28 August–2 September 2016; pp. 1608–1612. [CrossRef]
66. Lagunas, E.; Rugini, L. Performance of Compressive Sensing Based Energy Detection. In Proceedings of the IEEE 28th Annual International Symposium on Personal, Indoor, and Mobile Radio Communications (PIMRC), Montreal, QC, Canada, 8–13 October 2017. [CrossRef]
67. Maleki S.; Chepuri S.P.; Leus, G. Energy and Throughput Efficient Strategies for Cooperative Spectrum Sensing in Cognitive Radios. In Proceedings of the IEEE 12th International Workshop on Signal Processing Advances in Wireless Communications (SPAWC), San Francisco, CA, USA, 26–29 June 2011; pp. 71–75. [CrossRef]

68. Chatterjee, S.; Maity, S.P.; Acharya, T. Energy-Spectrum Efficiency Trade-Off in Energy Harvesting Cooperative Cognitive Radio Networks. *IEEE Trans. Cognitive Commun. Netw.* **2019**, *5*, 295–303. [CrossRef]
69. Chatterjee, S.; De, S. QoE-Aware Cross-Layer Adaptation for D2D Video Communication in Cooperative Cognitive Radio Networks. *IEEE Syst. J.* **2022**, *16*, 2078–2089. [CrossRef]

sensors

MDPI

Article

Respecting Partial Privacy of Unstructured Data via Spectrum-Based Encoder

Qingcai Luo [1,*] and Hui Li [2]

[1] School of Cyber Engineering, Xidian University, Xi'an 710126, China
[2] School of Computer Science and Technology, Xidian University, Xi'an 710071, China; hli@xidian.edu.cn
* Correspondence: luoqc@inspur.com

Abstract: Since the popularity of Machine Learning as a Service (MLaaS) has been increasing significantly, users are facing the risk of exposing sensitive information that is not task-related. The reason is that the data uploaded by users may include some information that is not useful for inference but can lead to privacy leakage. One straightforward approach to mitigate this issue is to filter out task-independent information to protect user privacy. However, this method is feasible for structured data with naturally independent entries, but it is challenging for unstructured data. Therefore, we propose a novel framework, which employs a spectrum-based encoder to transform unstructured data into the latent space and a task-specific model to identify the essential information for the target task. Our system has been comprehensively evaluated on three benchmark visual datasets and compared to previous works. The results demonstrate that our framework offers superior protection for task-independent information and maintains the usefulness of task-related information.

Keywords: spectrum-based encoder; latent code; machine learning; privacy preserving

1. Introduction

Machine learning has demonstrated impressive performance in several areas, such as natural language processing [1] and computer vision [2]. However, training an effective machine learning model requires proper model design, massive computing resources, and large datasets that may be beyond the reach of many individuals. In addition, deploying and running the model requires significant storage and computing resources that are also unfriendly to edge devices such as smartphones or sensors [3]. One promising approach is Machine Learning as a Service (MLaaS) [4], which supports the outsourcing of prediction. Well-trained models can be deployed by vendors in the cloud. This is attractive because it offloads the user's local computing and storage requirements and eliminates the cost of training new models. However, the outsourced data consist of not only task-related information, but also task-independent information [5], which does not significantly affect the inference results, but exposes users to unwanted risks of misuse or theft. Recently, China's Personal Information Protection Law has prompted information processors to prevent unauthorized access to personal information. Therefore, it is of paramount importance to protect unauthorized information while ensuring the usefulness of the data.

Previous works addressing privacy concerns have been devoted to balancing the trade-offs between privacy and utility. An obvious and widely adopted solution is to extract task-oriented features and upload them to servers instead of raw data, such as Google Now [6] and Google Cloud [7]. Although the mere transmission of features avoids direct disclosure of raw data, recent developments in model inversion attacks show that adversaries can use intermediate features to reconstruct the input and infer privacy attributes [8–10]. Ossia et al. [11] apply dimensionality reduction and noise injection to defend against adversaries before uploading features to the servers, but the cost is a non-negligible loss in utility. Inspired by Generative Adversarial Networks (GANs) [12], PAN [13], DeepObfuscator [3], and TIPRDC [14] propose to obtain an encoder through adversarial training to extract partial privacy-preserving features that keep a subset of the attributes available while specifying the attributes anonymously. However, these schemes artificially simulate proxy adversaries during the training phase, leading to dangers from potential attack models. This suspicion is also supported by the results of the potential adversary detection experiments in Section 4.2.

Therefore, we propose a partial privacy-preserving framework to preserve data utility while protecting task-independent attributes. An intuitive phenomenon is that not all data information is useful for inference. Some of the recent literature shows that the task model pays more attention to a part of regions [15–18], which becomes evidence that the data can be regarded as composed of task-related and task-independent information. Inspired by these works, our framework focuses on selecting the information relevant to the target task. This is feasible for structured data, but difficult for unstructured data. Taking Figure 1 as an example, users can flexibly select the attributes necessary for the task in Table (a) due to the naturally independent entries, while it is impractical for image (b) because different attributes are entangled and expressed in the same region. An intuitive approach is to express unstructured data in a structured form. However, naturally occurring data are often accompanied by redundant information, which hinders structured expression. Therefore, we introduce Fourier transform as a pre-processing method to reduce data redundancy, and propose the spectrum-based encoder to disentangle the unstructured data into a latent space [19]. We then propose a universally interpretable model, called an indicator, which marks the information necessary for the target task in the latent representation.

Attribute	Value
Hair Color	Brown
Age	31
Gender	Male
Race	White
Mood	Smile

(**a**)

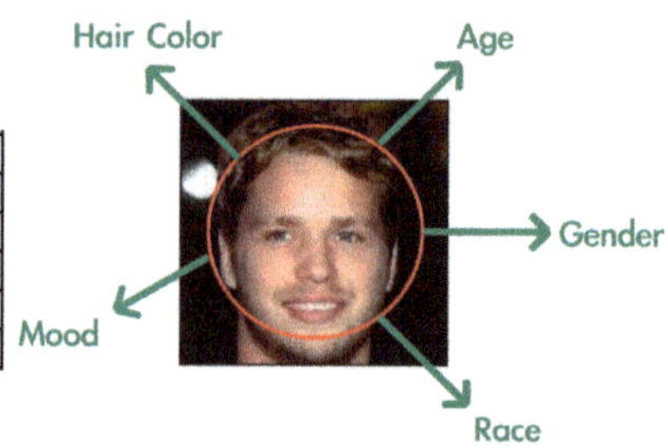

(**b**)

Figure 1. (**a**) Structured data and (**b**) unstructured data.

As shown in Figure 2, our framework consists of three parts: a spectrum encoder E, an indicator I, and a decoder D. The encoder E is intended to be used on the user side to extract the disentangled representation from unstructured data. Indicator I is also used on the user side, recommending task-related information by marking representation dimensions. The marked dimensions indicate the information required by the target task model, and the corresponding anonymized transform is designed. Specifically, the values of the marked dimensions are retained, while the values of the ignored dimensions are discarded and reassigned as default values. The decoder D runs on the server to reconstruct the data based on the transformed representation uploaded by the users. The classifier (green) is considered the target task model, and the reconstruction data will strive to maintain its usefulness. At the same time, the reconstruction data are expected to prevent adversaries (red) from inferring unauthorized attributes.

Discarding task-independent information according to Indicator's recommendations has four advantages. First, interpretable indicators provide interpretability for anonymized transformation. Second, target-task-driven attribute retention avoids unconscious utility loss and sensitive information leakage. Third, disentangled representation-based information selection provides an explicit and controllable balance for privacy–utility trade-offs. Finally, this allows our framework to withstand potential attack models. Furthermore, Indicator and encoder–decoder pairs of our framework are trained separately in two phases.

Compared to existing end-to-end adversarial training methods, our framework can adapt to the changes in the target tasks and adjust the retained attribute information more flexibly.

In summary, our key contributions are as follows:

- We introduce a novel interpretable model called Indicator, which can effectively indicate the critical information required for a specific target task within unstructured data.
- We present a partial privacy-preserving framework that utilizes the designed Indicator to restrict the access of undesired task-independent attacks while preserving the utility of target tasks.
- We fully implement our framework and demonstrate its wide applicability by performing experiments on several standard datasets. The evaluation results show that our framework can achieve sweet trade-offs between privacy and utility, and is resistant to potential attackers.

The rest of this paper is organized as follows. Section 2 introduces the preliminaries and reviews the related work. Section 3 describes the framework overview and the details of core modules. Section 4 reports the evaluation results. Section 5 concludes and discusses this paper.

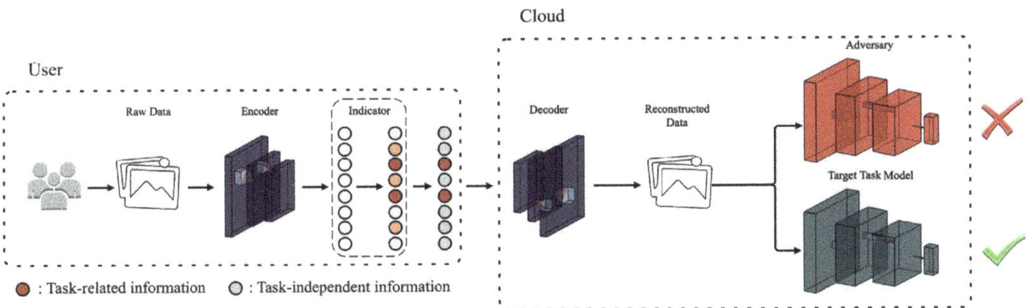

Figure 2. The VAE-based encoder maps the raw data to the latent space, and the proposed indicator points out the relevance of the latent code to the target task and removes irrelevant elements. The subsequent decoder reconstructs the data from the filtered code, with the target attributes being preserved while the remaining attributes are obfuscated.

2. Preliminaries and Related Work

In this section, we first introduce the work involved in this article. Then, we briefly review the most relevant work on privacy.

2.1. Disentangled Representation Learning

In general, disentangled representation learning aims to isolate different attributes into non-overlapping sub-dimensions in the latent space. As shown in Figure 3, different colours represent different attributes in the raw data, and the ball represents the factor containing attribute information. In the raw data, these factors are messy and entangled, and it is difficult to filter all the factors corresponding to a certain attribute in a common way. At the same time, the latent code obtained by the disentangled representation learning can express attributes regularly and independently. In other words, different attributes in the raw data can be determined by the different representation sub-dimensions in the latent space.

Existing works about disentangled representation can be roughly divided into three categories: (1) based on Variational Autoencoders (VAE) [19–21], (2) based on GAN [22] and (3) based on the flow model [23]. Among them, the VAE-based model is attractive due to its lower cost and stability in the training phase.

VAE is an unsupervised generative network based on variational bayes inference, consisting of an encoder and a decoder. Given a sample x, VAE determines a distribution

z in the latent space as the encoding result. The optimization objective of VAE consists of two parts. The first part is to maximize the Evidence Lower Bound (ELBO) so that the variational distribution is close to the isotropic Gaussian prior $p(z)$, and the second part is to minimize the pixel-level metrics of the generated data and the original data:

$$L_{VAE} = -\mathbb{E}_{q(z|x)}[log(p(x|z)] + D_{KL}(q(z|x)||p(z)) \quad (1)$$

β-VAE [19] modified the objective function as:

$$L_{\beta VAE} = -\mathbb{E}_{q(z|x)}[log(p(x|z)] + \beta D_{KL}(q(z|x)||p(z)) \quad (2)$$

Compared to the original VAE, the hyperparameter $\beta > 1$ encourages the variational distribution to be closer to the Gaussian prior, thereby producing a disentangled latent code. Kim et al. [20] and Chen et al. [21] believed that the total correlation term obtained by decomposing the KL divergence plays a crucial role and proposed Factor-VAE and β-TCVAE, respectively.

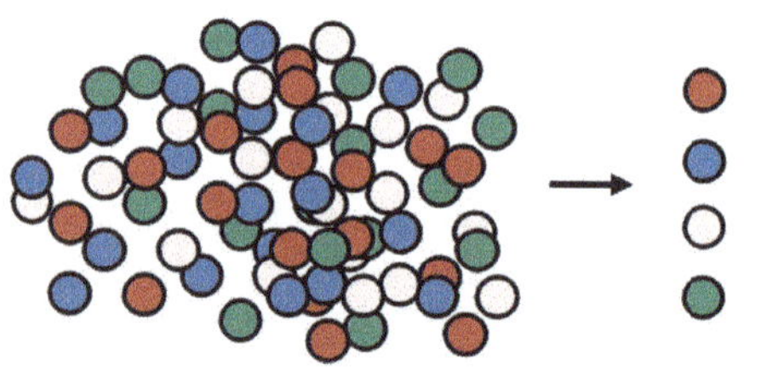

Figure 3. Different colours represent different attributes in the unstructured data, and the balls represent the factors that affect the attributes.

2.2. Data Privacy Protection

Several methods have been proposed to protect privacy. k-anonymity [24], l-diversity [25], and t-closeness [26] have been proposed as desensitization criteria. However, these methods are only designed for structured data and are difficult to scale to unstructured data. Differential privacy [27–29] and random noise injection [30,31] are common methods that are widely used to protect sensitive information in structured and unstructured data. Although security guarantees are provided, these methods often significantly reduce the usefulness of the data. Homomorphic encryption (HE) [32,33] and Secure Multi-Party Computation (MPC) [34–36] support the manipulation of encrypted data, but the computation of non-linear functions is always accompanied by unrealistic computational and communication complexity, leading to much lower efficiency than plaintext inference. iPrivacy [37] focuses on visual tasks by constructing a multi-task learning model to detect and blur objects that may leak sensitive information in the image. The types of these objects are preset. RAE [38] follows the same idea but is time-series-oriented. This scheme proposes to replace the features of each section corresponding to sensitive inferences with the values corresponding to non-sensitive inferences. Using GAN, RAE provides the security guarantee that it is almost impossible to detect the nature of sensitive inferences.

2.3. Representation Privacy–Utility Trade-Offs

Aloufi et al. [39] focused on the disentanglement of voice for the Voice User Interfaces (VUIs). VQ-VAE [40] was introduced to construct independent representations of emotion, identity, and semantics, while WaveRNN [41] was employed to reconstruct voice information. Gong et al. [42] are concerned about attributes preserving face de-identification and propose R^2VAEs to obfuscate identity-related information so as to achieve a balance between facial privacy protection and data utilities. Wu et al. [43] jointly proposed a securely recoverable visual information transformation and steganography PECAM based on deep learning. They used this technology to design a more general VSA privacy enhancement architecture and system implementation. PECAM can effectively transform the original data to other domains to hide sensitive information. At the same time, authorized users can inversely transform and restore the original data to complete detailed investigations. This secure reversible transformation relies on a security-enhanced generative adversarial network. Also, it introduces a key mechanism to ensure that attackers cannot restore the data protected by PECAM. The adversary and the defender are given the conflicting utility–privacy optimization goal, and the game between them is simulated. AttriGuard [44] proposed a two-phase practical framework to resist private attribute inference attacks. In phase I, existing evasion attacks in adversarial learning are adopted to find the minimum noise for each attribute value. In phase II, the attribute values are sampled with a certain probability distribution, and the minimum noise found in phase I is added to the dataset. Therefore, finding the probability distribution is formulated as a constrained convex optimization problem. Liu et al. proposed PAN to protect the privacy of a specific attribute while maintaining the data utility for a certain task. The representation obtained by PAN will remain anonymous, and the adversaries cannot launch reconstruction attacks or privacy attributes inference attacks. Wu et al. [45] designed an adversarial training framework to obtain the degradation transform of video inputs to resist privacy attribute attacks. Considering the diversity of attack models, and that it is impossible to enumerate all adversary models to enhance the features privacy, Budget Model Restarting and Budget Model Ensemble are enabled to enrich potential adversaries. TIPRDC is a task-independent privacy-respecting data crowdsourcing framework but following the same idea. Unlike the above works, the data utility maintained by TIPRDC does not limit to specific tasks but is effective for arbitrary tasks by maximizing mutual information. In a sense, our work is diametrically opposed to the idea of TIPRDC: TIPRDC struggles to **retain** all information in the data, except for privacy attributes, while our framework is expected to **remove** all information, except for the target task required.

3. Design of Framework

In this section, we introduce the VAE-based disentanglement method and propose the model called Indicator for filtering the factors related to the target task.

3.1. Overview

Because models do not need all the information in the uploaded data to make credible inferences, users tend to share only task-relevant details in a controlled manner. This is practical for structured data with naturally independent attribute records but is difficult for unstructured data. Thus, our framework is proposed to sift task-related information from the unstructured data while confusing task-independent information. Figure 2 shows that our framework addresses this problem in three stages. In the first stage, the encoder in the VAE family model is used to obtain the disentangled representation, from which different attribute information can be independently selected. Although the disentangled representation is similar in form to the structured data, users are still confused about which dimensions are necessary due to the lack of semantic interpretation. Therefore, in the second step, we propose a model called Indicator that provides suggestions for explicit user control over the information. In the representation, the dimensions marked by Indicator are frozen, while the values of the remaining dimensions are discarded and refilled. In the third stage, the transformed representation is fed to the decoder that corresponds to the encoder in the first stage for data reconstruction. The task-related information in the reconstructed data is preserved, while the task-independent information is unreliable.

3.2. Unstructured Data Disentanglement

The information of different attributes in unstructured data is often intertwined and almost impractical to select independently. By disentangling different attributes, it is possible to preserve the task-related factors of unstructured data while obfuscating the task-independent factors. As shown in Figure 4, we employ the VAE family models (β-VAE, Factor-VAE, and β-TCVAE) in the training phase to obtain an encoder–decoder pair. The encoder is used to extract the disentangled representation, and the decoder is used to reconstruct the data. However, the common problem is that the data generated by VAEs is always ambiguous. One view is that the pixel-wise reconstruction error metric causes the generated data to be too smooth [46]. In contrast, the main idea of GAN is to provide a game between the generator and the discriminator. During this game, the discriminator judges the original data as true and the generated data as false at each iteration. Meanwhile, the generator tries to fool the discriminator into judging the generated data as true in the same iteration. Therefore, the decoder can be considered as the generator and a discriminator is introduced to improve the quality of the generated data. To avoid affecting the disentanglement of the representation, in each iteration, the training of the GAN is carried out after the training of the VAEs, which means that the encoder and the discriminator are not end-to-end. Formally, the loss function can be defined as:

$$L_{VAEs}(\theta_{End}, \theta_{Dec}) = L_{VAEs} \tag{3}$$

$$L_{GAN}(\theta_{Dec}) = E_x[log(Dis(x)) + log(1 - Dis(x'))] \tag{4}$$

$$L_{GAN}(\theta_{Dis}) = -E_x[log(log(Dis(x')))] \tag{5}$$

where L_{VAEs} represents different loss functions in the VAE family and θ_* indicate the parameters to be updated.

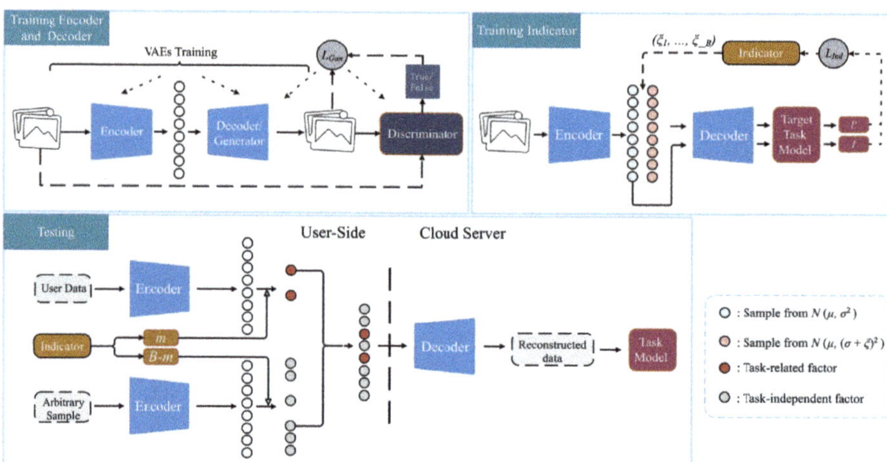

Figure 4. The workflow of our framework. The top line is the training stage, including the training of the encoder–decoder pair and Indicator. The bottom line is the test stage. An indicator is introduced to recommend the indexes of the representation dimensions that need to be retained. At the same time, an arbitrary sample is used as a carrier to supplement the remaining dimensions.

In the testing phase, the encoder is deployed on the user side while the corresponding decoder runs on the cloud server.

3.3. Representation Oriented Indicator

After the encoder and decoder training, the encoder can standard express different attributes in the latent space. Such a disentangled representation allows us to obfuscate the task-independent factors without changing task-related factors. However, data contain many factors, and it is impractical to enumerate all task-independent attributes. In addition, whether a factor is related to the task depends on the specific task model. Different classifiers may focus on different associated attributes for the same classification task. For example, one classifier will concentrate on hair when judging the gender of a face image, while another classifier may focus on beards. The tendency of the classifier depends on the training set and model structure, which is uncontrollable for the user. If the factors to be obfuscated are rashly determined based on human perception, it will inevitably affect the effectiveness of the primary task. For this consideration, the task-adaptive Indicator is proposed to mark the attributes that the specific task model focuses on.

Different dimensions in the disentangled representation are considered disjoint, and a set of sub-dimensions can only express a particular data attribute. Meanwhile, the task model does not view all the information to make credible inferences but pays more attention to specific attributes. This is equivalent to that only one set of sub-dimensions in the disentangled representation contributes to the task model inference while discarding the values of the remaining dimensions has almost no effect. Following the idea, the proposed Indicator is designed to search this set of sub-dimensions. Indicator is expected to have both fidelity and interpretability. Fidelity means that Indicator can accurately mark the representation dimensions necessary for the task model. The interpretability signifies that the decision-making process is consistent with the human perspective.

Figure 5 reviews the paradigm of the VAE family. Each original datapoint $x^{(i)}$ is encoded into a multivariate gaussian distribution $\mathcal{N}(\mu^{(i)}, (\sigma^{(i)})^2)$, and the decoding results $x'^{(i)}$ of all samples in $\mathcal{N}(\mu^{(i)}, (\sigma^{(i)})^2)$ are similar to the original data $x^{(i)}$. Given an original datapoint $x^{(i)}$, its disentangled representation $z^{(i)}$ can be represented by $z^{(i)} \sim \mathcal{N}(\mu^{(i)}, (\sigma^{(i)})^2)$, and $z^{(i)} \in \mathbb{R}^B$. For the inference of a certain task model, there are m necessary dimensions in $z^{(i)}$, whose value fluctuation will significantly affect the result of the inference, while the change in the remaining $B - m$ dimensions can hardly have impacts. This demonstrates that under the premise of not affecting the inference confidence, the larger variance is not tolerated by the m dimensions, but is acceptable for the $B - m$ dimensions. Therefore, the ultimate goal of Indicator can be expressed as finding a variance bias ξ as large as possible and encoding the data $x^{(i)}$ into the new distribution $\mathcal{N}(\mu^{(i)}, (\sigma^{(i)} + \xi)^2)$, as shown in Figure 5. Among them, the decoding result $\widetilde{x}'^{(i)}$ of the sample on $\mathcal{N}(\mu^{(i)}, (\sigma^{(i)} + \xi)^2)$ and the decoding result $x'^{(i)}$ of the sample on $\mathcal{N}(\mu^{(i)}, (\sigma^{(i)})^2)$ show the same confidence in the task model. In general, the optimization goal of ξ can be formulated as:

$$
\begin{aligned}
&minL(\xi) = minE_{p(x)}[E_{q(z|x),\, \widetilde{q}(\widetilde{z}|x)}[L2(TM(Dec(z)), TM(Dec(\widetilde{z})))]] - \lambda \sum_{i=1}^{B} \xi_i, \\
&q(z|x) = \mathcal{N}(\mu, \sigma^2),\ \widetilde{q}(\widetilde{z}|x) = \mathcal{N}(\mu, (\sigma + \xi)^2)
\end{aligned} \tag{6}
$$

where TM represents the target task model.

After training, the representation dimensions corresponding to the smaller ξ_i cannot support the larger sampling ranges while maintaining effectiveness for the task model, which means that the task model will pay more attention to these dimensions. Conversely, the dimensions corresponding to a larger ξ_i contribute less to the task model inference. To determine ξ, an intuitive method is to set ξ as trainable parameters. However, there are two problems with this method. First, since Indicator aims to explore the tolerance of different dimensions to the larger variance, ξ_i is expected to be non-negative. Also, too large a variance $\sigma + \xi$, which leads to meaningless sampling, can cause training to collapse. Therefore, ξ_i should be restricted to the interval $[0, \delta)$. Second, ξ is the variable in the distribution $\mathcal{N}(\mu, (\sigma + \xi)^2)$ and the sampling process is not differentiable.

For the first problem, we design a function $\xi_i = f(\beta_i),\ \beta_i \in R,\ \xi_i \in [0, \delta)$ to eliminate the constraint on ξ, where β is Indicator parameters. Considering the $\lambda \sum_{i=1}^{B} \xi_i$ term in Equation (6), $f()$ should also satisfy monotonicity. If ξ_i can take the minimum value of 0 when $\beta_i = 0$, the training of Indicator will benefit from the sparse parameters. Formally, $f()$ can be defined as:

$$\xi_i = f(\beta_i) = \delta \cdot (\frac{1 - e^{-\beta_i^2}}{e^{-\beta_i^2} + 1}) \tag{7}$$

Among them, $f()$ is monotonically increasing in $[0, +\infty)$, monotonically decreasing in $(-\infty, 0]$, and the minimum value is 0 at $\beta_i = 0$. In order to solve the second problem, we borrow the reparameterization trick to convert $\tilde{z}^{(i)} \sim \mathcal{N}(\mu^{(i)}, (\sigma + \xi)^{2(i)})$ to $\tilde{z}^{(i)} = \mu^{(i)} + (\sigma + \xi)^{(i)}\epsilon$, $\epsilon \sim \mathcal{N}(0, I)$ to make ξ differentiable. In summary, the formal loss function $L_{Indicator}$ is expressed as follows:

$$\begin{aligned} L_{indicator}(\beta) =& E_{p(x)}[E_{q(z|x),\, \tilde{q}(\tilde{z}|x)}[L2(TM(Dec(z)), TM(Dec(\tilde{z})))]] - \lambda \sum_{i=1}^{B} f(\beta_i) + \|\beta\|_2, \\ & q(z|x) = \mathcal{N}(\mu, \sigma^2),\ \tilde{q}(\tilde{z}|x) = \mathcal{N}(\mu, (\sigma + f(\beta))^2) \end{aligned} \tag{8}$$

Finally, the representation dimensions corresponding to the parameters satisfying $|\beta_i| < \psi$ are considered more relevant by the task model and their indices are recorded, where ψ is the threshold. The entire training process of the proposed Indicator is shown in Algorithm 1.

Algorithm 1 Indicator Training

Input: Encoder $Enc()$, Conversion function $f()$, Decoder $Dec()$, task model $Tm()$, disentangled representation size B, threshold ψ.
Output: Dimension index to be retained $Index$.
$\beta \leftarrow$ random initialize Indicator parameters
for epochs **do**
    Random mini-batch $X = \{x^{(1)}, x^{(2)}, \ldots, x^{(n)}\}$,
    $\mu, \sigma = Enc(X), \mu, \sigma \in R^{n \times B}$
    $z = \mu + \sigma \times \epsilon,\ \epsilon \sim \mathcal{N}(0, I)$
    $\tilde{z} = \mu + (\sigma + f(\beta)) \times \epsilon,\ \epsilon \sim \mathcal{N}(0, I)$
    $l = Tm(Dec(z)),\ \tilde{l} = Tm(Dec(\tilde{z}))$
    $\beta \xleftarrow{update} min(L2(l, \tilde{l}) - \sum_{i=1}^{B} f(\beta_i)) + \|\beta\|_2$
end for
$Index = \{\}$
for $\beta_i \in \beta$ **do**
    if $|\beta_i| < \psi$ **then**
        $Index.append(i)$
    end if
end for

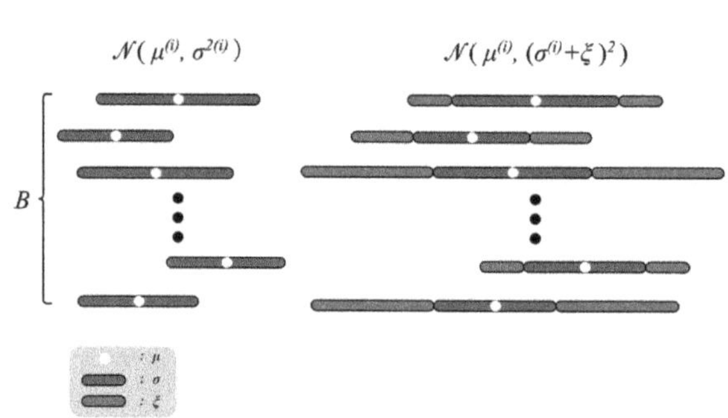

Figure 5. Illustration of how the indicator works. Indicator searches for the maximum allowable oscillation range that remains utility for the task model in the B representation dimensions.

3.4. Data Reconstruction

The disentangled representation encoded by the encoder allows obfuscation of the task-independent features without changing the task-related features. It is necessary to preserve the m dimensions marked by Indicator because the task model pays more attention to them. At the same time, the remaining $B - m$ dimensions, which contribute less to task inference but contain excessive task-independent information, should be discarded. Theoretically, it is possible to replace the original values of the $B - m$ dimensions with arbitrary values. In practice, however, completely random values will make it easier for the transformed representation to decode ambiguous data, resulting in task-relevant information not being correctly expressed. Even though the $B - m$ dimensions have nothing to do with the task model, it is still necessary to be careful when choosing their replacement values. As shown in the test phase of Figure 4, our method uses an arbitrary sample as a carrier. It concatenates the $B - m$ dimensions in the carrier representation with the m dimensions in the original data representation. By reconstructing the data from such a representation, only the factors that the task model focuses on are credible, while others are confusing.

4. Experimental Study

In this section, we first qualitatively evaluate the proposed Indicator and report the experimental results. Then, we quantify the privacy–utility trade-offs of our framework and present a comparison with other popular methods. The following experiments involve three datasets: dSprites [47], MNIST [48], and CelebA [49]. dSprites contains 737,280 2D synthesis samples with 6 attributes. We randomly divide 589,824 samples for training VAE family models and Indicators and 147,456 samples for testing. MNIST contains grayscale images of 10 classes of handwritten digits, including 60,000 training samples and 10,000 testing samples. CelebA includes 202,599 face images labeled with 40 binary attributes, of which 162,770 images are divided for training and 39,829 images for testing. The experiments are conducted on Nvidia GTX 3080Ti GPU in Pytorch.

4.1. Indicator Evaluation

To qualitatively demonstrate the effectiveness of the proposed Indicator, we conduct experiments on dSprites and MNIST from three perspectives. (a) Versatility: whether the proposed Indicator can be effectively combined with the various VAEs models. (b) Reliability: whether the task model considers the dimensions marked by Indicator. (c) Stability: whether the Indicator can make the same decision under different initial conditions and training subsets.

4.1.1. Versatility

To illustrate the versatility, the following experiments are performed on β-VAE, Factor-VAE, and β-TCVAE, respectively.

4.1.2. Reliability

The verification of the reliability is studied by two experiments. For dSprites and MNIST, the dimension B of the disentangled representation is set to 10, and the threshold δ is set to 0.5. For dSprites, a classifier focusing on the X-position is used as the target task model. For β-VAE, Factor-VAE, and β-TCVAE, Indicator finds 3, 4, and 2 dimensions on which the target task model focuses. For MNIST, a classifier that distinguishes digits is the target task model. In the above three VAE models, Indicator finds 3, 4, and 3 task-related dimensions in the representation.

The first experiment is to interpolate the dimensions marked by Indicator while freezing the remaining dimensions. Figure 6 visualizes the reconstructed image traversing the marked dimensions. The first line is the original data, and the second to fifth lines are the reconstruction of the interpolated representation. On the one hand, the "X-position" that the task model focuses on changes with the change in the marked representation dimension. On the other hand, the experimental results show the difference in disentangling performance of different VAE schemes. In the second experiment, we fix the dimensions marked by Indicator and replace the values of the remaining dimensions with 0. Figure 7 shows the reconstruction of the processed representation. The attribute focused by the target task model is preserved, while the others become irrelevant to the original data. The above two experiments show that the task-related dimensions determined by the proposed Indicator are consistent with the human view, which confirms the reliability of Indicator to a certain extent. The quantitative measure of reliability can be decomposed into target task accuracy and the availability of task-independent attributes. Target task accuracy reflects whether the task-related dimensions are fully selected. The availability of task-independent attributes is directly proportional to the redundancy of the selected dimensions. Therefore, reliability is equivalent to the privacy–utility trade-offs of our framework, which will be discussed on the CelebA dataset in Section 4.2.

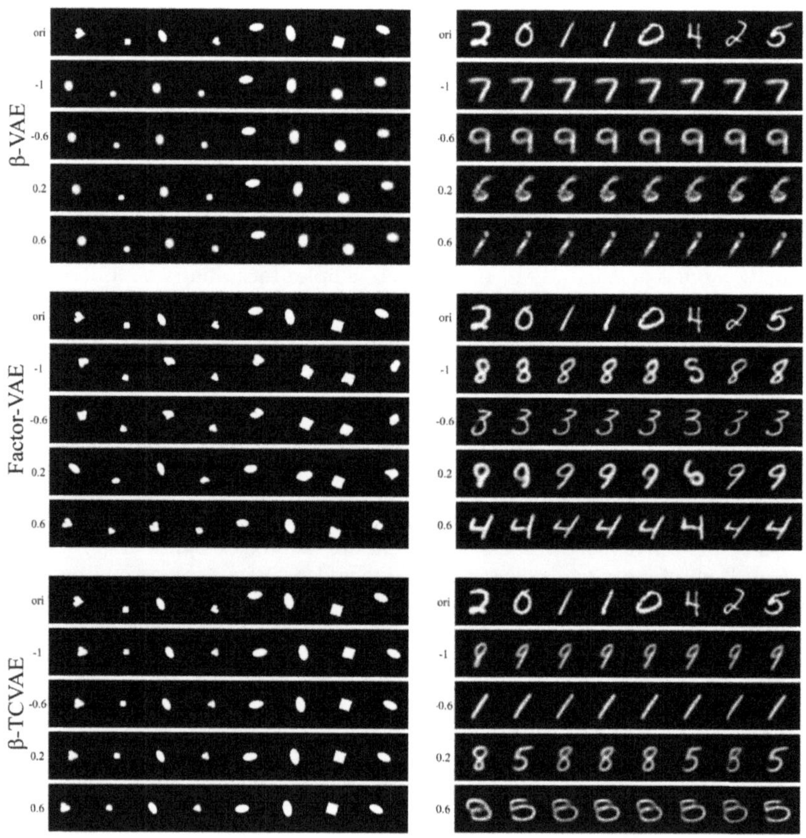

(**a**) dSprites (**b**) MNIST

Figure 6. Reconstructed image visualization of traversing the representation dimensions marked by the indicator.

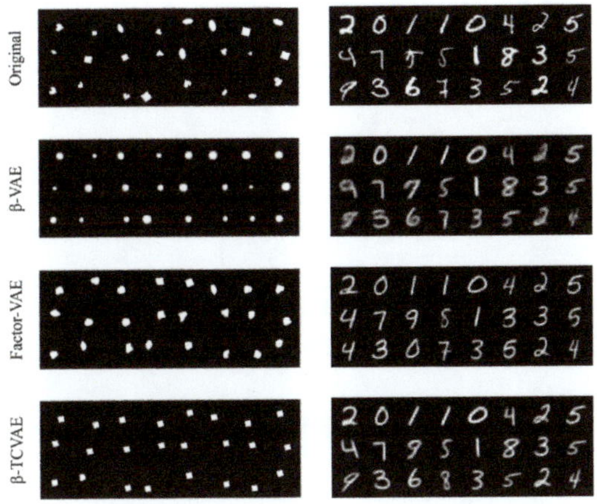

(**a**) dSprites (**b**) MNIST

Figure 7. The representation dimensions marked by Indicator are fixed, while the values of the remaining dimensions are replaced with 0. The above illustrations are reconstructed images based on these processed representations.

4.1.3. Stability

To illustrate the stability of the proposed Indicator, we perform experiments on dSprites and MNIST with the same settings as in Section 4.1.2. The train sets of dSprites and MNIST are divided into 3 subsets, and then Indicator searches for task-related dimensions on each subset with random initial parameters. Figure 8 provides a visualization of Indicator parameters changing with epochs. In Figure 8, the dimensions that fall into the yellow area are considered more concerned by the task model. In rows 1 and 2, the indicators mark the disentangled representations generated from β-VAE. The Indicators in lines 3 and 4 mark the disentangled representations generated using Factor-VAE. Lines 5 and 6 are Indicator marking the disentangled representations generated using β-TCVAE. Taking the three subfigures in the first row as an example, the parameters corresponding to dimensions 4, 8, and 10 in the Indicator eventually converge to the yellow region, while the rest diverge. This represents that Indicator considers dimensions 4, 8 and 10 as being attended to by the target task. Under different dataset slices and random initial parameters, the tendency of Indicators to represent dimensions shows the same trend. This demonstrates the stability of the Indicator, where the marking process is not affected by the initial parameters and the division of the dataset. Moreover, the experimental results also support the conclusion that the same dimensions of the latent code of different samples correspond to the same information.

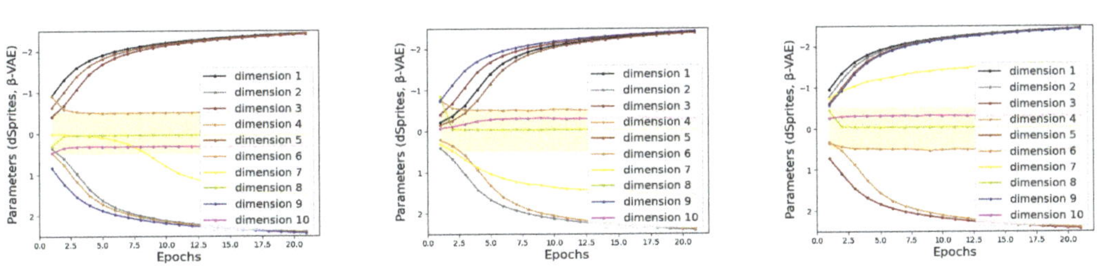

Figure 8. *Cont.*

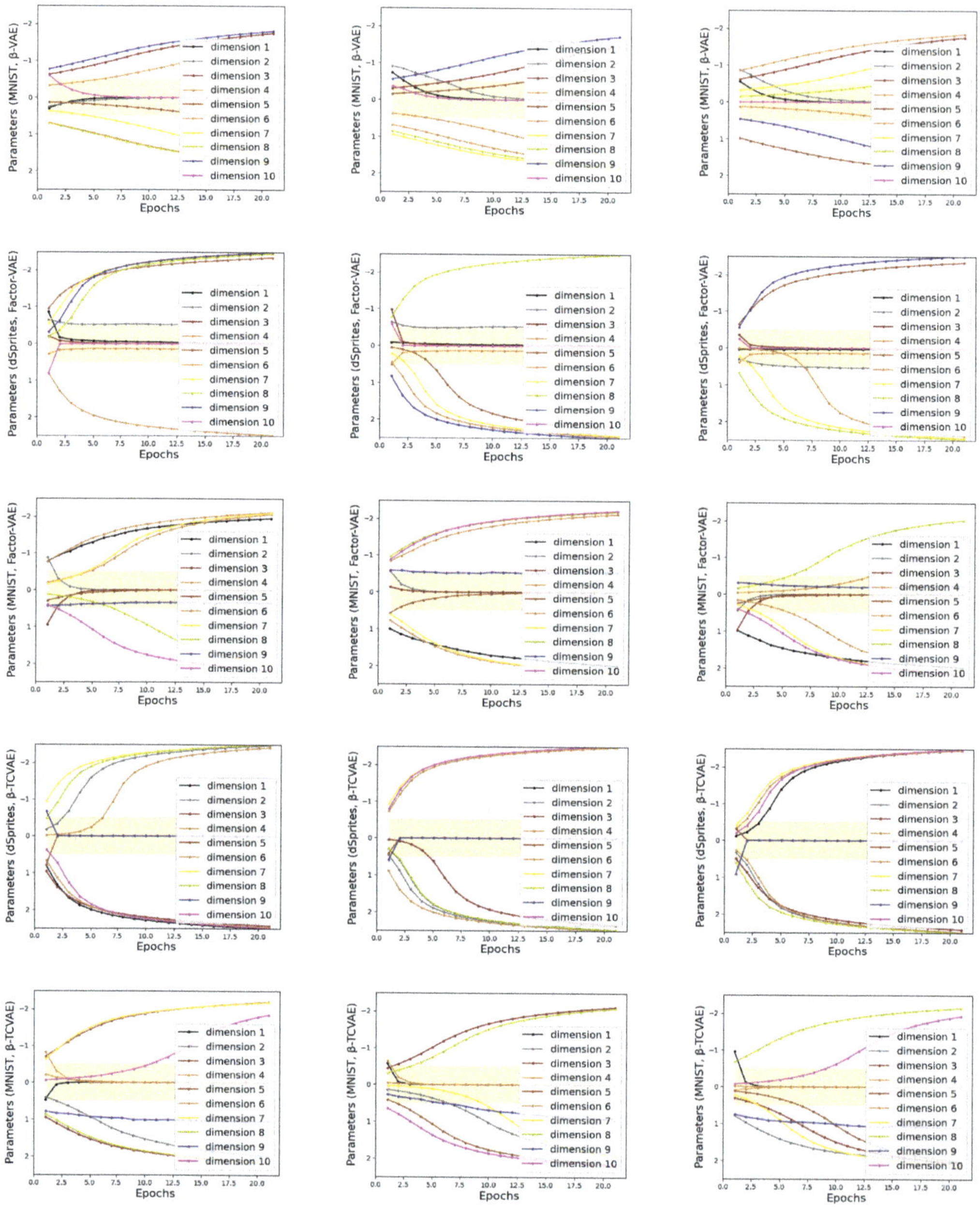

Figure 8. The training sets of dSprites and MNIST are divided into 3 subsets, respectively. The above illustration is the parameter curve obtained by Indicator training on these subsets. In the illustration, the dimensions that fall into the yellow area are considered more concerned by the task model. In rows 1 and 2, the indicators mark the disentangled representations generated from β-VAE. Indicators in lines 3 and 4 mark the disentangled representations generated using Factor-VAE. Lines 5 and 6 are Indicator marking the disentangled representations generated using β-TCVAE.

4.2. *Privacy–Utility Trade-Offs Evaluation and Comparison*

4.2.1. Setup

We design experiments to verify the effectiveness of our framework's utility–privacy trade-offs on the real-world dataset CelebA. The images are normalized and resized to $3 \times 64 \times 64$ for preprocessing. Due to the better disentanglement of β-TCVAE, the β-TCVAE optimized by GAN is chosen to construct our partial privacy-preserving framework. The encoder and decoder are optimized using RmSprop, with $alpha$ and eps set to 0.9 and 1×10^{-8}, respectively. The discriminator is trained using an SGD optimizer, with $momentum$ and $weight\_decay$ set to 0.9 and 1×10^{-4}, respectively. We train these three components for 40 epochs with a fixed learning rate of 3×10^{-4}, and the batch_size is set to 128. The dimension size of the disentangled representation is set to 128, which represents the output of the encoder, including 128 means and 128 variances. Indicator is trained using the SGD optimizer with 0.9 $momentum$ and 1×10^{-4} $weight\_decay$ for 20 epochs with batch_size 256. The learning rate is set to 1×10^{-4}. Empirically, we set the hyperparameter λ to 2 and δ to 1. The classifier trained on the original data with the standard ResNet18 architecture is considered the task model.

In our experiments, the accuracy of the task model is used to quantify utility. Several attack models designed to infer privacy attributes are introduced, and we propose two new metrics as privacy measures. The data processed by the task model and the attack models come from the reconstruction of the decoder. Despite the introduction of GAN, the reconstructed data still inevitably loses details. In order to avoid exaggerating the protective effect of our privacy attribute framework due to the fuzziness of the reconstruction, we set the easily recognizable attributes as the platform for the privacy utility measurement. Specifically, we set "Eyeglasses" and "Gender" as the target attributes of the task model, while enumerating "Wearing_Hat" and "Bald" as the target for the attack models.

4.2.2. Baselines

We choose three classical privacy-preserving schemes that are widely used in the literature as a baseline against which to compare our framework. A brief description of these schemes is given below. Gaussian noise obfuscates the raw data by adding Gaussian noise $\mathcal{N}(0, \sigma^2)$, where σ is set to 0.5 and 1, respectively. Because Gaussian noise can provide rigorous differential privacy guarantees with less local noise, it is widely used in federated learning scenarios [30,50]. Laplacian noise is also a classic differential privacy method that injects Laplacian noise into the raw data according to the privacy budget $\{0.3, 0.9\}$. PAN is a representative framework for adversarial training methods [13]. In the training phase, PAN simulates adversaries interested in private information to obtain an encoder that can extract the representation with good utility–privacy trade-offs. In the comparison phase, the objective function adopts two sets of coefficients $\{0.1, 0.7, 0.2\}$, $\{0.5, 0.3, 0.2\}$ to show its performance under different privacy budgets.

4.2.3. Evaluation and Comparison

To quantify the data utility maintained by different schemes, the classification accuracy of the target attribute is measured. Specifically, the two noise injection methods and our framework use classifiers trained on the raw dataset, while PAN uses the utility discriminator generated in the adversarial training. In terms of privacy measurement, the accuracy of the adversarial model's inference of privacy attributes is not convincing. This is because the model's decision is biased and the test set samples may be uneven, which means that lower accuracy does not necessarily mean better privacy protection. Taking the "bald" attribute as an example, the uniform random noise with a value in the range $[0, 1]$ will be 100% judged as not bald by the adversary model with an average confidence of 0.99. Although these noises are unrelated to the raw data, they will still achieve 97.88% inference accuracy when considered as a processed private image. In addition, the confidence difference in the attack models in inferring private images will reveal additional information compared to inferring random noise. Therefore, we propose the average confidence difference *Con-Diff* and the distribution shift *Dis-Shift* as the privacy quantification. The two formulas are defined as follows:

$$Con\_Diff = \frac{1}{N}\sum_{i=1}^{N}(AM^{l}(x'^{(i)})^{(0)} - AM^{l}(noise^{(i)})^{(0)}) \tag{9}$$

$$Dis\_Shift = \frac{1}{N}\sum_{i=1}^{N}|AM(x'^{(i)}) - AM(noise^{(i)})| \tag{10}$$

Among them, AM is the attack model with l layers, x' represents private images generated by different methods, and N is the total number of samples in the test set. $AM^{l}(\cdot)^{(0)}$ indicates the first element output by the AM, and $AM(\cdot)$ represents class 0 or 1. The lower *Con-Diff* and *Dis-Shift* represent that the attack model's inference on the processed data privacy attributes is closer to a non-priority guess, which demonstrates better privacy considerations.

In Figure 9, we use t-SNE [51] to visualize the features learned by the attack model at layer l-1 to analyze the effectiveness of our framework. The first column is the t-SNE plot of the original image facing the attack model. The second and third columns show the t-SNE plot for anonymously transformed reconstruction with "Eyeglasses" and "Gender" as the task-related attributes. The original data features show significant clustering for the two task-independent attributes, while the features of anonymously transformed reconstruction are indistinguishable.

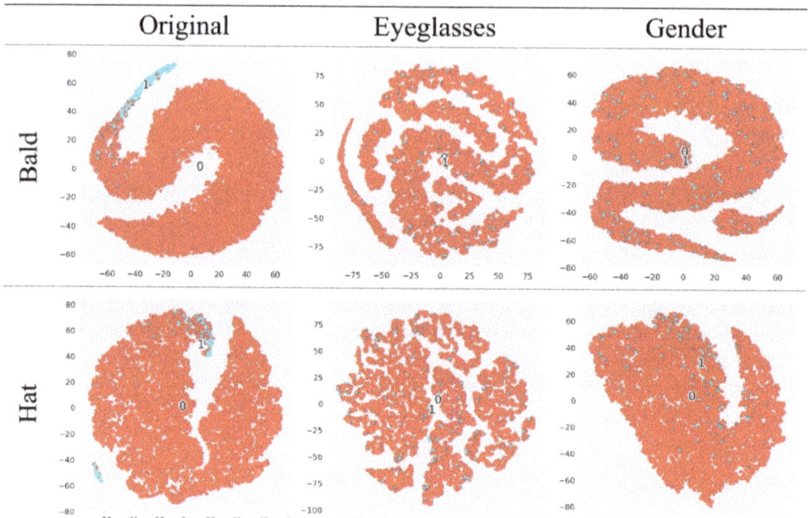

Figure 9. The t-SNE visualization of the AM^{l-1} output. The first column represents the performance of the original data in the face of the attack model. The second and third columns are the anonymously transformed reconstruction performance facing the attack model, with "Eyeglasses" and "Gender" as the task-related attributes, respectively.

Table 1 shows the evaluation and comparison of different methods on the utility–privacy trade-offs. It also includes L2 distance to measure the similarity between the processed image and the original data. "Target Attribute #1" and "Target Attribute #2" represent "Eyeglasses" and "Gender", while "Privacy Attribute #1" and "Privacy Attribute #2" represent "Wearing_Hat" and "Bald". Injecting Gaussian noise and Laplacian noise are general methods that will affect all attributes indiscriminately. Therefore, the privacy protection of these methods will significantly sacrifice its utility. In addition, L2-DIS indicate that the processed image still has a high similarity to the original data, which will also lead to the risk of privacy leakage. For the evaluation of PAN, we follow its recommendation on the encoder structure and design 4 convolutional layers, 4 normalization layers, 2 Maxpooling, and 2 upsampling layers. After 15 epochs of training, PAN achieves the ideal utility, but more discussion about privacy is necessary. In our experiment, the attack model for privacy attribute #1 will classify uniform random noise as class "1" with 100% probability, and 3.3% of the samples in the test set are class "1". At the same time, the attack model for privacy attribute #2 will infer uniform random noise as class "0" with 100% probability, and 97.9% of the test set are "0" samples. Under different experimental settings, some samples fool the attack model's judgment on privacy attribute #1, but have little effect on privacy attribute #2. Similarly, for these two privacy attributes, the attack model's judgments on the encrypted data are closer to its judgments on random noise. The mechanism of classifiers based on neural networks can be simply described as being oriented to data distribution. The nonlinear transformation of the original data by the encoder in PAN essentially causes a distribution shift. From our point of view, this is the privacy guarantee of PAN. However, there are still a large number of samples that reveal the privacy attributes. The evaluation of our framework uses the inference results of the attack model on the carrier as a benchmark. While the reconstructed image retains utility, there is almost no difference in confidence and distribution shift compared to the carrier. This shows that the reconstructed image produces a low-level information gain for the attack model, demonstrating the privacy of our framework. In order to compare different frameworks more intuitively, we further describe the evaluation results in Figure 10. The visualization of the reconstructed images shown in Figure 11 supports the evaluation results. The upper part of each sub-region in Figure 9 are the original images, and the lower part are the reconstructed images, which retain the target attribute while others still belong to the carrier. Since the reconstructed images are also facial images, the structure is similar to the original image. Another advantage of our framework is its flexibility. When the target or privacy attributes change, our framework needs to retrain 128 parameters, but PAN needs to retrain 22.44 M.

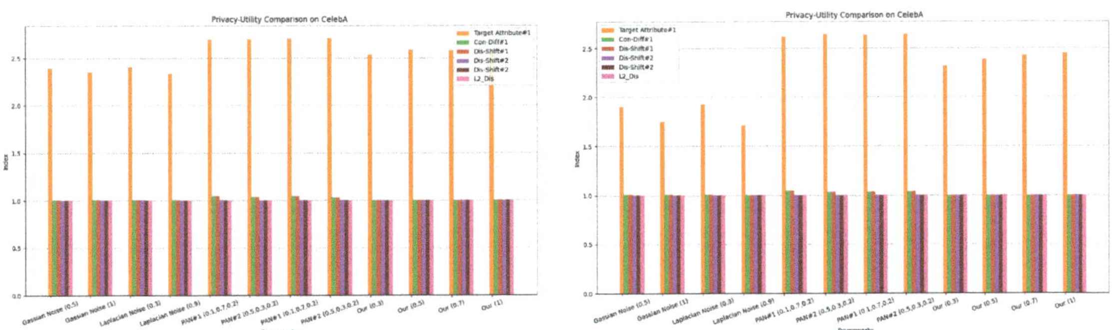

(**a**) Comparison on Target Attribute #1 (**b**) Comparison on Target Attribute #1

Figure 10. Privacy–utility comparison on CelebA. Among them, the y-axis takes the $exp(\cdot)$ of the evaluation result.

Using the privacy attribute #1 as a platform, we further explore potential attackers against the baseline methods and our framework. More powerful attack models are trained using the privacy-edited data as input and combining the ground truth. It should be noted that these attacks may not be feasible in real-world scenarios, and we aim to explore whether the above methods can effectively confuse the original data. The experimental results are reported in Table 2. The new attack models do not perform more effective attacks against the two noise injection methods and our framework, which shows that the topological space of the original data is broken. On the contrary, PAN is vulnerable to new attacks, supporting the suspicion mentioned in Section 1.

Table 1. Privacy–utility comparison on CelebA.

Methods	Target Attribute #1	Privacy Attribute #1		Privacy Attribute #2		L2-DIS
		Dis-Shift	Con-Diff	Dis-Shift	Con-Diff	
Gaussian noise ($\sigma^2 = 0.5$)	87.3%	2.45%	6.11×10^{-2}	0.01%	1.90×10^{-2}	6.19×10^{-4}
Gaussian noise ($\sigma^2 = 1$)	85.5%	0.31%	5.05×10^{-2}	0.02%	1.85×10^{-2}	8.04×10^{-4}
Laplacian noise ($\lambda = 0.3$)	87.8%	12.8%	5.51×10^{-2}	0.03%	1.91×10^{-2}	5.82×10^{-4}
Laplacian noise ($\lambda = 0.9$)	85.0%	3.53%	3.49×10^{-2}	0.01%	1.80×10^{-2}	8.06×10^{-4}
PAN (#1, (0.1, 0.7, 0.2))	99.1%	50.7%	4.72×10^{-1}	0.05%	1.33×10^{-4}	1.52×10^{-3}
PAN (#1, (0.5, 0.3, 0.2))	99.3%	37.3%	3.53×10^{-1}	0.02%	1.89×10^{-4}	1.95×10^{-3}
PAN (#2, (0.1, 0.7, 0.2))	99.4%	41.4%	4.12×10^{-1}	0.02%	3.49×10^{-4}	1.26×10^{-3}
PAN (#2, (0.5, 0.3, 0.2))	99.5%	16.1%	2.73×10^{-1}	0.05%	3.18×10^{-4}	1.69×10^{-3}
Our framework ($\psi = 0.3$)	93.1%	0%	$\mathbf{1.32 \times 10^{-6}}$	0%	**0**	$\mathbf{9.86 \times 10^{-4}}$
Our framework ($\psi = 0.5$)	95.0%	0%	$\mathbf{1.81 \times 10^{-6}}$	0%	**0**	$\mathbf{8.96 \times 10^{-4}}$
Our framework ($\psi = 0.7$)	94.6%	0%	$\mathbf{1.44 \times 10^{-6}}$	0%	**0**	$\mathbf{8.44 \times 10^{-4}}$
Our framework ($\psi = 1$)	94.6%	0%	$\mathbf{3.15 \times 10^{-6}}$	0%	$\mathbf{2.60 \times 10^{-10}}$	$\mathbf{7.52 \times 10^{-4}}$
Methods	**Target Attribute #2**	**Privacy Attribute #1**		**Privacy Attribute #2**		**L2-DIS**
		Dis-Shift	**Con-Diff**	**Dis-Shift**	**Con-Diff**	
Gaussian noise ($\sigma^2 = 0.5$)	64.0%	2.45%	6.11×10^{-2}	0.01%	1.90×10^{-2}	6.19×10^{-2}
Gaussian noise ($\sigma^2 = 1$)	55.8%	0.31%	5.05×10^{-2}	0.02%	1.85×10^{-2}	8.04×10^{-4}
Laplacian noise ($\lambda = 0.3$)	65.4%	12.8%	5.51×10^{-2}	0.03%	1.91×10^{-2}	5.82×10^{-4}
Laplacian noise ($\lambda = 0.9$)	53.7%	3.53%	3.49×10^{-2}	0.01%	1.80×10^{-2}	8.06×10^{-4}
PAN (#1, (0.1, 0.7, 0.2))	96.1%	46.9%	4.47×10^{-1}	0.15%	2.20×10^{-4}	1.05×10^{-3}
PAN (#1, (0.5, 0.3, 0.2))	97.0%	43.3%	3.34×10^{-1}	0.22%	3.69×10^{-4}	1.05×10^{-3}
PAN (#2, (0.1, 0.7, 0.2))	96.8%	28.5%	3.45×10^{-1}	0.27%	2.28×10^{-4}	1.06×10^{-3}
PAN (#2, (0.5, 0.3, 0.2))	97.0%	35.6%	3.93×10^{-1}	0.15%	2.45×10^{-4}	1.05×10^{-3}
Our framework ($\psi = 0.3$)	84.0%	**0.02%**	$\mathbf{1.65 \times 10^{-6}}$	0%	**0**	$\mathbf{7.41 \times 10^{-4}}$
Our framework ($\psi = 0.5$)	86.9%	**0.03%**	$\mathbf{1.60 \times 10^{-6}}$	0%	**0**	$\mathbf{5.56 \times 10^{-4}}$
Our framework ($\psi = 0.7$)	88.6%	**0.05%**	$\mathbf{4.09 \times 10^{-6}}$	**0.12%**	$\mathbf{1.03 \times 10^{-7}}$	$\mathbf{3.29 \times 10^{-4}}$
Our framework ($\psi = 1$)	89.5%	**0.78%**	$\mathbf{1.61 \times 10^{-4}}$	**0.40%**	$\mathbf{2.92 \times 10^{-5}}$	$\mathbf{2.21 \times 10^{-4}}$

Figure 11. The above illustrations are facial images whose task-independent attributes are confused. The upper part takes "Eyeglasses" as the task-related attribute, and the bottom part, "Gender" is regarded as the task-related attribute.

To further show the effectiveness of our framework, we also use SVM for experiments. We choose the RBF kernel function, use libsvm to set the hyperparameters and let the latent code be the input. Evaluation results are shown in Table 3.

Table 2. Search for potential attacker.

Methods	Dis-Shift (Original Attacker)	Dis-Shift (Potential Attacker)
Gaussian noise ($\sigma^2 = 0.5$)	2.45%	3.76%
Gaussian noise ($\sigma^2 = 1$)	0.31%	3.51%
Laplacian noise ($\lambda = 0.3$)	12.8%	4.44%
Laplacian noise ($\lambda = 0.9$)	3.53%	3.14%
PAN $(0.5, 0.3, 0.2)$	43.3%	95.52%
PAN $(0.1, 0.7, 0.2)$	46.9%	96.74%
Our framework ($\psi = 0.5$)	**0.03%**	**0.05%**
Our framework ($\psi = 1$)	**0.78%**	**0.91%**

Table 3. Evaluating privacy–utility on CelebA via SVM.

Method	Target Attribute #1	Privacy Attribute #1	Privacy Attribute #2
Our framework ($\psi = 0.3$)	84.3%	43.1%	54.2%
Our framework ($\psi = 0.5$)	85.9%	46.3%	55.6%
Our framework ($\psi = 0.7$)	87.2%	48.4%	56.9%
Our framework ($\psi = 1$)	87.5%	50.3%	58.4%
Method	**Target Attribute #1**	**Privacy Attribute #1**	**Privacy Attribute #2**
Our framework ($\psi = 0.3$)	72.5%	44.5%	54.6%
Our framework ($\psi = 0.5$)	73.2%	47.8%	55.8%
Our framework ($\psi = 0.7$)	73.7%	47.7%	57.2%
Our framework ($\psi = 1$)	74.1%	48.4%	59.1%

Based on the above experimental results, on the one hand, it can be observed that our framework can be more effective against potential attackers compared to adversarial training based PAN. On the other hand, our framework maintains better data availability as well as privacy of task-independent attributes compared to the noise adding approach.

5. Discussion and Conclusions

In this work, we design an Indicator to indicate the region of interest of the target task model on the disentangled representation. By retaining the information necessary for the target task through Indicator, we further construct a privacy-preserving prediction framework that respects the task-independent attributes. Evaluations on multiple standard datasets show that our framework achieves competitive utility–privacy trade-offs.

However, our framework has not yet reached the ideal situation of preserving all utility and protecting all privacy. On the one hand, our framework partially loses accuracy in the target task. On the other hand, the attacker's accuracy in inferring privacy attributes is higher than the guess without prior knowledge. We speculate that there are two reasons: (a) the quality of the reconstructed image limits the utility; (b) there is information overlap between the different representation dimensions, leading to sensitive information leakage. These are also problems that we hope to solve in the future.

Author Contributions: Conceptualization, H.L. and Q.L.; methodology, H.L.; software, Q.L.; validation, H.L.; formal analysis, Q.L.; investigation, Q.L.; resources, H.L.; data curation, Q.L.; writing—original draft preparation, Q.L.; writing—review and editing, Q.L.; visualization, Q.L.; supervision, H.L.; project administration, H.L.; funding acquisition, H.L. All authors have read and agreed to the published version of the manuscript.

Funding: This research received no external funding.

Institutional Review Board Statement: Not applicable.

Informed Consent Statement: Not applicable.

Data Availability Statement: Our experiments use the public datasets MNIST, dSprites and CelebA.

Conflicts of Interest: The authors declare no conflicts of interest.

References

1. Devlin, J.; Chang, M.; Lee, K.; Toutanova, K. BERT: Pre-training of Deep Bidirectional Transformers for Language Understanding. In Proceedings of the 2019 Conference of the North American Chapter of the Association for Computational Linguistics: Human Language Technologies, NAACL-HLT 2019, Minneapolis, MN, USA, 2–7 June 2019; Burstein, J., Doran, C., Solorio, T., Eds.; Long and Short Papers; Association for Computational Linguistics: Stroudsburg, PA, USA, 2019; Volume 1, pp. 4171–4186. [CrossRef]
2. Dosovitskiy, A.; Beyer, L.; Kolesnikov, A.; Weissenborn, D.; Zhai, X.; Unterthiner, T.; Dehghani, M.; Minderer, M.; Heigold, G.; Gelly, S.; et al. An Image is Worth 16x16 Words: Transformers for Image Recognition at Scale. *arXiv* **2020**, arXiv:2010.11929. Available online: http://arxiv.org/abs/2010.11929 (accessed on 6 December 2023).
3. Li, A.; Guo, J.; Yang, H.; Salim, F.D.; Chen, Y. DeepObfuscator: Obfuscating Intermediate Representations with Privacy-Preserving Adversarial Learning on Smartphones. In Proceedings of the IoTDI'21: International Conference on Internet-of-Things Design and Implementation, Charlottesville, VA, USA, 18–21 May 2021; ACM: New York, NY, USA, 2021; pp. 28–39. [CrossRef]
4. Ribeiro, M.; Grolinger, K.; Capretz, M.A. MLaaS: Machine Learning as a Service. In Proceedings of the 2015 IEEE 14th International Conference on Machine Learning and Applications (ICMLA), Miami, FL, USA, 9–11 December 2015; pp. 896–902. [CrossRef]
5. Achille, A.; Soatto, S. Emergence of Invariance and Disentanglement in Deep Representations. *J. Mach. Learn. Res.* **2018**, *19*, 1947–1980.
6. Google. Google Now Launcher. 2018. Available online: https://en.wikipedia.org/wiki/Google_Now (accessed on 6 December 2023).
7. Google. Data Preparation. 2018. Available online: https://cloud.google.com/ml-engine/docs/tensorflow/data-prep (accessed on 6 December 2023).
8. Fredrikson, M.; Jha, S.; Ristenpart, T. Model Inversion Attacks That Exploit Confidence Information and Basic Countermeasures. In Proceedings of the CCS'15, 22nd ACM SIGSAC Conference on Computer and Communications Security, Denver, CO, USA, 12–16 October 2015; pp. 1322–1333. [CrossRef]
9. Mahendran, A.; Vedaldi, A. Understanding deep image representations by inverting them. In Proceedings of the 2015 IEEE Conference on Computer Vision and Pattern Recognition (CVPR), Boston, MA, USA, 7–12 June 2015; pp. 5188–5196. [CrossRef]
10. Hidano, S.; Murakami, T.; Katsumata, S.; Kiyomoto, S.; Hanaoka, G. Model Inversion Attacks for Prediction Systems: Without Knowledge of Non-Sensitive Attributes. In Proceedings of the 2017 15th Annual Conference on Privacy, Security and Trust (PST), Calgary, AB, Canada, 28–30 August 2017; pp. 115–11509. [CrossRef]
11. Osia, S.A.; Shahin Shamsabadi, A.; Sajadmanesh, S.; Taheri, A.; Katevas, K.; Rabiee, H.R.; Lane, N.D.; Haddadi, H. A Hybrid Deep Learning Architecture for Privacy-Preserving Mobile Analytics. *IEEE Internet Things J.* **2020**, *7*, 4505–4518. [CrossRef]
12. Goodfellow, I.; Pouget-Abadie, J.; Mirza, M.; Xu, B.; Warde-Farley, D.; Ozair, S.; Courville, A.; Bengio, Y. Generative Adversarial Nets. In *Advances in Neural Information Processing Systems*; Ghahramani, Z., Welling, M., Cortes, C., Lawrence, N., Weinberger, K.Q., Eds.; Curran Associates, Inc.: Red Hook, NY, USA, 2014; Volume 27.
13. Liu, S.; Du, J.; Shrivastava, A.; Zhong, L. Privacy Adversarial Network: Representation Learning for Mobile Data Privacy. *Proc. Acm Interact. Mobile, Wearable Ubiquitous Technol.* **2019**, *3*, 144. [CrossRef]
14. Li, A.; Duan, Y.; Yang, H.; Chen, Y.; Yang, J. TIPRDC: Task-Independent Privacy-Respecting Data Crowdsourcing Framework for Deep Learning with Anonymized Intermediate Representations. In Proceedings of the KDD'20, 26th ACM SIGKDD International Conference on Knowledge Discovery & Data Mining, Virtual, 6–10 July 2020; pp. 824–832. [CrossRef]
15. Zhou, B.; Khosla, A.; Lapedriza, A.; Oliva, A.; Torralba, A. Learning Deep Features for Discriminative Localization. In Proceedings of the 2016 IEEE Conference on Computer Vision and Pattern Recognition (CVPR), Las Vegas, NV, USA, 27–30 June 2016; pp. 2921–2929. [CrossRef]
16. Chattopadhay, A.; Sarkar, A.; Howlader, P.; Balasubramanian, V.N. Grad-CAM++: Generalized Gradient-Based Visual Explanations for Deep Convolutional Networks. In Proceedings of the 2018 IEEE Winter Conference on Applications of Computer Vision (WACV), Lake Tahoe, NV, USA, 12–15 March 2018; pp. 839–847. [CrossRef]
17. Zhang, Q.; Rao, L.; Yang, Y. Group-CAM: Group Score-Weighted Visual Explanations for Deep Convolutional Networks. *arXiv* **2021**, arXiv:2103.13859. Available online: http://arxiv.org/abs/2103.13859 (accessed on 6 December 2023).
18. Zhang, Q.; Wang, X.; Wu, Y.N.; Zhou, H.; Zhu, S.C. Interpretable CNNs for Object Classification. *IEEE Trans. Pattern Anal. Mach. Intell.* **2021**, *43*, 3416–3431. [CrossRef] [PubMed]
19. Higgins, I.; Matthey, L.; Pal, A.; Burgess, C.; Glorot, X.; Botvinick, M.; Mohamed, S.; Lerchner, A. beta-VAE: Learning Basic Visual Concepts with a Constrained Variational Framework. In Proceedings of the 5th International Conference on Learning Representations, ICLR 2017, Toulon, France, 24–26 April 2017; Conference Track Proceedings. Available online: http://OpenReview.net (accessed on 6 December 2023).
20. Kim, H.; Mnih, A. Disentangling by Factorising. In Proceedings of the Proceedings of the 35th International Conference on Machine Learning, ICML 2018, Stockholmsmässan, Stockholm, Sweden, 10–15 July 2018; Dy, J.G., Krause, A., Eds.; Proceedings of Machine Learning Research: Breckenridge, CO, USA, 2018; Volume 80, pp. 2654–2663.
21. Chen, T.Q.; Li, X.; Grosse, R.B.; Duvenaud, D. Isolating Sources of Disentanglement in Variational Autoencoders. In *Advances in Neural Information Processing Systems 31, Proceedings of theAnnual Conference on Neural Information Processing Systems 2018, NeurIPS 2018, Montréal, QC, Canada, 3–8 December 2018*; Bengio, S., Wallach, H.M., Larochelle, H., Grauman, K., Cesa-Bianchi, N., Garnett, R., Eds.; Neural Information Processing Systems: La Jolla, CA, USA, 2018; pp. 2615–2625.

22. Chen, X.; Duan, Y.; Houthooft, R.; Schulman, J.; Sutskever, I.; Abbeel, P. InfoGAN: Interpretable Representation Learning by Information Maximizing Generative Adversarial Nets. In *Advances in Neural Information Processing Systems 29, Proceedings of theAnnual Conference on Neural Information Processing Systems 2016, Barcelona, Spain, 5–10 December 2016*; Lee, D.D., Sugiyama, M., von Luxburg, U., Guyon, I., Garnett, R., Eds.; Neural Information Processing Systems: La Jolla, CA, USA, 2016; pp. 2172–2180.
23. Kingma, D.P.; Dhariwal, P. Glow: Generative Flow with Invertible 1x1 Convolutions. In *Advances in Neural Information Processing Systems 31, Proceedings of the Annual Conference on Neural Information Processing Systems 2018, NeurIPS 2018, Montréal, QC, Canada 3–8 December 2018*; Bengio, S., Wallach, H.M., Larochelle, H., Grauman, K., Cesa-Bianchi, N., Garnett, R., Eds.; Neural Information Processing Systems: La Jolla, CA, USA, 2018; pp. 10236–10245.
24. Sweeney, L. k-Anonymity: A Model for Protecting Privacy. *Int. J. Uncertain. Fuzziness Knowl. Based Syst.* **2002**, *10*, 557–570. [CrossRef]
25. Machanavajjhala, A.; Gehrke, J.; Kifer, D.; Venkitasubramaniam, M. L-diversity: Privacy beyond k-anonymity. In Proceedings of the 22nd International Conference on Data Engineering (ICDE'06), Atlanta, GA, USA, 3–7 April 2006; pp. 24–24. [CrossRef]
26. Li, N.; Li, T.; Venkatasubramanian, S. t-Closeness: Privacy Beyond k-Anonymity and l-Diversity. In Proceedings of the 2007 IEEE 23rd International Conference on Data Engineering, Atlanta, GA, USA, 3–7 April 2006; pp. 106–115. [CrossRef]
27. Dwork, C.; Roth, A. The Algorithmic Foundations of Differential Privacy. *Found. Trends Theor. Comput. Sci.* **2014**, *9*, 211–407. [CrossRef]
28. Mironov, I. Rényi Differential Privacy. In Proceedings of the 2017 IEEE 30th Computer Security Foundations Symposium (CSF), Santa Barbara, CA, USA, 21–25 August 2017; pp. 263–275. [CrossRef]
29. Abadi, M.; Chu, A.; Goodfellow, I.; McMahan, H.B.; Mironov, I.; Talwar, K.; Zhang, L. Deep Learning with Differential Privacy. In Proceedings of the CCS'16, 2016 ACM SIGSAC Conference on Computer and Communications Security, Vienna, Austria, 24–28 October 2016; pp. 308–318. [CrossRef]
30. Papernot, N.; Song, S.; Mironov, I.; Raghunathan, A.; Talwar, K.; Erlingsson, Ú. Scalable Private Learning with PATE. In Proceedings of the 6th International Conference on Learning Representations, ICLR 2018, Vancouver, BC, Canada, 30 April–3 May 2018; Conference Track Proceedings. Available online: http://OpenReview.net (accessed on 6 December 2023).
31. Oh, S.J.; Benenson, R.; Fritz, M.; Schiele, B. Faceless Person Recognition: Privacy Implications in Social Media. In *Computer Vision—ECCV 2016*; Leibe, B., Matas, J., Sebe, N., Welling, M., Eds.; Springer: Cham, Switzerland, 2016; pp. 19–35.
32. Dowlin, N.; Gilad-Bachrach, R.; Laine, K.; Lauter, K.; Naehrig, M.; Wernsing, J. CryptoNets: Applying Neural Networks to Encrypted Data with High Throughput and Accuracy. In Proceedings of the ICML'16 33rd International Conference on International Conference on Machine Learning, New York, NY, USA, 19–24 June 2016; Volume 48, pp. 201–210. Available online: http://JMLR.org (accessed on 6 December 2023).
33. Li, J.; Kuang, X.; Lin, S.; Ma, X.; Tang, Y. Privacy preservation for machine learning training and classification based on homomorphic encryption schemes. *Inf. Sci.* **2020**, *526*, 166–179. [CrossRef]
34. Riazi, M.S.; Weinert, C.; Tkachenko, O.; Songhori, E.M.; Schneider, T.; Koushanfar, F. Chameleon: A Hybrid Secure Computation Framework for Machine Learning Applications. In Proceedings of the ASIACCS'18, 2018 on Asia Conference on Computer and Communications Security, Incheon, Republic of Korea, 4 June 2018; pp. 707–721. [CrossRef]
35. Liu, J.; Juuti, M.; Lu, Y.; Asokan, N. Oblivious Neural Network Predictions via MiniONN Transformations. In Proceedings of the CCS'17, Proceedings of the 2017 ACM SIGSAC Conference on Computer and Communications Security, Dallas, TX USA, 30 October–3 November 2017; pp. 619–631. [CrossRef]
36. Mohassel, P.; Zhang, Y. SecureML: A System for Scalable Privacy-Preserving Machine Learning. In Proceedings of the 2017 IEEE Symposium on Security and Privacy (SP), San Jose, CA, USA, 22–26 May 2017; pp. 19–38. [CrossRef]
37. Yu, J.; Zhang, B.; Kuang, Z.; Lin, D.; Fan, J. iPrivacy: Image Privacy Protection by Identifying Sensitive Objects via Deep Multi-Task Learning. *IEEE Trans. Inf. Forensics Secur.* **2017**, *12*, 1005–1016. [CrossRef]
38. Malekzadeh, M.; Clegg, R.G.; Haddadi, H. Replacement AutoEncoder: A Privacy-Preserving Algorithm for Sensory Data Analysis. In Proceedings of the 2018 IEEE/ACM Third International Conference on Internet-of-Things Design and Implementation (IoTDI), Orlando, FL, USA, 17–20 April 2018; pp. 165–176. [CrossRef]
39. Aloufi, R.; Haddadi, H.; Boyle, D. Privacy-Preserving Voice Analysis via Disentangled Representations. In Proceedings of the CCSW'20, 2020 ACM SIGSAC Conference on Cloud Computing Security Workshop, Virtual, 9 November 2020; pp. 1–14. [CrossRef]
40. van den Oord, A.; Vinyals, O.; kavukcuoglu, K. Neural Discrete Representation Learning. In *Advances in Neural Information Processing Systems*; Guyon, I., Luxburg, U.V., Bengio, S., Wallach, H., Fergus, R., Vishwanathan, S., Garnett, R., Eds.; Curran Associates, Inc.: Red Hook, NY, USA, 2017; Volume 30.
41. Kalchbrenner, N.; Elsen, E.; Simonyan, K.; Noury, S.; Casagrande, N.; Lockhart, E.; Stimberg, F.; van den Oord, A.; Dieleman, S.; Kavukcuoglu, K. Efficient Neural Audio Synthesis. In Proceedings of the 35th International Conference on Machine Learning, ICML 2018, Stockholmsmässan, Stockholm, Sweden, 10–15 July 2018; Dy, J.G., Krause, A., Eds.; Proceedings of Machine Learning Research: Breckenridge, CO, USA, 2018; Volume 80, pp. 2415–2424.
42. Gong, M.; Liu, J.; Li, H.; Xie, Y.; Tang, Z. Disentangled Representation Learning for Multiple Attributes Preserving Face Deidentification. *IEEE Trans. Neural Netw. Learn. Syst.* **2020**, *33*, 244–256. [CrossRef]

43. Wu, H.; Tian, X.; Li, M.; Liu, Y.; Ananthanarayanan, G.; Xu, F.; Zhong, S. PECAM: Privacy-Enhanced Video Streaming and Analytics via Securely Reversible Transformation. In Proceedings of the MobiCom'21, 27th Annual International Conference on Mobile Computing and Networking, New Orleans, Louisiana, 25–29 October 2021; pp. 229–241. [CrossRef]
44. Jia, J.; Gong, N.Z. Attriguard: A Practical Defense against Attribute Inference Attacks via Adversarial Machine Learning. In Proceedings of the SEC'18, 27th USENIX Conference on Security Symposium, Baltimore, MD, USA, 15–17 August 2018; pp. 513–529.
45. Wu, Z.; Wang, Z.; Wang, Z.; Jin, H. Towards Privacy-Preserving Visual Recognition via Adversarial Training: A Pilot Study. In *Computer Vision—ECCV 2018*; Ferrari, V., Hebert, M., Sminchisescu, C., Weiss, Y., Eds.; Springer: Cham, Switzerland, 2018; pp. 627–645.
46. Larsen, A.B.L.; Sønderby, S.K.; Larochelle, H.; Winther, O. Autoencoding beyond Pixels Using a Learned Similarity Metric. In Proceedings of the ICML'16, 33rd International Conference on International Conference on Machine Learning, New York, NY, USA, 19–24 June 2016; Volume 48, pp. 1558–1566.
47. Matthey, L.; Higgins, I.; Hassabis, D.; Lerchner, A. dSprites: Disentanglement Testing Sprites Dataset. 2017. https://github.com/deepmind/dsprites-dataset/ (accessed on 6 December 2023).
48. Lecun, Y.; Bottou, L.; Bengio, Y.; Haffner, P. Gradient-based learning applied to document recognition. *Proc. IEEE* **1998**, *86*, 2278–2324. [CrossRef]
49. Liu, Z.; Luo, P.; Wang, X.; Tang, X. Deep Learning Face Attributes in the Wild. In Proceedings of the International Conference on Computer Vision (ICCV), Santiago, Chile, 7–13 December 2015.
50. Truex, S.; Baracaldo, N.; Anwar, A.; Steinke, T.; Ludwig, H.; Zhang, R. A Hybrid Approach to Privacy-Preserving Federated Learning. In Proceedings of the AISec'19, 12th ACM Workshop on Artificial Intelligence and Security, London, UK, 15 November 2019; pp. 1–2.
51. van der Maaten, L.; Hinton, G. Viualizing data using t-SNE. *J. Mach. Learn. Res.* **2008**, *9*, 2579–2605.

sensors

MDPI

Article

Algorithm Design and Convergence Analysis for Coexistence of Cognitive Radio Networks in Unlicensed Spectrum

Yuan Zhang [1,†], Weihua Wu [2], Wei He [3] and Nan Zhao [1,*]

1 State Key Laboratory of ISN, School of Telecommunications Engineering, Xidian University, Xi'an 710126, China; zhangyuan031400@hollysys.com
2 School of Physics and Information Technology, Shaanxi Normal University, Xi'an 710119, China; whwu@snnu.edu.cn
3 College of Electronic Countermeasures, National University of Defense Technology, Changsha 410073, China; billfovy@hotmail.com
* Correspondence: zhaonan@xidian.edu.cn
† Current address: School of Telecommunications Engineering, Xidian University, North Campus, No. 2 South Taibai Road, Xi'an 710071, China.

Abstract: This paper focuses on achieving the low-cost coexistence of the networks in an unlicensed spectrum by making them operate on non-overlapping channels. For achieving this goal, we first give a universal convergence analysis framework for the unlicensed spectrum allocation algorithm. Then, a one-timescale iteration-adjustable unlicensed spectrum allocation algorithm is developed, where the step size and timescale parameter can be jointly adjusted based on the system performance requirement and signal overhead concern. After that, we derive the sufficient condition for the one-timescale algorithm. Furthermore, the upper bound of convergence error of the one-timescale spectrum allocation algorithm is obtained. Due to the multi-timescale evolution of the network states in the wireless network, we further propose a two-timescale iteration-adjustable joint frequency selection and frequency allocation algorithm, where the frequency selection iteration timescale is set according to the slow-changing statistical channel state information (CSI), whereas the frequency allocation iteration timescale is set according to the fast-changing local CSI. Then, we derive the convergence condition of two-timescale algorithms and the upper bound of the corresponding convergence error. The experimentalresults show that the small timescale adjustment parameter and large step size can help decrease the convergence error. Moreover, compared with traditional algorithms, the two-timescale policy can achieve throughput similar to traditional algorithms with very low iteration overhead.

Keywords: cognitive radio networks; convergence analysis; continuous-time Lyapunov drift

1. Introduction

As high-speed data services continue to grow, the unlicensed spectrum has become an option for cellular operators to increase their service capabilities [1]. In total, there is over 500 MHz of spectrum bandwidth for public (e.g., (5.15, 5.25) GHz and (5.47, 5.48) GHz in the US). Compared with the licensed spectrum, the unlicensed spectrum is a completely open spectrum resource open to different operators as long as they comply with the relevant regulations. Through sharing the completely open unlicensed spectrum, operators no longer need to deploy its network resources with over-provisioning according to the peak load. In addition, it is very cost-effective to use the unlicensed spectrum. Depending on the free unlicensed spectrum resource, operators can efficiently cope with the diminishing effective revenue per GB.

Currently, WiFi is the most popular and successful wireless technology on the unlicensed bandwidths. However, the spectrum efficiency of the WiFi system is pretty low, especially when the number of subscribers is large. Hence, deploying cellular technology over the unlicensed spectrum has become a popular research topic in recent years. However, in the coexistence scenario of 5G and the WiFi network, the decision of one network can impair the decision of another markedly if they share the same unlicensed spectrum band. For example, the LTE unlicensed network (LTE-U) does not implement CSMA, whereas WiFi does. As a result, the WiFi network always senses the channel is busy until the LTE network does not have any traffic to transmit. Therefore, knowing how to achieve harmonious coexistence between LTE and incumbent systems that are already operating in unlicensed spectrum is a key challenge.

The spectrum-sharing and coexistence scenarios can be categorized into two categories, i.e., uncoordinated and coordinated schemes [2]. In the former, networks achieve coexistence based on their local information, whereas in the latter, networks should directly or indirectly exchange signals with each other. In the coordinated schemes, coordination can be implemented centralized by a cloud-based control plane via a direct communication channel [3]. With the coordination of the cloud-based control plane, heterogeneous technologies can be allocated with its corresponding clean channels. Hence, each of them can use its own resource segments without cross-tier interference. Moreover, the cloud-based control plane can also consider the load dynamics and failure scenarios of the network for adopting an adaptive coexistence scheme to avoid the underutilization of the unlicensed spectrum [4]. Due to the fact that each network can identify the existence and operation parameters of another network, coordinated schemes can provide higher performance than uncoordinated schemes. However, this performance efficiency is achieved at the expense of infrastructure/protocol complexity and coordination signal overhead. Especially when the coordination logic is deployed in a wider coverage region beyond one-hop or two-hop nodes, huge communication overhead will take place through the backhaul links.

Therefore, in this paper, we focus on achieving the low-cost coexistence of networks in the unlicensed spectrum by making them operate on separate non-overlapping channels. The use of an unlicensed spectrum allocation algorithm is critical to achieving this goal. To avoid the over-provisioning of the spectrum resource on each network, the spectrum allocation algorithm should have adaptive ability for solving the optimal solution after network state changes. However, if the spectrum allocation algorithm iteration evolves much faster than the network state dynamics, a lot of signal exchange between the coordination logic and networks will be incurred. On the contrary, if the spectrum allocation algorithm iteration evolves in a comparative timescale as the network state dynamics, the network states may have changed after a few iterations, and the existing convergence results failed to apply in this case of network states. Hence, we provide a theoretical framework in this paper for studying the convergence of the adaptive unlicensed spectrum allocation algorithm with adjustable iteration timescale. It should be noted that the theoretical framework of convergence analysis in this paper not only can be used for the coexistence of 5G and WiFi networks but also for the IoT networks with high heterogeneities.

The main contributions of this paper are summarized as follows:

- A universal convergence analysis framework is developed for the unlicensed spectrum allocation algorithm. Specifically, the algorithm iteration evolutions are modeled by a system of stochastic differential equations as a virtual dynamic system. We further reveal that the stability of the dynamic system is equivalent to the convergence of the algorithm iteration.
- A one-timescale iteration-adjustable unlicensed spectrum allocation algorithm is developed, where the step size and timescale parameter can be jointly adjusted based on the system performance requirement and signal overhead consideration. After that, we obtain the sufficient condition for the convergence of the one-timescale iteration-adjustable algorithm and the upper bound of the convergence error.
- We further propose a two-timescale iteration-adjustable joint frequency selection and frequency allocation algorithm, where the frequency selection iteration timescale is set according to the slow-changing statistical channel state information (CSI), whereas

the frequency allocation iteration timescale is set according to the fast-changing local CSI. Then, we derive the convergence condition of the two-timescale algorithm and the upper bound of the corresponding convergence error under mixed timescale network states.

We organize the paper as follows. Section 2 summarizes the related work. Section 3 outlines the system model and problem formulation. Section 4 presents the static spectrum allocation iteration algorithm. Section 5 includes the one-timescale dynamic spectrum allocation algorithm. Then, in Section 6, we outline the two-timescale network state structure and develop the two-timescale dynamic spectrum allocation algorithm. In Section 7, we conduct the simulation. Section 8 concludes the paper.

2. Related Work

Many approaches have been proposed in these years for achieving the coexistence between WiFi and LTE-U in the unlicensed bands. These methods can be classified into the power domain approach, time domain approach and frequency domain approach.

A joint user association and power control algorithm was developed in [5] for maximizing the number of QoS-preferred users supported by LAA-LTE while protecting the WiFi users. The work in [6] proposed a power control algorithm based on Q-Learning to realize the coexistence of licensed users and unlicensed users. A stochastic optimization framework was developed in [7] for minimizing the average power consumption in the LAA-enabled SBSs and WiFi networks. To realize coexistence, the works in [8] regarded WiFi users as protected users, and then proposed a power control scheme to ensure that the interference caused by cellular users to WiFi users is less than a certain threshold. A distributed solution based on Lagrangian relaxation was proposed in [9] to assist the LTE-unlicensed network in making decisions on transmit power. Although the power domain scheme can theoretically realize the coexistence of LTE and WiFi, it is very complex to enact fine-grained control on the transmit power in real networks. On the other hand, in time-varying networks, the power control algorithm will bring huge algorithm iteration overhead and signaling transmission overhead.

The time domain scheme relies on LTE's scheduling to periodically turn off its transmission so that WiFi users can have adequate access time [10]. Hence, the time domain scheme can be also considered as an uncoordinated scheme. The listen before talk (LBT)-based access mechanisms have been proposed in [11–13] for adjusting the backoff window size based on WiFi traffic load to enhance the service capacity of an LTE-U. Different from the LBT mechanism, the devices in [3,14] transmitted data on different timeslots by the polling method. However, there are many issues for using these approaches. Under the current LTE service, the time domain transmission approach degrades the rate guarantee that devices have been accustomed to. As a result, it is difficult to know why a device would select LTE service instead of using free WiFi directly. The time domain scheme usually requires all of the devices to be synchronized with each other, which would involve a major change to some LTE protocols.

Different from the time-domain scheme, in the frequency domain, the coexistence between LTE and WiFi can be achieved by having the two operate on separate, non-overlapping channels [15]. In order to achieve the frequency domain coexistence, Refs. [16–18] developed the unlicensed spectrum splitting strategy between WiFi and the femtocell network. Note that all of these schemes need a central server to coordinate the spectrum sharing between WiFi and LTE BS, so these schemes can be considered as coordinated schemes. It should be pointed out that these schemes have no adaptive ability; the light-loaded networks cannot deliver its superfluous spectrum resource to the heavy-loaded ones. As a result, the capacity and robustness of the entire network will be seriously restrained. With this consideration, ref. [15] developed the adaptive spectrum allocation schemes for achieving the significant improvement in network performance. However, this scheme needs to reformulate a new optimization problem under the updated network states. Solving the optimization problem often needs an algorithm to be iterated many times. Hence, huge signal overhead over both the air interface and the backhaul links will be incurred. Therefore, we need to design the low-cost adaptive spectrum allocation schemes.

3. System Model

LTE-U and the WiFi network are representative technologies deployed in the unlicensed spectrum. In this paper, our goal is how to realize the coexistence of LTE and WiFi. We assume that there is a cloud server deployed at the backend to implement the coordination of LTE-U and WiFi [1]. As shown in Figure 1, the cloud server connected with LTE-U and WiFi by backhaul links, and the coordinating networks have an infrastructure to exchange information about themselves or their demands. We represent the devices covered by LTE-U as sDevice, and we represent the devices covered by WiFi as wDevice. We assume that the system model in this paper consists of one WiFi, N LTE-U and K devices. Let $\mathcal{N} = \{1, 2, \cdots, N\}$ denote the sets of LTE-U. Furthermore, we use $\mathcal{K} = \mathcal{M} \cup \mathcal{L}$ to denote the devices set, where $\mathcal{M}$ and $\mathcal{L}$ denote the sets of wDevices and sDevices, respectively. Let M and L denote the number of devices in sets $\mathcal{M}$ and $\mathcal{L}$. We assume that the total available bandwidth for the unlicensed spectrum is B.

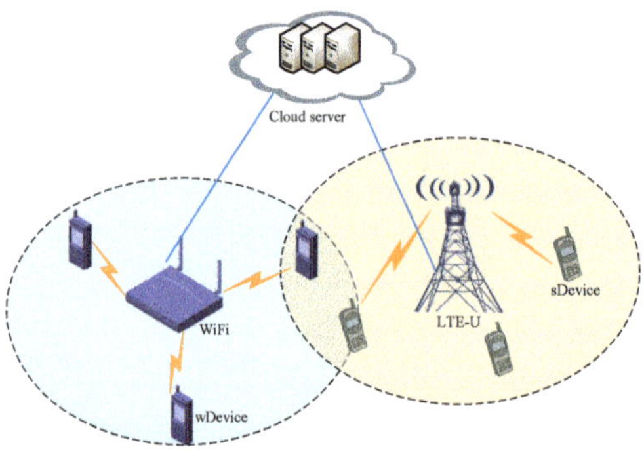

Figure 1. System model. The cloud server connects with LTE-U and WiFi by backhaul links. LTE and WiFi operate in different frequency bands that do not overlap each other.

The throughput of CSMA networks has been investigated by [19]. However, this throughput model cannot reveal the relationship between the spectrum bandwidth, channel state information and the network throughput. Different from [19], we focus more on investigating the relationship between throughput, dynamic CSI and the allocated spectrum resource. To this end, we adopt the Ideal CSMA Network (ICN) model to compute the WiFi throughput [20]. For the wDevices in set $\mathcal{M}$, they compete to access the channel through CSMA protocol. We assume that each wDevice always has packets to transmit; i.e., the network operates in the saturation state. Then, the wDevice needs to spend some time on waiting, confirming and retransmitting when transmitting packets. We use T_{tr} to represent the total time for transmitting a packet, which consists of packet duration, distributed Interframe Spacing, acknowledgement and short Interframe Spacing. In addition, the period of the random backoff countdown process is T_{cd}. Thus, the countdown overhead can be represented as $c = \mathbb{E}[T_{cd}]/\mathbb{E}[T_{tr}]$. The transmission state of wDevice m is expressed as s_m, where $s_m = 1$ and $s_m = 0$ indicate that the device is transmitting and waiting, respectively. After that, we use $s = s_1 s_2 \cdots s_M$ to represent the state of the network. Based on the state transition equation in [20], the stationary distribution of the state $s = 0 \cdots 0$ and $s = s_1 s_2 \cdots s_M$ can be given as

$$P_{0 \cdots 0} = (1 + M/c)^{-1}, P_{s_1 s_2 \cdots s_M | s : s_m = 1} = (c + M)^{-1}.$$

The normalized transmission rate of wDevice m is computed as

$$x_m = P_{s_1 s_2 \cdots s_M | s: s_m = 1} = \frac{1}{M+c}.$$

Then, multiplying the normalized throughput by the achievable uplink transmission rate of wDevice m, we can obtain its actual throughput as

$$r_m = \frac{\alpha_w}{M+c} \log_2\left(1 + \frac{h_m}{\alpha_w}\right)$$

where α_w denotes the fraction of the unlicensed spectrum for WiFi, $h_m = p_m |g_m|^2 / N_0$, p_m denotes the transmit power for wDevice m, N_0 is the power spectral density of the Gaussian background noise, and g_m is the channel gain from wDevice m to WiFi. For the modest-size network, the effect of c on approximating the real throughput is negligible. Therefore, we give the overall throughput of the WiFi as

$$R_{WiFi} = \sum_{m=1}^{M} \frac{\alpha_w}{M} \log_2\left(1 + \frac{h_m}{\alpha_w}\right) \tag{1}$$

In the LTE-U network, we use $\mathcal{L}_n$ to denote the set of sDevices associated with LTE-U n. In the proposed model, LTE-U n will allocate a dedicated channel ρ_k^n to sDevice $k \in \mathcal{L}_n$. Then, the uplink transmission rate for sDevice k can be computed as

$$R_k^n = \rho_k^n \log_2\left(1 + \frac{h_k^n}{\rho_k^n}\right), \forall n \in \mathcal{N}, k \in \mathcal{L}_n. \tag{2}$$

The definition of h_k^n is similar to h_m.

Generally, evaluating the throughput maintained by each network is still a common approach to ensure that all collocated networks can sustain a certain level of performance [1]. Therefore, the optimization problem can be formulated as

$$\begin{aligned} \max \quad & F(\alpha_w, \boldsymbol{\rho}; \mathbf{h}) \\ \text{s.t.} \quad & \alpha_w + \sum_{n=1}^{N} \sum_{k \in \mathcal{L}_n} \rho_k^n \leq B, \\ & \alpha_w > 0, \rho_k^n > 0, \forall k \in \mathcal{L}_n, n \in \mathcal{N}, \end{aligned} \tag{3}$$

where $F(\alpha_w, \boldsymbol{\rho}; \mathbf{h}) =$

$$w_1 \sum_{m=1}^{M} \frac{\alpha_w}{M} \log_2\left(1 + \frac{h_m}{\alpha_w}\right) + w_2 \sum_{n=1}^{N} \sum_{k \in \mathcal{L}_n} \rho_k^n \log_2\left(1 + \frac{h_k^n}{\rho_k^n}\right),$$

$\mathbf{h} = [h_k]_{k \in \mathcal{K}}$, $\boldsymbol{\rho} = [\rho_k^n]_{k \in \mathcal{L}}$, w_1 and w_2 are weights of throughput of sDevices and wDevices, respectively, which provide the preference of operators on wDevices or sDevices. The constraint indicates that the partitioned spectrum can't be more than the total bandwidth.

4. Spectrum Allocation Iteration Algorithm

This section considers that the CSI remains unchanged before the corresponding algorithm-solving problem (3). At this case, problem (3) is a concave optimization problem. We will propose a spectrum allocation algorithm for the LTE-U/WiFi networks and then analyze the convergence of this algorithm.

4.1. Spectrum Allocation Iteration Algorithm

For problem (3), we construct the following Lagrange function

$$L(\alpha_w, \boldsymbol{\rho}, \lambda; \mathbf{h}) = F(\alpha_w, \boldsymbol{\rho}; \mathbf{h}) - \lambda\left(\alpha_w + \sum_{n=1}^{N} \sum_{k \in \mathcal{L}_n} \rho_k^n - B\right),$$

where λ is the Lagrange multiplier. Then, the following Karush-Kuhn-Tucker (KKT) conditions for problem (3) are given as

$$\frac{\partial L(\alpha_w, \boldsymbol{\rho}, \lambda; \mathbf{h})}{\partial \alpha_w} = 0, \tag{4}$$

$$\frac{\partial L(\alpha_w, \boldsymbol{\rho}, \lambda; \mathbf{h})}{\partial \rho_k^n} = 0, \forall n \in \mathcal{N}, k \in \mathcal{L}_n, \tag{5}$$

$$\lambda\left(\alpha_w + \sum_{n=1}^{N} \sum_{k \in \mathcal{L}_n} \rho_k^n - B\right) = 0. \tag{6}$$

For convenience, we express the spectrum allocation variables α_w,$\boldsymbol{\rho}$ and λ as $\mathbf{x} = (\alpha_w, \boldsymbol{\rho}, \lambda) \in \mathbb{R}_+^{L+2}$. Therefore, at the n_s-th solt, $\mathbf{x}$ is iterated as

$$\mathbf{x}_{n_s+1} = \Gamma_+[\mathbf{x}_{n_s} + \gamma_{n_s} \mathbf{G}(\mathbf{x}_{n_s}; \mathbf{h}_{n_s})], \tag{7}$$

where γ_{n_s} is the step size, and Γ_+ denotes the projection onto the non-negative domain. The mapping of $\mathbf{G}(\cdot)$ is given as

$$\mathbf{G}(\cdot) = \begin{pmatrix} \frac{\partial F(\alpha_w, \boldsymbol{\rho}; \mathbf{h})}{\partial \alpha_w} - \lambda \\ \frac{\partial F(\alpha_w, \boldsymbol{\rho}; \mathbf{h})}{\partial \boldsymbol{\rho}} - \lambda \\ \alpha_w + \sum_{n=1}^{N} \sum_{i \in \mathcal{L}_n} \rho_i^n - B \end{pmatrix} \in \mathbb{R}_+^{L+2}. \tag{8}$$

Although many works have proved the convergence of the iteration algorithm [21,22], it is difficult to apply them directly in the scenario where the CSI is time-varying. In a dynamic wireless network, when the network state changes, the optimal spectrum allocation solution will also change, which may result in existing convergence results failling to apply. Hence, in the following, we will explore the convergence performance of the spectrum allocation algorithm in static LTE/WiFi networks. This study will provide a basis for analyzing the convergence of spectrum partitioning algorithms in dynamic network environments.

4.2. Convergence Analysis

If we regard the spectrum allocation solution $\mathbf{x}$ as a dynamic system evolving with the change in network state $\mathbf{h}$, then the state trajectory $\mathbf{x}$ can be considered as the solution to the following equation:

$$d\mathbf{x} = \gamma \mathbf{G}(\mathbf{x}; \mathbf{h}) dt, \tag{9}$$

where the subscript of γ is omitted for the sake of convenience.

Then, the equilibrium point of a specific dynamic system can be defined as

Definition 1. *(Equilibrium Point): The solution $\mathbf{x}^*$ satisfying $\mathbf{G}(\mathbf{x}^*; \mathbf{h}) = 0$ is defined as the equilibrium point.*

In static wireless networks, in the iterative process of the algorithm, CSI $\mathbf{h}$ is always unchanged. Hence, it should be noted that the equilibrium point $\mathbf{x}^*$ is fixed. In order to analyze the convergence performance of the algorithm, we give the definition of convergence error.

Definition 2. *(Convergence Error): Convergence error of the spectrum allocation algorithm is defined as the error between the iterative output* $\mathbf{x}$ *and the optimality* $\mathbf{x}^*$*, i.e.,* $\mathbf{x}_e = \mathbf{x} - \mathbf{x}^*$.

Then, based on this definition, the differential equation of convergence error is defined as

$$d\mathbf{x}_e = d\mathbf{x} - d\mathbf{x}^* = \gamma \mathbf{G}(\mathbf{x}; \mathbf{h})dt. \tag{10}$$

Equation (10) can be also considered as a dynamic system. Therefore, the convergence of the algorithm in Equation (7) is equivalent to the asymptotic stability of the dynamic system defined in Equation (10).

Then, we use Lyapunov theory to study the stability of the system. First, we define a Lyapunov function as $V = \frac{1}{2}\mathbf{x}_e^T \mathbf{x}_e$. Motivated from the definition of Lyapunov drift in a discrete-time system, the Lyapunov drift in a continuous-time dynamic system is defined as

$$LV(\mathbf{x}_e) = \lim_{\delta \to 0} \frac{\mathbb{E}[V(\mathbf{x}_e(t+\delta)) - V(\mathbf{x}_e(t)) | \mathbf{x}_e(t)]}{\delta}. \tag{11}$$

Lyapunov drift indicates the growth direction of the Lyapunov function $V(\mathbf{x}_e)$ over time, which can be also considered as an energy function. Obviously, the negative Lyapunov drift contributes to the stability of the system. However, Equation (11) is not easy to calculate. In order to obtain an accurate expression, the Lyapunov drift $LV(\mathbf{x}_e)$ can also be calculated by the following lemma.

Lemma 1. *(Continuous-Time Lyapunov Drift [23]): Suppose that there is a stochastic process* $\mathbf{z}$ *described as*

$$d\mathbf{z} = f(\mathbf{z})dt + g(\mathbf{z})d\mathbf{W}, \tag{12}$$

the stochastic Lyapunov drift of $V(\mathbf{z})$ *can be written as*

$$LV(\mathbf{z}) = \frac{\partial V(\mathbf{z})}{\partial \mathbf{z}} f(\mathbf{z}) + tr\left[g(\mathbf{z})^T \frac{\partial^2 V(\mathbf{z})}{\partial \mathbf{z} \partial \mathbf{z}^T} g(\mathbf{z})\right],$$

where $\mathbf{W}$ *is a standard complex Wiener process.*

Lemma 1 builds a bridge between the dynamic system in Equation (10) and the Lyapunov drift. Therefore, in the following, we will study the stability of dynamic systems by exploring the characteristics of Lyapunov drift.

Lemma 2. *Suppose that the Lyapunov drift of* $V(\mathbf{z})$ *satisfies*

$$V(\mathbf{z}) \le -a\|\mathbf{z}\| + g(\mathbf{s}),$$

then $\mathbf{z}(t)$ *is stable and satisfies*

$$\limsup_{t \to \infty} \frac{1}{t} \int_0^t E[\mathbf{z}(\tau)]d\tau \le \frac{d}{a}.$$

where $\mathbf{s}(t)$ *satisfies* $\limsup_{t\to\infty} \frac{1}{t} \int_0^t E[g(\mathbf{s}(\tau))]d\tau \le d$ *for* $d < \infty$ *and* a *is a positive constant.*

Proof. The similar proof is shown in [24]. □

In Lemma 2, $\frac{d}{a}$ can be considered as an estimation on the upper bound of convergence error. The advantage of Lemma 2 is that it provides a qualitative analysis of the dynamic system. Then, based on Lemma 2, the Lyapunov drift of the dynamic system in Equation (10) can be computed as shown in Equation (13).

$$
\begin{aligned}
LV(\mathbf{x}_e) =& \mathbf{x}_e^T \gamma \mathbf{G}(\mathbf{x};\mathbf{h}) = \gamma(\alpha_w - \alpha_w^*)(\frac{\partial F(\cdot)}{\partial \alpha_w} - \lambda) + \gamma \sum_{n=1}^{N} \sum_{k \in \mathcal{L}_n} (\rho_k^n - \rho_k^{n,*})(\frac{\partial F(\cdot)}{\partial \rho_k^n} - \lambda) \\
&+ \gamma(\lambda - \lambda^*)(\alpha_w + \sum_{n=1}^{N} \sum_{k \in \mathcal{L}_n} \rho_k^n - B) \\
=& \gamma(\alpha_w - \alpha_w^*)(\frac{\partial F(\cdot)}{\partial \alpha_w} - \frac{\partial F(\cdot)}{\partial \alpha_w^*}) + \gamma(\alpha_w - \alpha_w^*)(\frac{\partial F(\cdot)}{\partial \alpha_w} - \lambda^*) \\
&+ \gamma \sum_{n=1}^{N} \sum_{k \in \mathcal{L}_n} (\rho_k^n - \rho_k^{n,*})(\frac{\partial F(\cdot)}{\partial \rho_k^n} - \frac{\partial F(\cdot)}{\partial \rho_k^{n,*}}) \\
&+ \gamma \sum_{n=1}^{N} \sum_{k \in \mathcal{L}_n} (\rho_k^n - \rho_k^{n,*})(\frac{\partial F(\cdot)}{\partial} \rho_k^n - \lambda^*) + (\lambda - \lambda^*)(\alpha_w^* + \sum_{n=1}^{N} \sum_{k \in \mathcal{L}_n} \rho_k^{n,*} - B) \\
\leq& \gamma(\alpha_w - \alpha_w^*)(\frac{\partial F(\cdot)}{\partial \alpha_w} - \frac{\partial F(\cdot)}{\partial \alpha_w^*}) + \gamma \sum_{n=1}^{N} \sum_{k \in \mathcal{L}_n} (\rho_k^n - \rho_k^{n,*})(\frac{\partial F(\cdot)}{\partial \rho_k^n} - \frac{\partial F(\cdot)}{\partial \rho_k^{n,*}}) \leq -\alpha_x \gamma \|\mathbf{x}_e\|^2.
\end{aligned}
\tag{13}
$$

In Equation (13), the first inequality is based on the KKT conditions in Equations (5) and (6), the second inequality is because $F(\cdot)$ is a concave function.

It can be easily seen that the Lyapunov drift in Equation (13) satisfies the stochastically stable condition in Lemma 2. Hence, we can conclude that the algorithm can iterate to the optimal value and that the convergence error is zero, i.e., $\lim\sup_{t\to\infty} \frac{1}{t}\int_0^t \mathbb{E}[\mathbf{x}_e(\tau)]d\tau = 0$.

However, due to the constant step size, the iteration algorithm definitely converges to within some range of the optimal value; i.e., we have $\lim_{n_s\to\infty} |F_{bset}^{n_s} - F^*| \leq \frac{\gamma\epsilon^2}{2}$, where F^* denotes the optimal value problem (3), $F_{bset}^{n_s}$ is the best objective value found in n_s iterations, and ϵ is the upper bound of the norm of $\mathbf{G}$ [22]. We refer to this part of error as the steady-state error. As a result, the convergence of the spectrum allocation iteration algorithm can be measured by the convergence error and steady-state error. The former determines whether the algorithm is convergent, and the latter shows the quality of the convergence result.

4.3. Signaling Overhead

The spectrum allocation algorithm should be executed with the coordination of the centralized cloud server. Although the coordinated scheme provides higher performance, the coordination overhead will be increased sharply. Especially for the heterogeneous LTE-U and WiFi network, a cross-technology communication channel is used for exchanging the corresponding information [25]. In order to implement the spectrum allocation, the cloud server needs to acquire all of the CSI at each algorithm iteration. Let c_1 and c_2 denote the signaling overheads on wireless links and backhaul links, respectively. In order to achieve δ-optimality, i.e., $|\lambda - \lambda^*| \leq \delta$, it needs the order of $O(\frac{1}{\delta^2})$ iterations to solve the system through a maximization problem. Hence, the amount of signaling overhead between the cloud server and the devices is $O(\frac{K(c_1+c_2)}{\delta^2})$.

Remark 1. *It is certainly true that the iteration algorithm can' converge to the optimal value. However, we can observe from Equation (11) that the spectrum partitioning decision of the WiFi network depends on the statistical global CSI. It is not difficult to understand that the statistical global CSI is affected by the large-scale channel fading, whose statistical properties change over a relatively large timescale. In contrast, the spectrum partitioning decision of LTE user is affected by the local CSI. Moreover, the local CSI changes on a small timescale. Therefore, if the spectrum partitioning decisions of the LTE and WiFi are iterated at the same time on a smaller timescale, although the optimal solution can be found, the iteration of WiFi will bring additional iteration overhead and signaling overhead. In contrast, if we iterate the two decisions on a large timescale, the spectrum partitioning decision of the LTE will waste many transmission opportunities on the* *small timescale. Based on this motivation, we will first explore the convergence of an one-timescale algorithm and extend its theoretical framework to the two-timescale algorithm.*

5. One-Timescale Dynamic Spectrum Allocation

Obviously, the static CSI assumption during the algorithm iteration in the above section is not practical. In fact, the CSI is dynamic and behaves as a stochastic process, along with the spectrum allocation algorithm iteration. In this situation, a fast iterative algorithm, as shown in Figure 2b, may converge to the moving optimal solution. However, the complexity and coordination overhead may be too large to undertake. As shown in Figure 2c, the iteration number and signal overhead will reduce with the increase in the algorithm iteration timescale. However, the convergence of the algorithm cannot be ensured. With this consideration, in this section, we will explore whether the algorithm can converge in dynamic environments and its convergence error.

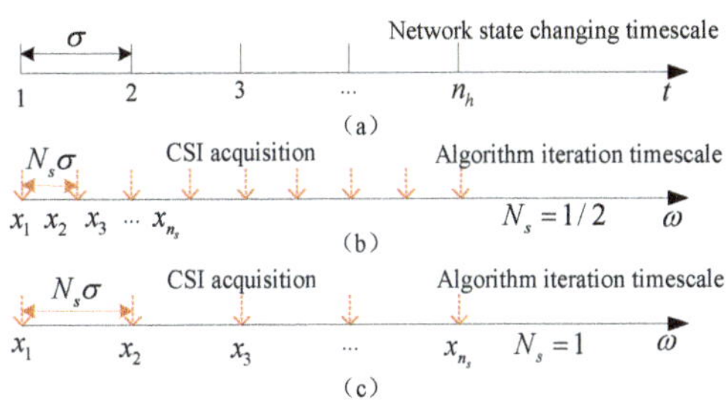

Figure 2. Network state changing timescale vs. one-timescale algorithm iteration

5.1. One-Timescale Algorithm

In dynamic wireless networks, the fading of wireless links satisfies Rayleigh distribution. The signal incoming to the receiver contains a large number of reflected radio waves, which are characterized by "multipath reception". Thus, the wireless channel behaves in a random-like fashion, and its random behavior can be described by the following stochastic equation [26],

$$d\mathbf{h} = -\frac{1}{2}a_h\mathbf{h}dt + a_h^{\frac{1}{2}}d\mathbf{W}_K, \tag{14}$$

where $\mathbf{W}_K$ denotes the K-dimensional Wiener process and a_h is the temporal correlation of $\mathbf{h}$.

Without loss of generality, we assume that the CSI is changing at the slotted time and the slot length is represented as sufficient constant σ. Then, the slot length of the iteration algorithm can be set as $N_s\sigma$, where N_s is the timescale adjustment parameter, and it can be continuously taken values in $(0, 1]$. As shown in Figure 2, a small N_s indicates more iterations of the spectrum allocation algorithm along with the CSI changing. Therefore, at the n_s-th slot, the spectrum allocation variable $\mathbf{x}$ can be also updated according to the following iterations

$$\mathbf{x}_{n_s+1} = \Gamma_+[\mathbf{x}_{n_s} + \gamma_{n_s}\mathbf{G}(\mathbf{x}_{n_s}; \mathbf{h}_{n_s})]. \tag{15}$$

Equation (15) characterizes a discrete iterative algorithm. To describe its iterative trajectory, we construct the following dynamic system

$$d\mathbf{x} = \gamma\mathbf{G}(\mathbf{x}; \mathbf{h})d\omega, \tag{16}$$

where ω is the timescale of the spectrum allocation algorithm.

In order to analyze the convergence of the algorithm in the network, we need to relate the algorithm iteration to the running time of the network. From the viewpoint of algorithm iteration, the length of each iteration is σ; thus, the length of n_s iterations can be expressed as the real number ω, i.e., $\omega = n_s\sigma$. From the viewpoint of the network running, the time of an single algorithm iteration is σN_s. Thus, after n_s iterations, the running time of the network is $t = n_s\sigma N_s$. After the above analysis, we found that $n_s = \omega/\sigma = t/(\sigma N_s)$. Hence, we can obtain the relation between ω and t as

$$t = \omega N_s, dt = N_s d\omega. \tag{17}$$

Based on the equation above, the dynamic system can be mapped from ω to t as

$$d\mathbf{x} = \gamma \mathbf{G}(\mathbf{x};\mathbf{h})d\omega = \gamma N_s^{-1}\mathbf{G}(\mathbf{x};\mathbf{h})dt. \tag{18}$$

5.2. Convergence Analysis

In a dynamic wireless network, the optimal solution, i.e., equilibrium point, is always disturbed by the external stochastic CSI processes. Therefore, the equilibrium point is also a stochastic process. Based on the optimality condition $\mathbf{G}(\cdot) = 0$ and the implicit theorem, the dynamic of the equilibrium point can be computed as

$$d\mathbf{x}^* = -\mathbf{G}_{\mathbf{x}^*}^{-1}\mathbf{G}_{\mathbf{h}}d\mathbf{h}, \tag{19}$$

where $\mathbf{G}_{\mathbf{x}^*} \triangleq \frac{\partial}{\partial \mathbf{x}^*}\mathbf{G}(\cdot)$, $\mathbf{G}_{\mathbf{h}} \triangleq \frac{\partial}{\partial \mathbf{h}}\mathbf{G}(\cdot)$.

Then, by subtracting Equation (19) from Equation (18), we can obtain the convergence error for the algorithm iteration.

$$d\mathbf{x}_e = d\mathbf{x} - d\mathbf{x}^* = \gamma N_s^{-1}\mathbf{G}(\cdot)dt - \mathbf{G}_{\mathbf{x}^*}^{-1}\mathbf{G}_{\mathbf{h}}d\mathbf{h}. \tag{20}$$

By substituting Equation (14) into Equation (20), the convergence error can be further computed as follows

$$d\mathbf{x}_e = \left(\gamma N_s^{-1}\mathbf{G}(\cdot) + \frac{1}{2}a\mathbf{G}_{\mathbf{x}^*}^{-1}\mathbf{G}_{\mathbf{h}}\mathbf{h}\right)dt - a^{\frac{1}{2}}\mathbf{G}_{\mathbf{x}^*}^{-1}\mathbf{G}_{\mathbf{h}}d\mathbf{W}_K. \tag{21}$$

We use $\mathbf{z} = (\mathbf{x}_e, \mathbf{h}) \in \mathbb{R}^{L+2+K}$ to represent the state of a joint system. Then, Equations (14) and (21) can be used to construct the following dynamic system

$$d\mathbf{z} = U_1(\mathbf{z})dt + U_2(\mathbf{z})d\tilde{\mathbf{W}}, \tag{22}$$

where

$$U_1 = \begin{bmatrix} \gamma N_s^{-1}\mathbf{G}(\cdot) + \frac{1}{2}a\mathbf{G}_{\mathbf{x}^*}^{-1}\mathbf{G}_{\mathbf{h}}\mathbf{h} \\ -\frac{1}{2}a\mathbf{h} \end{bmatrix}, U_2 = \begin{bmatrix} -a^{\frac{1}{2}}\mathbf{G}_{\mathbf{x}^*}^{-1}\mathbf{G}_{\mathbf{h}} \\ a^{\frac{1}{2}}\mathbf{I}_K \end{bmatrix},$$

$\tilde{\mathbf{W}}$ is a $(L+2+K)$-dimensional complex Wiener process.

Here, we find that the dynamic system is driven by the complex Wiener process $\tilde{\mathbf{W}}$. Hence, we can analyze the convergence of the spectrum allocation algorithm by studying the stability of the dynamic system. Motivated from this consideration, we first define a Lyapunov function of $\mathbf{z}$ as $V = \frac{1}{2}\mathbf{z}^T\mathbf{z}$. Similarly to the convergence analysis in Section 3, the Lyapunov drift of the dynamic system in (22) can be computed according to Lemma 1 as

$$\begin{aligned} LV(\mathbf{z}) &= \mathbf{z}^T U_1(\mathbf{z}) + \text{tr}\left[U_2(\mathbf{z})^T U_2(\mathbf{z})\right] \\ &\leq \mathbf{x}_e^T\left(\gamma \mathbf{G} N_s^{-1} + \frac{1}{2}a\mathbf{G}_{\mathbf{x}^*}^{-1}\mathbf{G}_{\mathbf{h}}\mathbf{h}\right) \\ &+ a\text{tr}\left[\left(\mathbf{G}_{\mathbf{x}^*}^{-1}\mathbf{G}_{\mathbf{h}}\right)^T\left(\mathbf{G}_{\mathbf{x}^*}^{-1}\mathbf{G}_{\mathbf{h}}\right)\right] - \frac{1}{2}\mathbf{h}^T\mathbf{h} + aK. \end{aligned} \tag{23}$$

In Equation (23), except for the network state parameters, such as a and $\mathbf{h}$, the step size γ and the timescale parameter N_s dominate the value of the Lyapunov drift. Based on

Lemma 2, the dynamic system will be stable if the Lyapunov drift can be upper bounded by a negative function. Hence, by making the Lyapunov drift satisfying the stability constraint in Lemma 2, we can obtain the following theorem.

Theorem 1. *(Convergence of the one-timescale algorithm): Suppose that there exists* $0 < v_{xh} < \infty$, *such that* $||G_{x^*}^{-1}G_h|| \leq v_{xh}$. *Then, if the step size parameter* γ *and timescale parameter* N_s *satisfy*

$$\frac{\gamma}{N_s} \geq \frac{av_{xh}^2}{8\alpha_x},$$

the one-timescale algorithm can converge to the moving optimal solution.

Proof. First of all, from the convergence analysis in Equation (13), we can obtain $\mathbf{x}_e^T\mathbf{G} \leq -\alpha_x||\mathbf{x}_e||^2$. Then, using this result and the assumption in **Theorem 1**, the upper bound of the Lyapunov drift can be computed as

$$\begin{aligned} LV(\mathbf{z}) &\leq -\alpha_x\gamma N_s^{-1}||\mathbf{x}_e||^2 - \frac{1}{2}a||\mathbf{h}||^2 + \frac{1}{2}av_{xh}||\mathbf{h}||\,||\mathbf{x}_e|| \\ &\quad + av_{xh}^2K + aK = -\chi^T\mathbf{A}\chi + av_{xh}^2N_h + aN_h, \end{aligned}$$

where $\chi = [||\mathbf{x}_e||, ||\mathbf{h}||]$ and

$$\mathbf{A} = \begin{bmatrix} \alpha_x\gamma N_s^{-1} & -\frac{1}{4}av_{xh} \\ -\frac{1}{4}av_{xh} & \frac{1}{2}a \end{bmatrix}.$$

Obviously, when the coefficient matrix $\mathbf{A}$ is positive, the dynamic system is stable. Therefore, by making the matrix positive definite, the condition in Theorem 1 can be obtained. □

In addition to the stability, the convergence error is another metric to show the convergence performance of the one-timescale algorithm.

Theorem 2. *(Upper Bound of Convergence Error): The convergence error of the one-timescale algorithm can be given as*

$$\eta \leq \left(\frac{N_s}{\gamma}\right)^2\left(\frac{av_{xh}\pi}{2\alpha_x}\right)^2 + \frac{N_s}{\gamma}\frac{2av_{xh}^2K}{\alpha_x}$$

Proof. Following from Equation (21), the Lyapunov drift of $\mathbf{x}_e$ can be calculated based on Lemma 1 as

$$\begin{aligned} LV(x_e) &\leq \mathbf{x}_e^T\left(\gamma\mathbf{G}N_s^{-1} + \frac{1}{2}a\mathbf{G}_{\mathbf{x}^*}^{-1}\mathbf{G}_{\mathbf{h}}\mathbf{h}\right) \\ &\quad + a\mathrm{tr}\left[\left(\mathbf{G}_{\mathbf{x}^*}^{-1}\mathbf{G}_{\mathbf{h}}\right)^T\left(\mathbf{G}_{\mathbf{x}^*}^{-1}\mathbf{G}_{\mathbf{h}}\right)\right] \\ &\leq -\alpha_x\gamma N_s^{-1}||\mathbf{x}_e||^2 + \frac{1}{2}av_{xh}\pi||\mathbf{x}_e|| + av_{xh}K \\ &\leq -\frac{1}{2}\alpha_x\gamma N_s^{-1}||\mathbf{x}_e||^2 + av_{xh}^2K + \frac{(av_{xh}\pi)^2}{8\alpha_x\gamma N_s^{-1}} \end{aligned}$$

where $\pi = \max_{k\in\mathcal{K}}\{h_k\}$ is the maximum channel gains among K transmission links. According to Lemma 2, the convergence error can be obtained as shown in Theorem 2. □

The above results motivate the following remark on the convergence and convergence error.

In Theorems 1 and 2, a specifies the changing speed of the wireless CSI, K can be interpreted as the network size, v_{xh} represents the sensitivity of the equilibrium point corresponding to the time-varying CSI and α_x is the convergence rate of the spectrum allocation problem. Therefore, for the given network situation, i.e., (a, K), we can increase the iteration number (N_s) or set a large step size (γ) for achieving the algorithm convergence and decreasing the convergence error for the certain spectrum allocation algorithm (α_x, v_{xh}). On the other hand, for the given spectrum allocation algorithm (α_x, v_{xh}) and adjustment parameters (γ, N_s), the CSI changing speed a not only affects the convergence but also determines the convergence error. However, the network size K only determines the convergence error, which comes from the accumulation of convergence error on each link.

5.3. Signaling Overhead

For the one-timescale spectrum allocation algorithm, it needs the order of $O(1/N_s)$ iterations at each slot of the CSI. Similar to the signal overhead in Section 4, at each iteration, the cloud server needs to acquire all of the CSI. Therefore, the signaling overhead amount between the cloud server and devices is $O(\frac{K(c_1+c_2)}{N_s})$, which is much smaller than the signaling overhead in Section 4.

6. Two-Timescale Dynamic Spectrum Allocation

In above section, we observe that the fraction of the unlicensed spectrum for WiFi α_w adapts to the CSI statistics $h_m, m \in \mathcal{M}$ with a global coordination, while the unlicensed spectrum allocation ρ_k^n should adapt to the instantaneous CSI h_k^n locally. In fact, In fact, obtaining real-time local CSI is practical, while obtaining real-time global CSI is extremely difficult [27]. Therefore, the one-timescale spectrum allocation algorithm not only is difficult to achieve in practice but also has very sensitive system performance due to signaling latency in acquiring the global CSI. In this section, we will decompose the spectrum allocation into a short-term control and a long-term control. Furthermore, by iterating them on different timescales, the spectrum allocation policy can not only exploit the instantaneous transmission opportunity at small timescale but also save a lot of signal overhead at the large timescale.

6.1. Two-Timescale Algorithm Dynamics

Motivated from the above discussion, we employ the mixed timescale CSI model, i.e., $h_k = h_k^l h_k^s$, where h_k^l is the large timescale fading and h_k^s is the small timescale fading [27,28]. Furthermore, the dynamics of h_k^l and h_k^s can be specified by the following stochastic differential equations:

$$dh_k^s = -\frac{1}{2} a h_k^s dt + a^{\frac{1}{2}} dW, \forall k \in \mathcal{K}, \tag{24}$$

$$dh_k^l = -c_0 D_k(t)^{-\iota-1} v_k(t) dt, \forall k \in \mathcal{K}, \tag{25}$$

where c_0 is an antenna-gain-related constant, $D_j(t) \geq D_{min}$ is the distance between device k and the corresponding network, $v_k(t)$ is the relative moving speed of device k, and ι is the path loss exponent.

In fact, in any form of wireless network, obtaining real-time global CSI from a central node is extremely difficult and expensive. Hence, we assume that only the network, WiFi or LTE-U, has the knowledge of the local CSI (h_k^l, h_k^s) of the connected devices at the small timescale. On the other hand, the cloud entity can only acquire the global CSI h_k^l for all of the devices at the large timescale. Therefore, the spectrum allocation can be achieved at two-timescales. At the large timescale, the cloud entity select the clean block of bandwidth for WiFi or LTE-U based the global CSI. On the small timescale, the frequency allocation can be executed in the selected frequency by LTE-U using the local CSI.

Let B_n denote the selected clean block of bandwidth for LTE-U n, and let ρ_k^n denote the allocated frequency for sDevice k. We can obtain the following constraint

$$\alpha_w + \sum_{n=1}^{N} B_n \leq B_{max}, \sum_{k \in \mathcal{L}_n} \rho_k^n \leq B_n,$$

where the first equation indicates that the selected bandwidth for WiFi and LTE-U should be smaller than the total unlicensed spectrum and the second equation indicates the frequency allocation constraint for LTE-U.

Then, the two-timescale spectrum allocation problem can be formulated as

$$\begin{aligned} \max \quad & \mathbb{E}[F(\alpha_w, \boldsymbol{\rho}, \mathbf{B}, \mathbf{h})] \\ \text{s.t.} \quad & \alpha_w + \sum_{n=1}^{N} B_n \leq B, \\ & \sum_{l \in \mathcal{L}_n} \rho_n^l \leq B_n, \\ & \alpha_w > 0, B_n > 0, \rho_k^n > 0, \forall n \in \mathcal{N}, k \in \mathcal{K}. \end{aligned} \tag{26}$$

In problem (26), the constraints are coupled by the frequency selection variable B_n. Hence, when the variable B_n is fixed, the problem would decouple. Therefore, it makes sense to decompose problem (26) into two levels of optimization. At the lower level, we have the frequency allocation subproblems for LTE-U n as follows

$$\begin{aligned} F_n(B_n, \mathbf{h}_n) = \max \ & w_2 \sum_{k \in \mathcal{L}_n} \rho_k^n \log_2\left(1 + \frac{h_k^n}{\rho_k^n}\right) \\ \text{s.t.} \ & \sum_{k \in \mathcal{L}_n} \rho_k^n \leq B_n, \rho_k^n > 0, \forall k \in \mathcal{L}_n, \end{aligned} \tag{27}$$

where $\mathbf{h}_n = [h_k^n]_{k \in \mathcal{L}_n}$. At the higher level, we have the master problem in charge of solving the frequency selection problem.

$$\begin{aligned} &\max \mathbb{E}\left[w_1 \sum_{m=1}^{M} \frac{a}{M} \alpha_w \log_2\left(1 + \frac{g_m}{\alpha_w}\right) + \sum_{n=1}^{N} F_n(B_n, \mathbf{h}_n)\right] \\ &\text{s.t.} \ \alpha_w + \sum_{n=1}^{N} B_n \leq B, \alpha_w > 0, B_n > 0, \forall n \in \mathcal{N}. \end{aligned} \tag{28}$$

Let $\mathbf{x} = (\boldsymbol{\rho}, \lambda) \in \mathbb{R}_+^{L+1}$ be the small timescale frequency allocation variable of the lower level subproblem. Similarly, let $\mathbf{y} = (\alpha_w, \mathbf{B}, \mu) \in \mathbb{R}_+^{N+2}$ be the large timescale frequency selection variable of the master problem. In $\mathbf{x}$ and $\mathbf{y}$, both λ and μ are the lagrange multiplier variables. As shown in Figure 3, the slot length of CSI is σ. The slot length of the small timescale iteration algorithm is set as $N_s\sigma$, where N_s is the small timescale adjustment parameter, and it can be continuously taken values in $(0, 1]$. On the other hand, the slot length of the large timescale iteration is $N_l\sigma$, where $N_l \geq 1$ is the large timescale adjustment parameter. Let n_s and n_l denote the slot index of the small timescale iteration and large timescale iteration, respectively. The frequency selection and frequency allocation can be updated according to the following iterations

$$\mathbf{y}_{n_l} = \Gamma_+\left[\mathbf{y}_{n_l-1} + \mu \mathbf{K}(\mathbf{y}_{n_l-1}; \mathbf{h}_{n_l-1}^l)\right], \tag{29}$$

$$\mathbf{x}_{n_s} = \Gamma_+\left[\mathbf{x}_{n_s-1} + \gamma \mathbf{G}(\mathbf{x}_{n_s-1}, \mathbf{y}_{n_l}; \mathbf{h}_{n_s-1}^s, \mathbf{h}_{n_l}^l)\right]. \tag{30}$$

Then, the trajectories of the algorithm iteration in Equations (29) and (30) can be characterized by the following dynamic system

$$d\mathbf{y} = \mu \mathbf{K}(\mathbf{y}; \mathbf{h}^l) d\kappa, d\mathbf{x} = \gamma \mathbf{G}(\mathbf{x}, \mathbf{y}; \mathbf{h}^s, \mathbf{h}^l) d\omega, \tag{31}$$

where κ and ω are the virtual timescales of frequency selection and frequency allocation, respectively. Under the continuous-time algorithm, the iteration indices of the two-timescale algorithm are represented as $n_s = \lfloor \frac{\omega}{N_s\sigma} \rfloor$, $n_l = \lfloor \frac{\kappa}{N_l\sigma} \rfloor$. Then, based on the relation with the CSI timescale, i.e., $n_h = \lfloor \frac{t}{\sigma} \rfloor$, the timescale relation of κ, ω and t is given as

$$t = \kappa N_l^{-1}, t = \omega N_s^{-1}, dt = N_l^{-1} d\kappa, dt = N_s^{-1} d\omega.$$

Then, the dynamics of the wireless network under frequency selection and frequency allocation algorithms are determined by the following coupled equations

$$d\mathbf{h} = -\frac{1}{2} a\mathbf{h}dt + a^{\frac{1}{2}} d\mathbf{W}, \tag{32}$$

$$d\mathbf{x} = \gamma N_s^{-1} \mathbf{G}(\mathbf{x}, \mathbf{y}; \mathbf{h}^s, \mathbf{h}^l) dt, \tag{33}$$

$$d\mathbf{y} = \gamma N_l^{-1} \mathbf{K}(\mathbf{y}; \mathbf{h}^l) dt. \tag{34}$$

The equilibrium point is a stochastic point driven by external conditions. Therefore, it is necessary to study whether the iteration of the algorithm can track the change in the above equilibrium point and its tracking error.

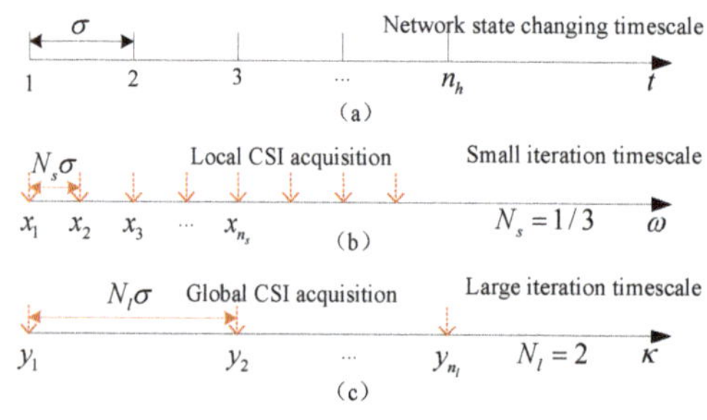

Figure 3. Network state changing timescale vs. two-timescale algorithm iteration.

6.2. Convergence Analysis

For the two-timescale iteration algorithms, the trajectory of equilibrium $\mathbf{y}^*(\mathbf{h}^l)$ is driven by the time-varying $\mathbf{h}^l$. Due to the slow variation of $\mathbf{h}^l$, the trajectory of equilibrium $\mathbf{x}^*(\mathbf{y}, \mathbf{h}^s, \mathbf{h}^l)$ is mainly driven by the time-varying $\mathbf{y}$ and $\mathbf{h}^s$. According to the implicit theorem and the optimality conditions $\mathbf{G}(\cdot) = 0, \mathbf{K}(\cdot) = 0$, we can obtain the following equations

$$d\mathbf{x}^* = -\mathbf{G}_{\mathbf{x}^*}^{-1} \mathbf{G}_{\mathbf{h}^s} d\mathbf{h}^s - \mathbf{G}_{\mathbf{x}^*}^{-1} \mathbf{G}_{\mathbf{y}} d\mathbf{y}, \tag{35a}$$

$$d\mathbf{y}^* = -\mathbf{K}_{\mathbf{y}^*}^{-1} \mathbf{K}_{\mathbf{h}^l} d\mathbf{h}^l, \tag{35b}$$

where $\mathbf{G}_{\mathbf{x}^*} \triangleq \frac{\partial}{\partial \mathbf{x}^*}\mathbf{G}(\cdot)$, $\mathbf{G}_{\mathbf{h}^s} \triangleq \frac{\partial}{\partial \mathbf{h}^s}\mathbf{G}(\cdot)$, $\mathbf{G}_{\mathbf{y}} \triangleq \frac{\partial}{\partial \mathbf{y}}\mathbf{G}(\cdot)$ and $\mathbf{K}_{\mathbf{y}^*} \triangleq \frac{\partial}{\partial \mathbf{y}^*}\mathbf{K}(\cdot)$, $\mathbf{K}_{\mathbf{h}^l} \triangleq \frac{\partial}{\partial \mathbf{h}^l}\mathbf{K}(\cdot)$. Then, the convergence error can be computed as

$$d\mathbf{x} - d\mathbf{x}^* = \gamma N_s^{-1} \mathbf{G}(\cdot) dt + \mathbf{G}_{\mathbf{x}^*}^{-1} \mathbf{G}_{\mathbf{h}^s} d\mathbf{h}^s + \mathbf{G}_{\mathbf{x}^*}^{-1} \mathbf{G}_{\mathbf{y}} d\mathbf{y}, \tag{36}$$

$$d\mathbf{y} - d\mathbf{y}^* = \gamma N_l^{-1} \mathbf{K}(\cdot) dt + \mathbf{K}_{\mathbf{y}^*}^{-1} \mathbf{K}_{\mathbf{h}^l} d\mathbf{h}^l. \tag{37}$$

Substituting Equations (24) and (25) into the above equations, we obtain

$$\begin{aligned} d\mathbf{x}_e = & \Big(\gamma N_s^{-1} \mathbf{G}(\cdot) + \gamma N_l^{-1} \mathbf{G}_{\mathbf{x}^*}^{-1} \mathbf{G}_{\mathbf{y}} \mathbf{K}(\cdot) \\ & -\frac{1}{2} a \mathbf{G}_{\mathbf{x}^*}^{-1} \mathbf{G}_{\mathbf{h}^s} \mathbf{h}^s \Big) dt + a^{\frac{1}{2}} \mathbf{G}_{\mathbf{x}^*}^{-1} \mathbf{G}_{\mathbf{h}^s} d\mathbf{W}, \end{aligned} \tag{38}$$

$$d\mathbf{y}_e = \Big(\gamma N_l^{-1} \mathbf{K}(\cdot) + \mathbf{K}_{\mathbf{y}^*}^{-1} \mathbf{K}_{\mathbf{h}^l} \mathbf{H}_L \Big) dt \tag{39}$$

where $\mathbf{H}_L$ is a diagonal matrix, and $\mathbf{H}_L(i,i) = c_0 D_i(t)^{-\iota-1} v_i(t)$. Let $\mathbf{u} = (\mathbf{x}_e, \mathbf{y}_e, \mathbf{h}^s)$ be a joint system state. Then, Equations (38) and (39) can construct the following dynamic system

$$d\mathbf{u} = U_1(\mathbf{u}) dt + U_2(\mathbf{u}) d\ddot{\mathbf{W}}, \tag{40}$$

where

$$U_1 = \begin{bmatrix} \gamma N_s^{-1}\mathbf{G} + \gamma N_l^{-1}\mathbf{G}_{\mathbf{x}^*}^{-1}\mathbf{G}_{\mathbf{y}}\mathbf{K} - \frac{1}{2}a\mathbf{G}_{\mathbf{x}^*}^{-1}\mathbf{G}_{\mathbf{h}^s}\mathbf{h}^s \\ \gamma N_l^{-1}\mathbf{K} - \mathbf{K}_{\mathbf{y}^*}^{-1}\mathbf{K}_{\mathbf{h}^l}\mathbf{H}_L \\ -\frac{1}{2}a\mathbf{h}^s \end{bmatrix}$$

and

$$U_2 = \begin{bmatrix} a^{\frac{1}{2}}\mathbf{G}_{\mathbf{x}^*}^{-1}\mathbf{G}_{\mathbf{h}^s} \\ \mathbf{0}_{(N+2)\times K} \\ a^{\frac{1}{2}}\mathbf{I}_K \end{bmatrix},$$

$\ddot{\mathbf{W}}$ is a $((L+1)+(N+2)+K)$-dimensional complex Wiener process.

We define the Lyapunov function of $\mathbf{u}$ as $V = \frac{1}{2}\mathbf{u}^T\mathbf{u}$. Based on Lemma 1, the Lyapunov drift of the dynamic system in (40) can be computed as

$$\begin{aligned} LV(\mathbf{u}) = {} & \gamma N_s^{-1}\mathbf{x}_e^T\mathbf{G} + \gamma N_l^{-1}\mathbf{x}_e^T\mathbf{G}_{\mathbf{x}^*}^{-1}\mathbf{G}_{\mathbf{y}}\mathbf{K} - \frac{1}{2}a\mathbf{x}_e^T\mathbf{G}_{\mathbf{x}^*}^{-1}\mathbf{G}_{\mathbf{h}^s}\mathbf{h}^s \\ & + \gamma N_l^{-1}\mathbf{y}_e^T\mathbf{K} - \mathbf{y}_e^T\mathbf{K}_{\mathbf{y}^*}^{-1}\mathbf{K}_{\mathbf{h}^l}\mathbf{H}_L - \frac{1}{2}a\mathbf{h}_s^T\mathbf{h}_s \\ & + \mathrm{tr}\left[a(\mathbf{G}_{\mathbf{x}^*}^{-1}\mathbf{G}_{\mathbf{h}_s}\mathbf{h}_s)^T(\mathbf{G}_{\mathbf{x}^*}^{-1}\mathbf{G}_{\mathbf{h}_s}\mathbf{h}_s) + a\mathbf{I}_K\right]. \end{aligned}$$

Based on the stability constraint in Lemma 2, we can obtain the following theorem for the dynamic system stability.

Theorem 3. *(Convergence of the two-timescale algorithm): Suppose there exists positive constants v_{xy}, v_{xh}, v_{yh} and l_y such that $\|\mathbf{G}_{\mathbf{x}}^{-1}\mathbf{G}_{\mathbf{y}}\| \le v_{xy}$, $\|\mathbf{G}_{\mathbf{x}}^{-1}\mathbf{G}_{\mathbf{h}^s}\| \le v_{xh}$, $\|\mathbf{K}_{\mathbf{y}}^{-1}\mathbf{K}_{\mathbf{h}^l}\| \le v_{yh}$ and $\mathbf{K} \le l_y\|\mathbf{y}_e\|$. If the step size parameter γ satisfies*

$$\gamma\left(\frac{\alpha_x}{N_s} - \frac{(v_{xy}l_l)^2}{4\alpha_y N_l}\right) \ge \frac{av_h^2}{8}$$

the two-timescale algorithm can converge to the moving equilibrium point.

Proof. The proof is shown in Appendix A. □

Then, we give the following remarks.

- The convergence of the small timescale variables and large timescale variables are coupled with each other. The term $\left(\frac{\alpha_x}{N_s} - \frac{(v_{xy}l_l)^2}{4\alpha_y N_l}\right) > 0$ is the premise condition for the convergence of two-timescale iterations. Recall that N_s and N_l are the adjustment parameters of the two-timescale iterations. Then, for the given certain spectrum allocation algorithm (α_x,α_y,v_{xy},l_l), the small timescale variables should be iterated much faster than the large timescale variables in order to achieve the convergence.
- By adjusting γ and N_s, the stability of the algorithm can be improved. However, when N_s is too small, the algorithm must iterate more times per unit of time, resulting in more iteration overhead. Excessive γ can also cause significant steady-state error $\mathcal{O}(\gamma)$ [22].

Then, the convergence error of the two-timescale algorithm is investigated in the following theorem.

Theorem 4. *(Upper Bound of Convergence Error): The convergence error of the two-timescale algorithm can be given as*

$$\eta_y \le \left(\frac{N_l}{\gamma}\right)^2\left(\frac{v_{yh}\varsigma}{\alpha_y}\right)^2$$

and

$$\eta_x \le \frac{N_s}{\gamma}\frac{2v_{xh}^2 Ka}{\alpha_x} + \left(\frac{N_s}{\gamma}\right)^2 \left(\frac{v_{xy} l_y v_{yh}^2 \varsigma^2}{\alpha_x \alpha_y^2}\frac{N_l}{\gamma} + \frac{a v_{xh} \pi}{2\alpha_x}\right)^2.$$

Proof. Please refer to Appendix B for the proof. □

From Theorem 4, we can give the following observations.

- For the convergence error, η_y is determined by the large timescale algorithm adjustment parameters (N_l, γ) and the global CSI parameter ς. However, the convergence error of the small timescale algorithm η_x is affected not only by the small timescale parameters N_s, γ, a and π but also by the large timescale parameters N_l, γ and ς.
- For the static CSI, where the CSI parameters $a = \varsigma = 0$, we have $\eta_x = \eta_y = 0$. Under static h^l, where $\varsigma = 0$, we obtain that $\eta_y = 0$ and that η_x is the same as Theorem 2. Hence, the overall convergence error only comes from the small timescale iterations.
- Under time-varying h^s and h^l, one can increase the step size γ and decrease the timescale adjustment parameter N_s and N_l to reduce the convergence error at the price of larger steady-state error $\mathcal{O}(\gamma)$, larger signaling overhead and computational complexity.

6.3. Signaling Overhead

For the one-timescale spectrum allocation algorithm, it needs the order of $O(1/N_s)$ iterations at the small timescale and $O(1/N_l)$ iterations at the large timescale during each slot of the CSI. The spectrum allocation has been decomposed into two levels of problems, which are separately solved by the cloud server and network (WiFi/LTE-U), respectively. Hence, the cloud server needs to acquire the global CSI at the large timescale, and the network needs to acquire the local CSI at the small timescale. With the average signaling overheads c_1 and c_2 on the backhaul links and wireless links, the signaling overhead amount of the network is $O(K(\frac{c_2}{N_s} + \frac{c_1}{N_l}))$.

7. Simulation Results

In this simulation, we consider one WiFi and two LTE-U nodes. We assume that the devices are moving based on the widely adopted *Levy walk mobility model* [29]. From $t = 0$, the device randomly chooses a destination and moves at a constant speed in $(0, v_{max}]$. Upon reaching the destination, the device randomly choose a new destination and speed to go on. We assume that the bandwidth of the whole unlicensed spectrum is 20 MHz. The maximum moving speed of the device is set as $v_{max} = 2$ m/s. The minimum distance between the device and the network is set as $D_{min} = 20$ m and the path loss exponent is set as $\iota = 1.8$ [30]. We compare our proposed algorithm with the following baseline schemes: **one-timescale optimal algorithm**: The cloud sever collects the real-time global CSI and computes the optimal spectrum allocation solution; this algorithm can be also considered as the case of an one-timescale algorithm under sufficient small N_s. **one-timescale based on statistical CSI**: In this algorithm, the optimal spectrum allocation solution is computed based on the statistical CSI at each slot of a large timescale. **Static algorithm**: The unlicensed spectrum is partitioned with static fraction.

In the first experiment, we compared the convergence performance of the vanishing step-size and constant step-size algorithms. From Figure 4, we observe that both step size rules can make the convergence errors converge to zero under sufficient iterations. However, due to the fact that a diminishing step-size will converge to zero along with the algorithm iterations, the output solution of the algorithm will not change with the dynamics of network state. Practical, when the network state changes, the algorithm iterations should continuously approach the time-varying optimal solution. Hence, for the spectrum allocation in time-varying wireless networks, we must consider a constant step size instead of diminishing step size. The figure also shows that a large steady-error will be incurred when the step size is excessively large. This is inevitable due to the discrete iteration of the spectrum allocation algorithm.

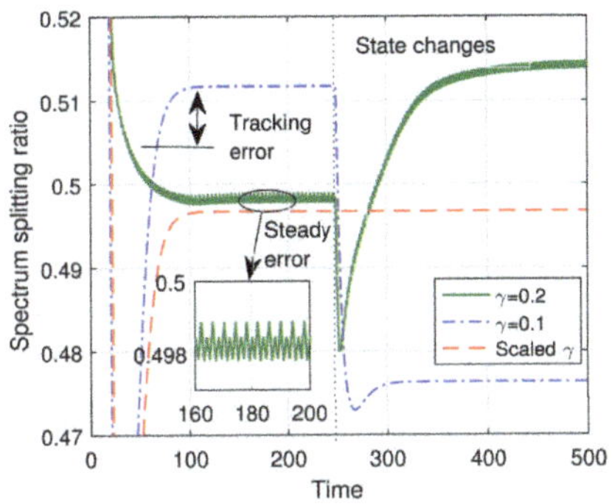

Figure 4. Convergence of the spectrum allocation iteration algorithm.

Figure 5 illustrates the convergence trajectory of the one-timescale spectrum allocation iteration algorithm. From the figure, we observe that the small timescale adjustment parameter N_s can help the spectrum allocation iteration algorithm converge to the optimal solution. However, as outlined in Section 4.3, the signal overhead will increase sharply with the decrease of N_s. Besides the effect of timescale adjustment parameter N_s, we investigate the relation between the step size γ and the system performance in Figure 6. The performance gap of Figure 6 is defined as the difference of weighted system throughput between the optimal algorithm and the one-timescale spectrum allocation iteration algorithm. From Figure 6, we observe that the performance gap decrease with the increase in step size parameter. However, when γ is larger than 0.13, the performance gap increase because of the steady error incurred by the constant step size. Therefore, it is not reasonable to increase the step size parameter blindly. The step size γ and timescale parameter N_s should be jointly adjusted based on the system performance requirement and signal overhead constraint.

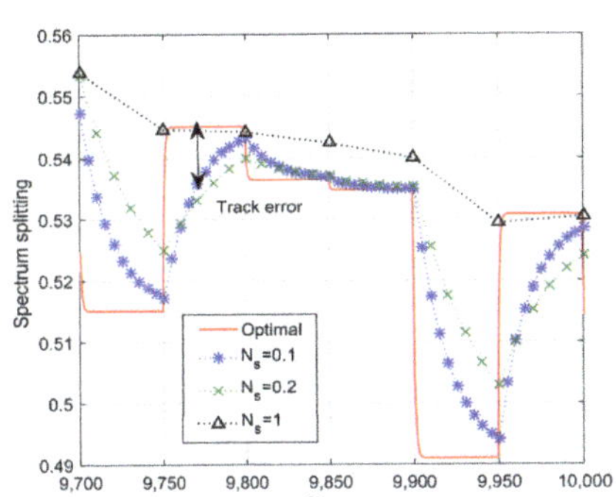

Figure 5. Convergence trajectory of the one-timescale spectrum allocation iteration algorithm ($\gamma = 0.01$).

Figures 7 and 8 illustrate the convergence trajectories of the two-timescale spectrum allocation algorithm. We observe that the small N_s and N_l can help decrease the convergence error, which validates the theoretical analysis in Theorems 3 and 4. Then, in Figures 9 and 10, we investigate the performance gap versus N_s, N_l. From these figures, we observe that the performance gap increases with the increase in N_s, N_l. However, the large N_s and N_l decrease the overhead of the two-timescale spectrum allocation algorithm. We also illustrate the relation between the step size γ and the system performance in Figure 10. Similar to the one-timescale algorithm, Figure 10 also shows that the performance gap decrease as the step size parameter increases. When the step size is larger than 0.85, a larger steady error will be incurred and the performance gap increase. Therefore, under the consideration of system performance requirement and signal overhead constraint, we can appropriately increase the step parameter γ to compensate for the performance gap of the larger N_s and N_l.

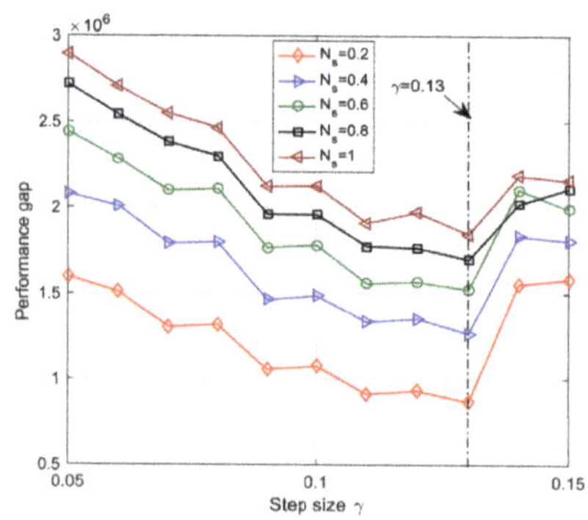

Figure 6. Performance gap vs. step size parameter γ.

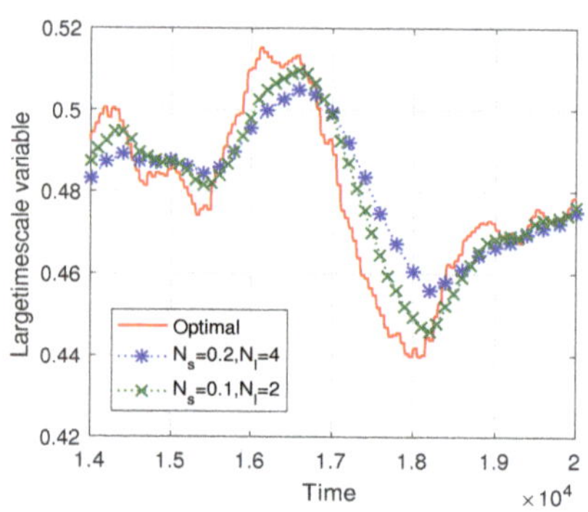

Figure 7. Convergence trajectory of the large timescale spectrum selection iteration algorithm.

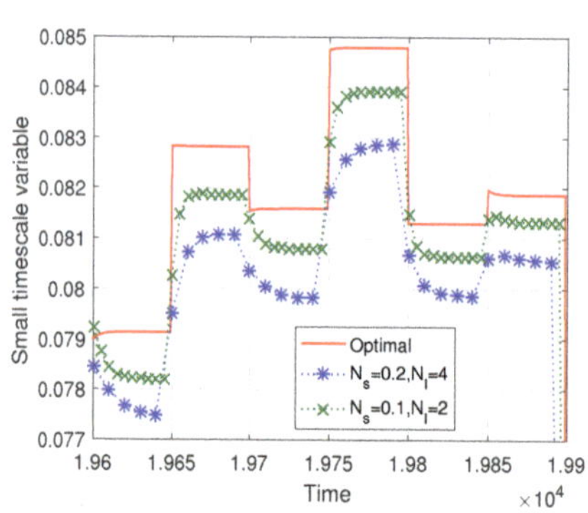

Figure 8. Convergence trajectory of the small timescale spectrum allocation iteration algorithm.

Figure 11 investigates the system performance versus the number of sDevices. The figure shows that the two-timescale optimal algorithm has the similar performance compared with the one-timescale optimal algorithm. This indicates the validity of decomposing the spectrum allocation into a two-timescale algorithm. That is, by decomposing the spectrum allocation into large timescale spectrum selection and small timescale spectrum allocation, the policy can achieve the maximum system throughput without the real-time global CSI. This can effectively reduce the signal overhead on both the backhaul links and wireless links. The figure also shows that although there is a little loss of system throughput of our proposed two-timescale algorithm, it performs much better than the static algorithm. Moreover, it performs better than the one-timescale dynamic spectrum allocation algorithm based on statistical CSI.

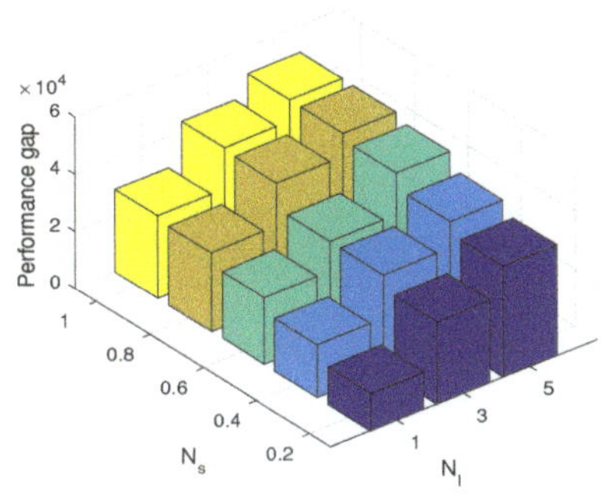

Figure 9. Performance gap versus N_s and N_l.

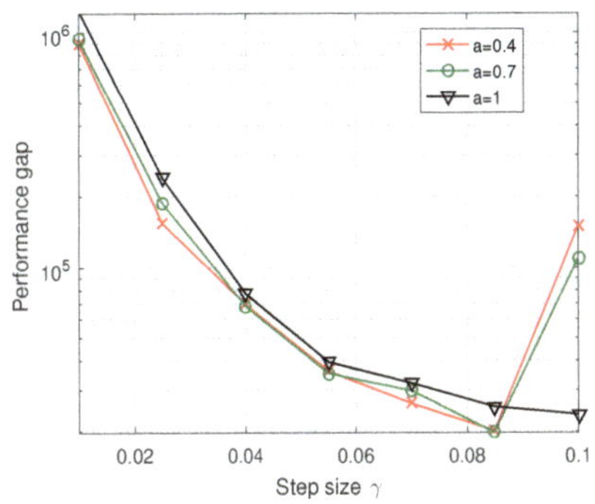

Figure 10. Performance gap versus a and γ.

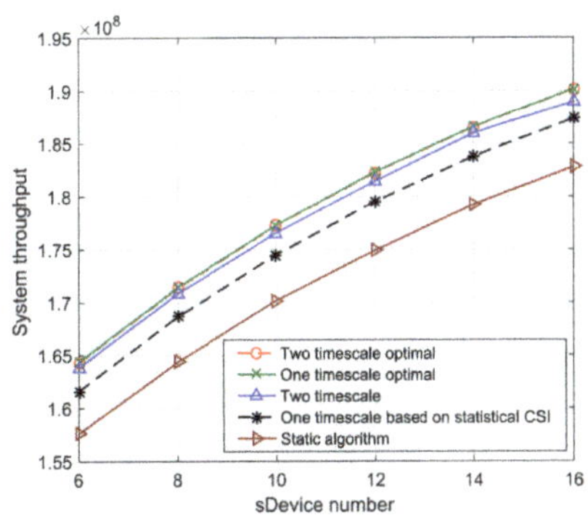

Figure 11. System throughput versus device number.

8. Conclusions

In this paper, we investigated the continuous-time control for the spectrum allocation in a wireless network. Specifically, we first proposed a universal convergence analysis framework for the unlicensed spectrum allocation algorithm. Then, we developed the one-timescale iteration-adjustable unlicensed spectrum allocation algorithm. For the networks with mixed timescale network sates, we proposed a two-timescale frequency allocation and derived its convergence condition and convergence error. When the user moves at high speed, the wireless channel will be affected by the Doppler shift, and it is difficult for the central server to obtain accurate CSI. In the future, we will model the uncertainty of CSI on different timescales. On the small timescale, the distribution of uncertainty is considered to be constant, but on the large timescale, the uncertainty is changing over time. According to this uncertainty model, we will design a multi-timescale algorithm for achieving the coexistence of LTE and WiFi networks.

Author Contributions: Conceptualization, Y.Z.; Methodology, Y.Z. and W.W.; Software, Y.Z. and W.H.; Validation, Y.Z.; Supervision, N.Z. All authors have read and agreed to the published version of the manuscript.

Funding: This research was funded in part by the NSF China (61801365), in part by the Natural Science Foundation of Shaanxi Province (2023-JC-YB-542).

Institutional Review Board Statement: Not applicable.

Informed Consent Statement: Not applicable.

Data Availability Statement: Data are contained within the article.

Conflicts of Interest: The authors declare no conflict of interest.

Appendix A

Proof of Theorem 3. The convergence analysis in Equation (13) indicates $\mathbf{x}_e^T\mathbf{G} \le -\alpha_x||\mathbf{x}_e||^2$. Similarly, we can obtain $\mathbf{y}_e^T\mathbf{K} \le -\alpha_y||\mathbf{y}_e||^2$. Then, the upper bound of the Lyapunov drift can be computed as

$$
\begin{aligned}
LV(\mathbf{z}) \le & -\alpha_x\gamma||\mathbf{x}_e||^2 N_s^{-1} + \gamma v_{xy} l_y N_l^{-1}||\mathbf{x}_e||\,||\mathbf{y}_e|| \\
& + \frac{1}{2} a v_{xh}||\mathbf{x}_e||\,||\mathbf{h}_s|| - \alpha_y N_l^{-1}\gamma||\mathbf{y}_e||^2 + ||\mathbf{y}_e|| v_{yh}\varsigma \\
& - \frac{1}{2}||\mathbf{h}_s||^2 + v_{xh} a K + aK \\
& = -\chi^T\mathbf{A}\chi + \mathbf{b}^T\chi + v_{xh} a K + aK
\end{aligned}
$$

where

$$
\mathbf{A} = \begin{bmatrix} \alpha_x\gamma N_s^{-1} & -\frac{\gamma v_{xy} l_y N_l^{-1}}{2} & -\frac{a v_h}{4} \\ -\frac{\gamma v_{xy} l_y N_l^{-1}}{2} & \gamma\alpha_y N_l^{-1} & 0 \\ -\frac{a v_h}{4} & 0 & \frac{1}{2}a \end{bmatrix}, \mathbf{b} = \begin{bmatrix} 0 \\ v_{yh}\varsigma \\ 0 \end{bmatrix}
$$

where $\varsigma = 2c_0\iota D_{min}^{-\iota-1} v_{max}$ is the upper bound of H_l. By calculating each of the leading principle minors of $\mathbf{A}$, and making them positive, Theorem 3 can be proved. □

Appendix B

Proof of Theorem 4. Firstly, we will show the convergence error of $\mathbf{y}_e$. Based on Lemma 1, the Lypounov drift of $\mathbf{y}_e$ can be computed as

$$\begin{aligned} LV \le & \gamma N_l^{-1}\mathbf{y}_e^T\mathbf{K} - \mathbf{y}_e^T\mathbf{K}_{\mathbf{y}^*}^{-1}\mathbf{K}_{\mathbf{h}_l}\varsigma \le -\alpha_y\gamma N_l^{-1}\|\mathbf{y}_e\|^2 \\ & + v_{yh}\varsigma\|\mathbf{y}_e\| \le -\frac{1}{2}\alpha_y\gamma N_l^{-1}\|\mathbf{y}_e\|^2 + \frac{(v_{yh}\varsigma)^2}{2\alpha_y\gamma N_l^{-1}}. \end{aligned}$$

According to Lemma 2, the convergence error of $\mathbf{y}_e$ is

$$\eta_y \le \left(\frac{v_{yh}\varsigma}{\alpha_y}\right)^2\left(\frac{N_l}{\gamma}\right)^2. \quad \text{(A1)}$$

For the convergence error $\mathbf{x}_e$, the Lypounov drift can be computed as

$$\begin{aligned} LV = & \gamma\mathbf{x}_e^T\mathbf{G}N_s^{-1} + \gamma N_l^{-1}\mathbf{x}_e^T\mathbf{G}_{\mathbf{x}^*}^{-1}\mathbf{G}_{\mathbf{y}^*}\mathbf{K} \\ & -\frac{1}{2}a\mathbf{x}_e^T\mathbf{G}_{\mathbf{x}^*}^{-1}\mathbf{G}_{\mathbf{h}_s}\mathbf{h}_s + \mathrm{tr}\left[a(\mathbf{G}_{\mathbf{x}^*}^{-1}\mathbf{G}_{\mathbf{h}_s})^T\mathbf{G}_{\mathbf{x}^*}^{-1}\mathbf{G}_{\mathbf{h}_s}\right] \\ \le & -\alpha_x\gamma N_s^{-1}\|\mathbf{x}_e\|^2 + \gamma v_{xy}l_yN_l^{-1}\|\mathbf{x}_e\|\|\mathbf{y}_e\| \\ & + \frac{1}{2}av_{xh}\pi\|\mathbf{x}_e\| + v_{xh}^2Ka. \end{aligned}$$

Substituting Equation (A1) into the equation above, we obtain

$$\begin{aligned} LV \le & -\alpha_x\gamma N_s^{-1}\|\mathbf{x}_e\|^2 + v_{xh}^2aK \\ & + \left(\frac{v_{xy}l_yv_{yh}^2\varsigma^2N_l}{\alpha_y^2\gamma} + \frac{1}{2}av_{xh}\pi\right)\|\mathbf{x}_e\| \\ \le & -\frac{1}{2}\alpha_x\gamma N_s^{-1}\|\mathbf{x}_e\|^2 + C_1, \end{aligned}$$

where

$$C_1 = v_{xh}^2aK + \frac{N_s}{2\alpha_x\gamma}\left(\frac{v_{xy}l_yv_{yh}^2N_l}{\alpha_y^2\gamma} + \frac{1}{2}av_{xh}\pi\right)^2.$$

Then, based on Lemma 1, the theorem can be proved. □

References

1. Zhang, H.; Leng, S.; Wei, Y.; He, J. A Blockchain Enhanced Coexistence of Heterogeneous Networks on Unlicensed Spectrum. *IEEE Trans. Veh. Technol.* **2022**, *71*, 7613–7624. [CrossRef]
2. Bhangale, K.; Mapari, R.; Randive, S.; Mapari, K. Cross Technology Communication between LTE-U and WiFi to Improve Overall QoS of 5G System. In Proceedings of the 2022 2nd Asian Conference on Innovation in Technology (ASIANCON), Ravet, India, 26–28 August 2022; pp. 1–6. [CrossRef]
3. Gopal, K.P.; Sambath, M.; Geetha, A.; Shekhar, H. Implementing Fast Router in Convergent LTE/Wifi Networks Using Software Defined Networks. In Proceedings of the 2022 IEEE 2nd Mysore Sub Section International Conference (MysuruCon), Mysuru, India, 16–17 October 2022; pp. 1–5. [CrossRef]
4. Challita, U.; Dong, L.; Saad, W. Proactive Resource Management for LTE in Unlicensed Spectrum: A Deep Learning Perspective. *IEEE Trans. Wirel. Commun.* **2018**, *17*, 4674–4689. [CrossRef]
5. Tan, J.; Xiao, S.; Han, S.; Liang, Y.C.; Leung, V.C.M. QoS-Aware User Association and Resource Allocation in LAA-LTE/WiFi Coexistence Systems. *IEEE Trans. Wirel. Commun.* **2019**, *18*, 2415–2430. [CrossRef]
6. Pei, E.; Zhu, B.; Li, Y. A Q-learning based Resource Allocation Algorithm for D2D-Unlicensed communications. In Proceedings of the 2021 IEEE 93rd Vehicular Technology Conference (VTC2021-Spring), Helsinki, Finland, 25–28 April 2021; pp. 1–6. [CrossRef]
7. Saha, G.; Abouzeid, A.A. Optimal Spectrum Partitioning and Licensing in Tiered Access under Stochastic Market Models. *IEEE/ACM Trans. Netw.* **2021**, *29*, 1948–1961. [CrossRef]
8. Girmay, G.G.; Pham, Q.V.; Hwang, W.J. Joint channel and Power Allocation for Device-to-Device Communication on Licensed and Unlicensed Band. *IEEE Access* **2019**, *7*, 22196–22205. [CrossRef]

9. LeAnh, T.; Tran, N.H.; Ngo, D.T.; Han, Z.; Hong, C.S. Orchestrating Resource Management in LTE-Unlicensed Systems with Backhaul Link Constraints. *IEEE Trans. Wirel. Commun.* **2019**, *18*, 1360–1375. [CrossRef]
10. Huang, Y.; Chen, Y.; Hou, Y.T.; Lou, W.; Reed, J.H. Recent Advances of LTE/WiFi Coexistence in Unlicensed Spectrum. *IEEE Netw.* **2018**, *32*, 107–113. [CrossRef]
11. Tuladhar, S.; Cao, L.; Viswanathan, R. Fair Coexistence of LAA and WiFi in Multi-Carrier LBT based on Joint Throughput and Airtime Fairness. In Proceedings of the 2021 IEEE International Symposium on Dynamic Spectrum Access Networks (DySPAN), Los Angeles, CA, USA, 13–15 December 2021; pp. 147–152. [CrossRef]
12. Kulkarni, V.P.; Joshi, R.D. An LBT Scheme with Adaptive Freezetime for LAA Outdoor Scenario. In Proceedings of the 2022 IEEE 7th International conference for Convergence in Technology (I2CT), Pune, India, 7–9 April 2022; pp. 1–5. [CrossRef]
13. Ren, Q.; Zheng, J.; Xiao, J.; Zhang, Y. Performance Analysis of an LAA and WiFi Coexistence System Using the LAA Category-4 LBT Procedure with GAP. *IEEE Trans. Veh. Technol.* **2021**, *70*, 8007–8018. [CrossRef]
14. Liu, R.; Yu, G.; Qu, F.; Zhang, Z. Device-to-Device Communications in Unlicensed Spectrum: Mode Selection and Resource Allocation. *IEEE Access* **2016**, *4*, 4720–4729.
15. Yuan, X.; Qin, X.; Tian, F.; Hou, Y.T.; Lou, W.; Midkiff, S.F.; Reed, J.H. Coexistence between Wi-Fi and LTE on Unlicensed Spectrum: A Human-Centric Approach. *IEEE J. Sel. Areas Commun.* **2017**, *35*, 964–977. [CrossRef]
16. Li, J.; Han, Y. Multi-RAT wireless network capacity optimization under optimal spectrum splitting in LTE-U. In Proceedings of the IEEE WCNC, Doha, Qatar, 3–6 April 2016; pp. 1–6.
17. Hajmohammad, S.; Elbiaze, H. Unlicensed spectrum splitting between Femtocell and WiFi. In Proceedings of the IEEE ICC, Budapest, Hungary, 9–13 June 2013; pp. 1883–1888.
18. Erturk, M.C.; Aki, H.; Guvenc, I.; Arslan, H. Fair and QoS-Oriented Spectrum Splitting in Macrocell-Femtocell Networks. In Proceedings of the IEEE GLOBECOM, Miami, FL, USA, 6–10 December 2010; pp. 1–6.
19. Bianchi, G. Performance analysis of the IEEE 802.11 distributed coordination function. *IEEE J. Sel. Areas Commun.* **2000**, *18*, 535–547. [CrossRef]
20. Liew, S.C.; Kai, C.H.; Leung, H.C.; Wong, P. Back-of-the-Envelope Computation of Throughput Distributions in CSMA Wireless Networks. *IEEE Trans. Mob. Comput.* **2010**, *9*, 1319–1331. [CrossRef]
21. Bertsekas, D.P. *Nonlinear Programming*; Athena Scientific: Belmont, MA, USA, 1999.
22. Boyd, S.; Xiao, L.; Mutapcic, A. Subgradient methods. *Lect. Notes Ee392o Stanf. Univ. Autumn Quart.* **2003**, *2004*, 2004–2005.
23. Kushner, H.J. Stochastic stability and control. In *Mathematics in Science & Engineering A*; Academic Press: New York, NY, USA, 1967.
24. Chen, J.; Lau, V.K.N. Delay Analysis of Max-Weight Queue Algorithm for Time-varying Wireless Adhoc Networks-Control Theoretical Approach. *IEEE Trans. Signal Process.* **2013**, *61*, 99–108. [CrossRef]
25. Gawlowicz, P.; Zubow, A.; Wolisz, A. Enabling Cross-technology Communication between LTE Unlicensed and WiFi. In Proceedings of the IEEE INFOCOM, Honolulu, HI, USA, 16–19 April 2018; pp. 144–152.
26. Feng, T.; Field, T.R.; Haykin, S. Stochastic Differential Equation Theory Applied to Wireless Channels. *IEEE Trans. Commun.* **2007**, *55*, 1478–1483. [CrossRef]
27. Mostofi, Y.; Gonzalez-Ruiz, A.; Gaffarkhah, A.; Li, D. Characterization and modeling of wireless channels for networked robotic and control systems—A comprehensive overview. In Proceedings of the IEEE/RSJ International Conference on Intelligent Robots and Systems, St. Louis, MO, USA, 10–15 October 2009; pp. 4849–4854.
28. Tse, D.; Viswanath, P. *Fundamentals of Wireless Communication*; Cambridge University Press: Cambridge, UK, 2005.
29. Rhee, I.; Shin, M.; Hong, S.; Lee, K.; Kim, S.J.; Chong, S. On the Levy-Walk Nature of Human Mobility. *IEEE/ACM Trans. Netw.* **2011**, *19*, 630–643. [CrossRef]
30. Chen, J.; Lau, V.K.N. Convergence analysis of mixed timescale cross-layer stochastic optimization. In Proceedings of the Asilomar Conference on Signals, Systems and Computers, Pacific Grove, CA, USA, 3–6 November 2013; pp. 225–229.

sensors

MDPI

Article

Multitask Learning-Based Deep Signal Identification for Advanced Spectrum Sensing

Hanjin Kim [1,†], Young-Jin Kim [2,†] and Won-Tae Kim [1,*]

1 Future Convergence Engineering Major, Department of Computer Science and Engineering, Korea University of Technology and Education, Cheonan 31253, Republic of Korea; gks359@koreatech.ac.kr

2 Department of Computer Science Engineering, Korea University of Technology and Education, Cheonan 31253, Republic of Korea; you359@koreatech.ac.kr

* Correspondence: wtkim@koreatech.ac.kr; Tel.: +82-41-560-1485

† These authors contributed equally to this work.

Abstract: The explosive demand for wireless communications has intensified the complexity of spectrum dynamics, particularly within unlicensed bands. To promote efficient spectrum utilization and minimize interference during communication, spectrum sensing needs to evolve to a stage capable of detecting multidimensional spectrum states. Signal identification, which identifies each device's signal source, is a potent method for deriving the spectrum usage characteristics of wireless devices. However, most existing signal identification methods mainly focus on signal classification or modulation classification, thus offering limited spectrum information. In this paper, we propose DSINet, a multitask learning-based deep signal identification network for advanced spectrum sensing systems. DSINet addresses the deep signal identification problem, which involves not only classifying signals but also deriving the spectrum usage characteristics of signals across various spectrum dimensions, including time, frequency, power, and code. Comparative analyses reveal that DSINet outperforms existing shallow signal identification models, with performance improvements of 3.3% for signal classification, 3.3% for hall detection, and 5.7% for modulation classification. In addition, DSINet solves four different tasks with a 65.5% smaller model size and 230% improved computational performance compared to single-task learning model sets, providing meaningful results in terms of practical use.

Keywords: deep signal identification; multitask learning; spectrum sensing; spectrum hyperspace; 5G-advanced

1. Introduction

The massive connectivity of wireless devices has caused a significant increase in spectrum demand, leading to a shortage in the available spectrum. To address this, wireless devices tend to (1) share the spectrum for opportunistic spectrum utilization and (2) apply various wireless technologies to densely use multidimensional spectrum space. The explosive demand for the spectrum is expected to increase further, even in 5G-advanced and 6G systems, and the network must accommodate heterogeneous devices and their complex access methods [1–3].

Spectrum sensing (SS), serving as the cornerstone of cognitive radio, has played a crucial role in mitigating spectrum scarcity by identifying unused resources within the spectrum and promoting their use by wireless devices [4]. Nonetheless, as spectrum dynamics become increasingly complex, wireless devices are demanding a more advanced level of detection from SS. To enhance spectrum efficiency with minimal interference amidst frequent spectrum sharing, it is essential for SS to not just detect 'empty holes' but also keep track of the activities of primary users [5]. Furthermore, to mitigate cross-technology interference (CTI) among devices utilizing diverse wireless technologies, there is a need to extract multidimensional spectrum state information, which is key in developing tailored interference mitigation strategies [6]. Given the functional limitations of SS, if wireless devices are confined to receiving fragmented information about the spectrum state, they may face frequent collisions and transmission delays [7]. Hence, it becomes critical for SS to evolve in a manner that enables the derivation of comprehensive spectrum state information, including time, frequency, power, and other relevant dimensions [8].

Signal identification [9] has the potential to be one of the key technologies facilitating the SS into advanced SS systems. It works by identifying signals from received radio frequency (RF) samples, enabling the SS system to infer spectrum usage characteristics of the identified signal, such as the channel in use, signal strength level, and modulation method. Numerous approaches have been explored for signal identification, including the use of cyclostationarity [10], maximum likelihood estimation [11], and the received signal strength indicator (RSSI) [12]. Recently, studies have focused on using machine learning or deep learning models, which are capable of capturing the characteristics of a signal source, and can replace static filters or statistical models that may not operate well in dynamic communication environments [13–15].

However, most signal identification studies conducted so far tend to align more with signal classification or modulation classification rather than the 'identification', which is defined as recognizing something and verifying its characteristics [16]. In other words, it is challenging to obtain comprehensive spectrum state information based only on the limited information about the presence of a specific signal within the spectrum. Of course, knowing the existence of certain signals allows us to estimate their spectrum usage characteristics, but it is difficult to derive their characteristics at the detected point. While the signal footprint left by signals using the multidimensional spectrum space, spectrum hyperspace [8], can be used for signal classification, leveraging it to derive used characteristics in each spectrum dimension can be seen as a more beneficial strategy. For clarity, we distinguish between deep signal identification, which adheres to the aforementioned definition, and shallow signal identification, which identifies only a single characteristic, such as signal classification or modulation classification.

The deep signal identification problem we are interested in is quite challenging, as it requires simultaneous extraction of not only the signal's classification but also its multidimensional usage characteristics. Considering the high adaptability of deep learning models, one might imagine creating multiple models for each individual identification task and carrying out deep signal identification. However, it is not that simple. As the number of signal properties to be identified increases, additional individual models need to be used, which might be difficult to accommodate in communication systems with limited resources, including access points and base stations [17]. Moreover, models biased towards individual tasks can overlook the correlation between signal footprints from the same signal source, making it difficult to capture multidimensional characteristics.

We posit that multitask learning (MTL), with its proven efficacy in computer vision [18], speech recognition [19], and natural language processing [20], is a prime solution to these challenges. Given MTL's capability to sufficiently learn a diverse array of characteristics from the signal source via a single model, it can yield substantial signal identification results. In this paper, we present DSINet, a multitask learning-based deep signal identification model. DSINet serves as a key component of an advanced spectrum sensing system for forthcoming 5G-advanced and 6G networks. The main contributions of this paper can be summarized as follows.

- We introduce an advanced spectrum sensing system designed to derive the multidimensional spectrum usage characteristics of wireless devices. The extraction process is formulated as a deep signal identification problem where each task derives usage characteristics for individual dimensions, such as time, frequency, power, and code.
- We propose DSINet, a multitask learning model explicitly tailored to address the deep signal identification problem. This model operates under the assumption that distinct wireless devices leave identifiable 'footprints' within a multidimensional spectrum

space. DSINet captures these 'footprints' as common features that directly contribute to the resolution of individual tasks.

- The DSINet is designed as a modular architecture, composed of three distinct types of blocks: hard sharing blocks, soft sharing blocks, and task-specific blocks. Hard sharing blocks and soft sharing blocks concentrate on common and task-dependent features, respectively, fostering effective information sharing. Conversely, task-specific blocks target unique features inherent to each task. The integration of these blocks enhances overall signal identification performance.
- We conduct three case studies to evaluate DSINet. Each study not only involves a comparative evaluation with existing shallow signal identification technology, but also includes a performance investigation based on task correlation and an evaluation of model size and computational efficiency. The comprehensive results of these studies emphasize DSINet's potential and practical utility for solving deep signal identification problems.

2. Background and Motivation

2.1. Background

The primary focus of SS is to detect opportunities for the unused spectrum that primary users (PUs) are not using in real time. To detect these gaps, various techniques, such as filter-based [21] and statistical model-based [22] methods, have been studied. However, given the inadequacies of static filters and statistical models in fully capturing actual communication environments, there has been a shift towards machine learning- and deep learning-based cognitive radio (CR) technologies [23,24]. These technologies adeptly extract unique characteristics of different communication environments, leading to significant performance improvements in terms of detection accuracy.

Recent studies have accorded considerable attention to estimating PU activity statistics and occupancy patterns, given their substantial impact on detection performance. For instance, one study [5] improved spectrum accuracy by detecting PU activity statistics in the time domain. Another study [25] enhanced the precision of channel occupancy estimation for wideband SS by considering the correlation between spectrum occupancy levels in adjacent channels. Given the robust performance of these studies, SS has evolved from mere hole detection [2] to a more nuanced concept that quantifies spectrum usage statistics and identifies signal interference across frequency bands at a specific location. As spectrum sharing expands into multidimensional spectral spaces, these new perspectives on spectral sensing will gain further importance.

SS is also highly important in unlicensed bands. With the growing popularity of unlicensed bands, wireless technologies such as WiFi, ZigBee, and Bluetooth are vying for spectrum access, thereby leading to CTI [26]. To attenuate interference among wireless devices in the spectrum, it is essential to consider the varying levels of interference among them and apply appropriate mitigation techniques in each situation. However, energy detectors [27] and simple filter-based detectors [28,29], which are extensively used in unlicensed bands, furnish only limited spectrum state information, proving inadequate for the application of appropriate mitigation methods [7].

In the impending 5G-advanced and 6G environments, scenarios of simultaneous spectrum sharing and CTI are expected to be prevalent. The role of SS is set to become increasingly important in these transitions. In essence, for more efficient utilization of the spectrum, it is crucial that SS evolves to adeptly extract the multidimensional spectrum usage characteristics of each signal within the spectrum.

2.2. Motivation

2.2.1. Deep Signal Identification

Signal identification is one of the technologies that can advance SS to function properly in complex spectrum dynamics. In [9], authors stated that understanding the use of spectrum resources to improve spectrum utilization would be an important asset for 5G, with the identification of radio signals at the core. Ali Gorcin [8] articulates that effective use of spectrum hyperspace dimensions demands more than just deciding the presence of primary users in the channel. It also necessitates the identification of primary user waveforms, such as air interface parameters, burst structures, chip rates, cyclic prefix sizes, preambles, and the radio access methods used, all without the need for demodulation.

Traditionally, signal identification has been explored through methods grounded in statistical and probabilistic approaches, such as cyclostationarity [10] and likelihood [11]. However, due to the limitations of these conventional techniques, including low accuracy and high computational complexity, research on feature-based signal identification methods has become more prevalent. The central premise of these methods revolves around utilizing the unique 'signal footprints' left in the spectrum during communication. These are treated as valuable features for discerning the corresponding signals. Some deep learning models are employed to comprehend these signal characteristics, including the type of modulation used and the frequency band occupied. Research evidence demonstrating the precise identification of target signals in complex communication environments through learning these features shows the superiority of feature-based methods over traditional approaches [13–15].

Despite improvements in the accuracy of signal identification technologies, the realization of advanced SS requires progress from different perspectives. The term 'Identification' refers to recognizing something and validation its properties [11], but in a strict sense, most studies fall under signal classification [17] and modulation classification [16]. Classified signal types or modulations can indicate the presence of specific signals in the current spectrum, but they cannot infer the spectrum usage characteristics of each signal. In other words, most research only performs shallow signal identification, which is insufficient information for executing spectrum sharing and CTI migration in complex spectrum dynamics.

As an alternative, the construction of multiple independent shallow identification models to derive multidimensional usage characteristics could be considered, but this introduces new problems, including increased memory size, processing speed, and system complexity. If shallow signal identification models are established for each dimension to extract the multidimensional spectrum usage characteristics required in advanced SS, the overall processing speed would increase linearly with each dimension to be detected, which is undesirable. Moreover, if separate preprocessing modules are additionally needed for each model, the complexity of chip configuration could significantly increase.

In the context of advanced SS, the importance of deep signal identification is prominent. The primary challenge resides in overcoming the limitations of multiple model-based approaches.

2.2.2. Multitask Learning

Traditionally, neural networks have been primarily focused on processing individual tasks. However, most of the problems around us are essentially multitasks, such as self-driving cars [30], intelligent advertising systems [31], etc. MTL is a learning paradigm [32] that aims to leverage useful information contained in multiple related tasks to help improve the generalization performance of all the tasks. It has shown remarkable performance compared to single-task learning in computer vision, speech recognition, and natural language processing. MTL offers three primary benefits: reduced memory footprint, increased inference speed, and most importantly, performance enhancement via complementary information and mutual regularization of each task [33].

Signal identification for SS is also a type of multitasking, as it involves recognizing signals and verifying their multiple attributes. The multidimensional spectrum utilization characteristics of a signal could be considered a collection of tasks, each deriving signal usage characteristics corresponding to individual dimensions, such as time, frequency, and power. A proper configuration of multitasks for the dimensions to be identified, coupled with the sophisticated design of an MTL model, can achieve ideal deep signal identification.

While the application of MTL for deep signal identification can be promising, it necessitates careful consideration before implementation. Challenges that need addressing include overall performance degradation due to an improperly configured multitasking setup, task interference due to low correlations, and dependencies on specific tasks at different levels of task difficulty [31]. In advanced SS scenarios where deep signal identification is required, there can be conflicting attributes among identification tasks, and an improper application of MTL may lead to undesirable results. Please see Section 4 for more details. In this paper, we aim to explore in depth the potential of MTL-based deep signal identification, taking into account the interrelations among signal identification tasks, through various case studies and experiments.

3. Deep Signal Identification in Advanced Spectrum Sensing Systems

3.1. Concept of Advanced Spectrum Sensing System

Figure 1 illustrates the operation of an advanced SS system. N wireless devices in the same spectrum band use W wireless technologies to communicate with each other. Some devices are secondary users performing spectrum sharing, which is considered one of the wireless technologies. Signals with similar characteristics can be grouped to reduce the number of types that need to be identified [34].

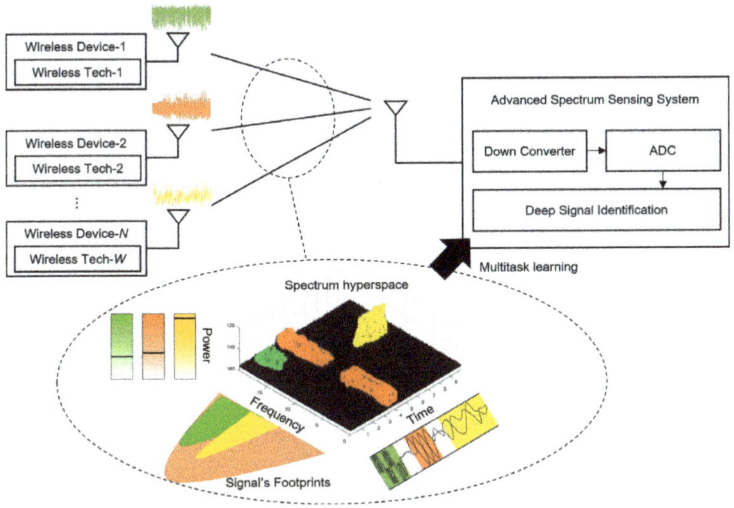

Figure 1. Advanced spectrum sensing scenario in the spectrum hyperspace.

The system comprises a down converter that lowers the signal band to the baseband, an analog–digital converter that digitizes analog signals, and a deep signal identification module that provides the spectrum usage characteristics of each signal to be sensed within the spectrum. We use the raw I/Q sample set to avoid system complexity instead of adding a preprocessing module, such as fast Fourier transform (FFT) or short-time Fourier transform (STFT), to obtain the transformation value.

As wireless devices communicate, they leave different footprints in the spectrum hyperspace, including time, frequency, and power footprints, as well as others [8]. This is formulated as follows:

$$\mathbb{S} \leftarrow \sum_{i=1}^{N} \hbar_i(g_i(s_i)). \quad (1)$$

where $\mathbb{S}$ is the spectrum hyperspace, s is the signal source of a wireless device, g is a function of the wireless technology used by the wireless device, and $\hbar$ is the wireless channel of each device. Since wireless devices can use the same wireless technology, N and W are not the same.

The goal of advanced SS is to derive the spectrum usage characteristics of wireless devices from $\mathbb{S}$. Specifically, it aims to identify a set of properties for signal sources that use a specific wireless technology within the spectrum band. By sensing the target band, advanced SS can obtain sample x and derive the spectrum usage characteristics through these samples, which can be expressed as follows:

$$\hat{Y} = \mathcal{F}(x), \forall x \in \mathbb{S}. \tag{2}$$

where $\hat{Y} = \{\hat{y}_p\}_{p \in \mathcal{P}}$ is the set of $\mathcal{P}$ pieces of spectrum state information to be estimated; x denotes the received RF signal samples, i.e., I/Q pairs; and $\mathcal{F} = \{f_p\}_{p \in \mathcal{P}}$ is a set of functions for estimating the $\mathcal{P}$ pieces of spectrum state information in SS.

3.2. Deep Signal Identification with Multitask Learning

Deep signal identification refers to the set of functions $\mathcal{F}$ that estimate $\hat{Y}$. One possible approach for estimating $\hat{Y}$ using $\mathcal{F}$ is to use individual machine learning model sets (independent shallow signal identification models), as shown below:

$$\hat{Y} = \begin{cases} \hat{y}_1 = f_1(x; \theta_1) \\ \quad \vdots \\ \hat{y}_p = f_p(x; \theta_p) \end{cases} \tag{3}$$

where θ is the weight parameter for each corresponding function f.

However, this approach is not optimal in terms of its system complexity, development costs, and dependence on a preprocessing module. Our proposed multitask learning-based deep signal identification approach derives $\hat{Y}$ from the same sample of x as that used by a single MTL model:

$$\hat{Y} = f_{MTL}(x; \theta_{MTL}) \tag{4}$$

This approach not only overcomes the limitations of individual model sets but also learns the correlations of each task.

3.3. MTL-Based Deep Signal Identification Network

3.3.1. Design Principles

Principle 1. Hybrid Shared Network Structure

MTL models have more weight parameters to be tuned during training than single-task learning models [35]. To establish effective learning strategies, we consider a hybrid network structure that combines hard sharing blocks (HBs) and soft sharing blocks (SBs). HBs extract common features shared by tasks through tightly coupled sharing layers, while an SB has a different layer for each task and shares information through a unit called a cross-stitch [36]. An HB uses unified shared layers, resulting in less memory usage and faster inference, but this may cause competition among tasks. In contrast, an SB consists of individual layers for each task, including the shared part, resulting in relatively less competition for each task.

Principle 2. Task-Specific Network Structure

In MTL, certain tasks may have different levels of difficulty or exhibit opposite characteristics compared to those of other tasks. These gaps and conflicts between tasks can impede the entire learning process, including specific tasks. To address this issue, we employ task-specific blocks (TBs).

3.3.2. Joint Optimization of Multiple Tasks

In multitask learning, MTL model training is performed by minimizing the losses of all tasks. In other words, the joint optimization of all tasks is completed as follows:

$$\mathcal{L}_{MTL} = \frac{1}{M} \sum_{t=1}^{M} \beta_t \mathcal{L}_t \tag{5}$$

where M is the number of tasks, $\mathcal{L}_t$ is the single loss associated with task t, and β_t is the weight for loss $\mathcal{L}_t$.

As we consider heterogeneous MTL, which consists of different types of supervised tasks, including classification and regression problems, the loss function for each individual task is one of the binary cross-entropy loss ($\mathcal{L}_{BCE}$), the cross-entropy loss ($\mathcal{L}_{CE}$), and the mean squared error loss ($\mathcal{L}_{MSE}$).

The binary cross-entropy is formulated as follows:

$$\mathcal{L}_{BCE} = \frac{1}{K}\sum_{i=1}^{K} y^{(i)} log(\hat{y}^{(i)}) + (1 - y^{(i)}) log(1 - \hat{y}^{(i)}) \tag{6}$$

where K represents the number of binary labels, and y and $\hat{y}$ are the vectors of ground truth and predicted probabilities, respectively; each vector consists of K elements with values between 0 and 1. When K is 1, it is used for binary classification tasks, such as determining the presence of a specific signal or the use of a single channel. When $K > 1$, the function is used for multilabel classification.

The $\mathcal{L}_{CE}$ is defined as

$$\mathcal{L}_{CE} = -\frac{1}{C}\sum_{i=1}^{C} y^{(i)} log(\hat{y}^{(i)}) \tag{7}$$

The cross-entropy loss is used for multiclass classification tasks, where C represents the number of classes. In this case, y denotes the ground truth vector in a one-hot encoded format, and $\hat{y}$ is the predicted probability distribution across C classes.

Due to the presence of complex spectrum states, advanced SS frequently encounters heterogeneous MTL problems in which both classification and regression tasks must be solved simultaneously. The loss function used for the regression task is as follows:

$$\mathcal{L}_{MSE} = -\frac{1}{R}\sum_{i=1}^{R} (y^{(i)} - \hat{y}^{(i)})^2 \tag{8}$$

where R represents the number of regression targets, and y and $\hat{y}$ are the vectors of the ground truth target and the model's prediction, respectively; each vector consists of R elements with continuous values. When R is 1, it is used for a single regression, and when $R > 1$, it is used for multiple regression.

3.3.3. Model Architecture

A set of tasks to be solved by DSINet is $\mathcal{T} = \{T_1, T_2, ..., T_i\}_{i=1}^{M}$, which can include signal classification, modulation classification, channel classification, and power estimation. The number of tasks, M, is determined based on the desired information, $\mathcal{P}$, for specific spectrum states. The training dataset consisting of K training samples for DSINet is $\mathcal{D} = \{(x_i, y_i)\}_{i=1}^{K}$, where $\forall x \in \mathbb{S}$. The data sample x is obtained by the advanced SS system when it senses the spectrum at a specific time, and the sample $x_i \in \mathbb{R}^{v \times 2}$ represents the I/Q samples as a vector of size $v \times 2$. The output $y_i = (y_i^{(1)}, y_i^{(2)}, ..., y_i^{(M)})$ corresponds to the labels for each of the M tasks.

Figure 2 illustrates that DSINet consists of three building blocks: HBs, SBs, and TBs, as we mentioned in the Section 3.3.1. The core aspect of multitask learning is acquiring shared representations from each signal identification task. We explain how each block learns essential shared features below.

Hard Sharing Blocks HBs are located at the front of DSINet, close to the input layer. They consist of weight parameters that are shared between all tasks. The HBs capture common features for all tasks, significantly reducing the risk of overfitting for each task. Since all tasks share weight parameters, the use of HBs can minimize the increase in memory usage and inference speed due to the increase in the number of tasks.

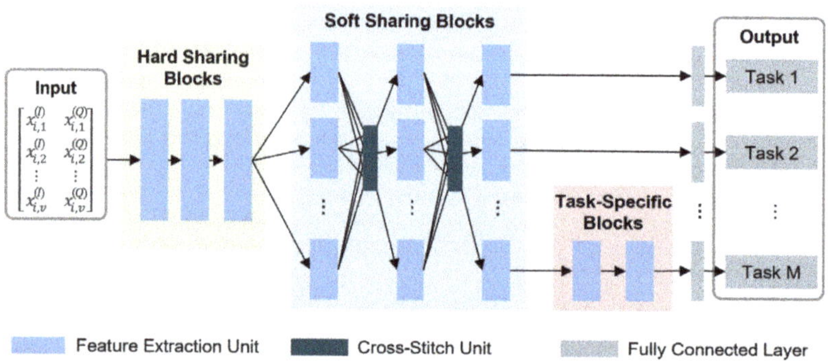

Figure 2. Structure of the DSINet.

There are two feature extraction units: a plain unit and a residual unit. The plain unit consists of two consecutive 1D convolutional (1D Conv) layers followed by a 1D max pooling (1D MaxPool) layer, as illustrated in Figure 3a. The parameters k, f, and $/n$ represent the kernel size, number of filters, and stride of the 1D Conv layers, respectively.

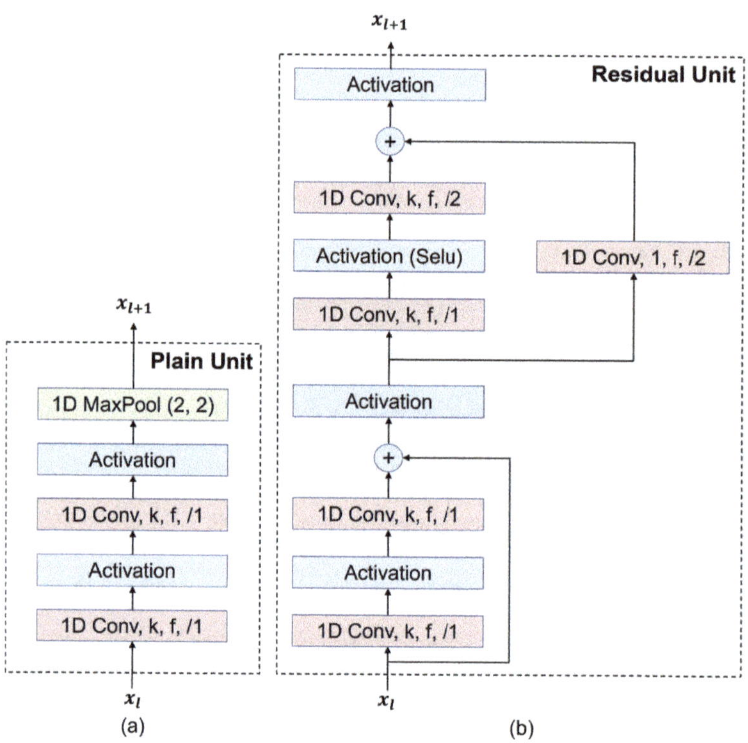

Figure 3. Structure of the (**a**) plain unit and (**b**) residual unit.

As the depths of deep neural networks increase, performance improvements are achieved, as has been proven in various fields, such as image processing, natural language processing, and signal processing [37]. However, deeper networks tend to suffer from vanishing gradient problems, leading to slow training. To mitigate this problem, we use the residual unit shown in Figure 3b as the second consideration for feature extraction. In the residual unit, two residual connections are employed to mitigate the effects of vanishing gradients and facilitate faster training processes [37].

Soft Sharing Blocks The use of HBs has some advantages in MTL, as mentioned above, but the overall performance may degrade due to competition issues. If the relatedness between tasks is low, each task trains the shared parameters of the HBs to improve its own performance, which results in performance degradations for other tasks. SBs mitigate this task competition problem by exploiting loosely coupled information sharing. In SBs, each task has its own weight parameters to capture independent features per task. In addition, each independent layer shares its information with the other layers by using the cross-stitch unit, as shown in Figure 4.

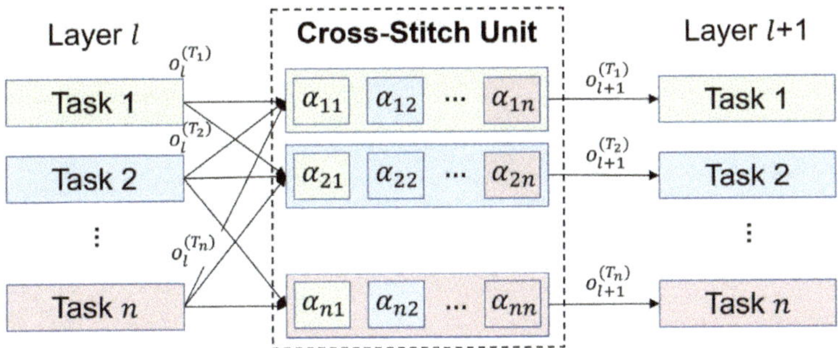

Figure 4. Structure of the cross-stitch unit.

The cross-stitch unit [36] is a transformation applied between independent layers, and it describes the relationships between different tasks with a linear combination of their activations, which is formulated by:

$$\begin{bmatrix} o_{l+1}^{(T_1)} \\ \vdots \\ o_{l+1}^{(T_n)} \end{bmatrix} = \begin{bmatrix} \alpha_{11} & \cdots & \alpha_{1n} \\ \vdots & \vdots & \vdots \\ \alpha_{n1} & \cdots & \alpha_{nn} \end{bmatrix} \begin{bmatrix} o_{l}^{(T_1)} \\ \vdots \\ o_{l}^{(T_n)} \end{bmatrix} \tag{9}$$

where $\left\{o_{l+1}^{(T_i)}\right\}_{i=1}^{n}$ is the transformed output associated with tasks T_1 to T_n, α_{ij} is a trainable parameter associated with tasks T_i and T_j, and $\left\{o_{l}^{(T_i)}\right\}_{i=1}^{n}$ is the input of the cross-stitch unit. By jointly training the parameters α_{ij}, the cross-stitch unit learns the relationship between tasks by itself.

Task-Specific Blocks In the MTL settings, the outputs corresponding to each task are highly heterogeneous. For example, the output of the classification task is discrete, while the output of the regression task is continuous. In this case, the cross-stitch unit in the SBs may result in undesired training effects due to the use of a linear combination of features in which one is related to a discrete output, while the other is related to a continuous output. In addition, it may be more difficult to learn features for some tasks than others, which may require more feature extraction layers. To mitigate these problems, we employ TBs, which have independent weight parameters without information-sharing mechanisms, such as cross-stitch units.

4. Case Studies

4.1. Methodology

In this section, we present three case studies to analyze the performance of DSINet: active primary user signal identification, CTI identification, and signal identification for 5G-advanced and 6G. For each case study, we evaluate the overall performance of DSINet by constructing realistic simulation scenarios that do not assume homogeneous devices, static access methods, or nearly noise-free environments. To ensure the objectivity of

DSINet's evaluations across multiple tasks, we compare its performance with that of other models for each task. The details of case studies are shown in Table 1.

Table 1. Details of Case Studies.

	Case Study 1 (CS-1) **Active PU Signal Identification**	**Case Study 2 (CS-2)** **CTI Identification**	**Case Study 3 (CS-3)** **Signal Identification for 5G-A/6G**
Purpose	Primary user in sub-6 GHz	Derive spectrum usage characteristics of Wireless devices in 2.4 GHz unlicensed band	Wireless devices in complex networks
Wireless Tech'	$W = \{\text{LTE, 5G, Wi-Fi}\}$	$W = \{\text{Wi-Fi, Bluetooth, Zigbee}\}$ Two or more wireless tech' can coexist	$W = \{\text{Wi-Fi, Bluetooth, Zigbee}\}$ Two or more wireless tech' can coexist
Tasks	$T_{sc} : x \to y^{(1)} \in \{0,1\}^{\lvert W \rvert}$ $T_{hd} : x \to y^{(2)} \in \{0,1\}^{\lvert hd \rvert}$ $T_{mc} : x \to y^{(3)} \in \{0,1\}^{\lvert mc \rvert}$	$T_{sc} : x \to y^{(1)} \in \{0,1\}^{\lvert W \rvert}$ $T_{pe} : x \to y^{(2)} \in [0,1]^{\lvert W \rvert}$ $T_{mc} : x \to y^{(3)} \in \{0,1\}^{\lvert mc \rvert}$	$T_{sc} : x \to y^{(1)} \in \{0,1\}^{\lvert W \rvert}$ $T_{cc} : x \to y^{(2)} \in \{0,1\}^{\lvert cc \rvert}$ $T_{pe} : x \to y^{(3)} \in [0,1]^{\lvert W \rvert}$ $T_{mc} : x \to y^{(4)} \in \{0,1\}^{\lvert mc \rvert}$
Data Gen'	$K = 6,881,175$, each with an SNR $\in [-20, 20]$ Training/validation/test set are randomly sampled at 80%, 10%, and 10% ratios $sc = W$ $\lvert hd \rvert = 16$ $mc = \{\text{BPSK, QPSK, 16QAM, 64QAM, 256QAM, 1024QAM, 4096QAM}\}$ $x \in \{\text{IQ, RSSI, STFT}\}$, (4096,2), (128,2), and (1024,2) I/Q vectors are generated for each task, (256,1) RSSI sample and (64,64) STFT image are generated for T_{sc} $y_{i,c}^{(1)} \in \{0,1\}$ for $c \in sc$, $\sum_{c \in sc} y_{i,c}^{(1)} = 1, \forall i$ $y_{i,c}^{(2)} \in \{0,1\}$ for $c \in hd$, $\sum_{c \in hd} y_{i,c}^{(2)} \leq \lvert hd \rvert, \forall i$ $y_{i,c}^{(3)} \in \{0,1\}$ for $c \in mc$, $\sum_{c \in mc} y_{i,c}^{(3)} = 1, \forall i$	$K = 6,300,000$, each with an SNR $\in [-20, 20]$ Training/validation/test set are randomly sampled at 80%, 10%, and 10% ratios $sc = \{$Wi-Fi, BT, ZB, Wi-Fi + BT, Wi-Fi+ZB, ZB + BT, Wi-Fi + BT + ZB$\}$ $mc = \{\text{GFSK, BPSK, QPSK, OQPSK, 16QAM, 64QAM, 256QAM, 1024QAM, 4096QAM}\}$ $x \in$ IQ, (1024,2) I/Q vectors are generated $y_{i,c}^{(1)} \in \{0,1\}$ for $c \in sc$, $\sum_{c \in sc} y_{i,c}^{(1)} = 1, \forall i$ $y_{i,c}^{(2)} \in [0,1]$ for $c \in W, \forall i$ $y_{i,c}^{(3)} \in \{0,1\}$ for $c \in mc$, $\sum_{c \in mc} y_{i,c}^{(3)} \leq \lvert W \rvert, \forall i$	$K = 6,300,000$, each with an SNR $\in [-20, 20]$ Training/validation/test set are randomly sampled at 80%, 10%, and 10% ratios $sc = \{$Wi-Fi, BT, ZB, Wi-Fi+BT, Wi-Fi + ZB, ZB + BT, Wi-Fi + BT + ZB$\}$ $\lvert cc \rvert = 20$ $mc = \{\text{GFSK, BPSK, QPSK, OQPSK, 16QAM, 64QAM, 256QAM, 1024QAM, 4096QAM}\}$ $x \in$ IQ, (1024,2) I/Q vectors are generated $y_{i,c}^{(1)} \in \{0,1\}$ for $c \in sc$, $\sum_{c \in sc} y_{i,c}^{(1)} = 1, \forall i$ $y_{i,c}^{(2)} \in \{0,1\}$ for $c \in cc$, $\sum_{c \in cc} y_{i,c}^{(2)} \leq \lvert cc \rvert, \forall i$ $y_{i,c}^{(3)} \in [0,1]$ for $c \in W, \forall i$ $y_{i,c}^{(4)} \in \{0,1\}$ for $c \in mc$, $\sum_{c \in mc} y_{i,c}^{(4)} \leq \lvert W \rvert, \forall i$
	Sense 20 MHz wideband spectrum Hole size: 1 MHz AWGN and Rayleigh fading	Sense 20 MHz wideband spectrum AWGN Pathloss model [38] with distance from 5 to 20	Sense 20 MHz wideband spectrum Min channel BW: 1 MHz AWGN and Rayleigh fading Pathloss model [38] with distance from 5 to 20
Models	T_{sc} :WTC-{CNNIQ, CNNRSSI, CNNSTFT} T_{hd} :DeepSense T_{mc} :RSC-{VGGNet, ResNet}	SSINet (sc) SSINet (mc) SSINet (pe)	* Comparison of DSINet's block configuration DSINet-HB3:TB$_{all}$1 DSINet-HB3:SB1 DSINet-HB3:SB1 + TB$_{pe}$ DSINet-HBx:SBy + TB$_{pe}$
	T_{sc} :DSINet-IQ4096 T_{hd} :DSINet-IQ128 T_{mc} :DSINet-IQ1024	* Task ablation study DSINet(all) DSINet(sc+pe) DSINet(sc+mc) DSINet(mc+pe)	

* Details of parameter settings. Adam optimizer ($\beta_1 = 0.9$, $\beta_2 = 0.999, \epsilon = 10^{(-7)}$) with a learning rate of 0.0005. Alpha dropout [39] in fully connected (fc) layer. Scaled exponential linear unit (SELU) [39] activation function is used to activate the functions of all hidden layers, including the 1D convolution, HB, SB, and fully connected layers.

In Case Study 1 (CS-1), we evaluate the performance of DSINet in spectrum sharing scenarios and compare it with the existing shallow signal identification approaches, WTC [40], DeepSense [41], RSC [42]. To ensure a fair comparison, we implement the DSINet variants with input shapes that match those of the compared methods. In Case Study 2 (CS-2), we investigate CTI scenarios by comparing DSINet with a single network (SSINet) that has the same architectural style as that of DSINet but with a single task. In both CS-1 and CS-2, we conduct task ablation studies to investigate the associations between the tasks and how each task affects DSINet. In Case Study 3 (CS-3), we consider complex heterogeneous network scenarios in 6G where both CS-1 and CS-2 co-occur. We study the effects of different DSINet configurations, including the selection of the feature extraction block, the proportions of HBs and SBs, and the addition of TBs. We also analyze the efficiency of DSINet in terms of memory usage and the number of operations. CS-1 and CS-2 focus on comparative evaluations, while CS-3 focuses on a performance evaluation to extract our best DSINet model.

4.2. Case Study 1: Active PU Signal Identification

The wireless technologies of primary users considered in CS-1 are LTE (LTE-U), 5G (NR-U), and Wi-Fi (Wireless LAN) operating in the sub-6 GHz frequency band. The fundamental concern in CS-1 is hole detection, T_{hd}, and we consider a wideband spectrum as DeepSense [41]. The waveforms of LTE, 5G, and Wi-Fi are generated using MATLAB's LTE Toolbox, 5G Toolbox, and WLAN Toolbox, respectively. We created 16 holes with approximately 1 MHz bandwidth, which is sufficient for wireless communication technologies, such as Bluetooth, ZigBee, and LoRa, while leaving room for guard bands. In this case, the LTE signals have a 180 kHz resource block size, while the 5G NR-U signals can have a resource block size of either 180 kHz, 360 kHz, or 720 kHz. The Wi-Fi signals, which can be generated based on either the IEEE 802.11ax or IEEE 802.11be standard , consist of a resource unit composed of multiple subcarriers, each with a bandwidth of 78.125 kHz. The comparison models in signal classification have different shapes and even different features, using a 256-size RSSI [40] and a 64 × 64-size STFT image [40]. For each task, the input vector size of the comparison models is different, so we constructed the models as follows for fair comparison.

- DSINet-IQ128:
  For the hole detection task T_{hd}, DSINet-IQ128, which is a DSINet variant, is implemented for comparison with DeepSense, which takes an input shape of (128, 2). DSINet-IQ128 has only one SB with the number of filters, kernel size, and stride equal to (32, 3, 2). For the task ablation study, DSINet-IQ128 (all), DSINet-IQ128 (hd + sc), and DSINet-IQ128 (hd + mc) are compared.
- DSINet-IQ1024:
  In the modulation classification task, the input of the comparison methods, including RSC-VGGNet and RSC-ResNet, is an IQ vector with a shape of (1024, 2). Therefore, DSINet-IQ1024 with a task combination of (all, mc + sc, mc + hd) is implemented with the same input as that of the comparison targets.
- DSINet-IQ4096:
  To conduct a comparison in the signal classification task, we implement DSINet-IQ4096, which has the same input shape as WTC-CNNIQ. DSINet-IQ4096 consists of five HBs and one SB. For the task ablation study, DSINet-IQ4096 variants with task combinations of (all, sc + hd, sc + mc) are implemented.

4.3. Case Study 2: CTI Identification

The purpose of models in CS-2 is an identification scheme that enables adaptive CCA to mitigate interference by varying the power levels assigned to different wireless technologies. So, we consider T_{pe}, estimating the average power of each signal, as the main task in CS-2. The waveforms of Bluetooth and Zigbee are generated using MATLAB's Bluetooth Toolbox and Communication Toolbox. The configuration of Wi-Fi signals is based on various WLAN standards, such as IEEE 802.11g (both 5 MHz and 20 MHz), IEEE 802.11n, IEEE 802.11ac, IEEE 802.11ax, and IEEE 802.11be, adding complexity to CTI identification.

In CS-2, DSINet is composed of three HBs, one SB, and one TB(pe). The TB(pe) is only applied for the power estimation task T_{pe} because task T_{pe} deals with a regression problem in which the outputs have different characteristics from those of other classification tasks. For the task ablation study, the four DSINet variants are implemented, which are related to the task combinations of (all, sc + mc, sc + pr, mc + pr). Note that even though T_{sc} is considered a multilabel classification task due to the mix of signals, we assigned a label of $y^{(1)}$ to multiclass classification for better performance through one-hot labeling.

4.4. Case Study 3: Signal Identification for 5G-A/6G

In this study, we leverage DSINet to estimate four tasks: the signals present in the spectrum, the channels in use, the average power utilized by each signal, and the modulations in use. The $\mathcal{T}$ for CS-3 involves adding a used channel classification task, T_{cc}, to

CS-2, which aims to identify PU behaviors for spectrum sharing while identifying CTI. The minimum channel size of wireless technology, Bluetooth, is 1 MHz within 20 MHz bandwidth. In CS-3, we implement a DSINet variant consisting of a block combination of {HB + TB, HB + SB, HB + SB + TB} to evaluate the effects of HBs, SBs, and TBs, which are the core components of DSINet. In addition, we implement and comparatively analyze five DSINet variants with HB:SB proportions of {0:4, 1:3, 2:2, 3:1, 4:0} to analyze the effect of this proportion.

- DSINet-HB3:TB$_{all}$1:
  DSINet-HB3:TB$_{all}$1 consists of three HBs and one TB$_{all}$. TB$_{all}$ indicates that all TBs are connected to each task so that DSINet-HB3:TB$_{all}$1 has exactly the same setting as traditional MTL approaches (hard parameter sharing) [33].
- DSINet-HB3:SB1:
  DSINet-HB3:SB1 is composed of three HBs and one SB. The SB is employed to maximize the information sharing between tasks by using the cross-stitch unit.
- DSINet-HB3:SB1 + TB$_{pe}$:
  DSINet-HB3:SB1 + TB$_{pe}$ has three HBs and one SB, similar to DSINet-HB3:SB1, but additionally uses a TB$_{pe}$ connected to the regression task T_{pe}.
- DSINet-HBx:SBy + TB$_{pe}$ variants for proportion analysis:
  To analyze the proportion of HB and SB, we implement DSINet-HBx:SBy + TB$_{pe}$ variants with HB:SB proportions of {0:4, 1:3, 2:2, 3:1, 4:0} are utilized.

5. Experimental Results and Analysis

5.1. Analysis of Case Study 1

5.1.1. Hole Detection Performance

Figure 5 presents the hole detection accuracies of four models: DeepSense [41] and three DSINet models. Our model, DSINet-IQ128(hd+sc), outperforms DeepSense by approximately 3.3% at an SNR of −4 dB, exhibiting the largest accuracy difference. Notably, the section with the most significant performance improvement due to MTL is from an SNR of −8 dB to −2 dB. Conversely, MTL's effect is relatively insignificant in sections where holes can be easily detected or where detection is almost impossible. All three of our models show promising results compared to DeepSense, which is a single-hole detection model, by sharing task features through MTL.

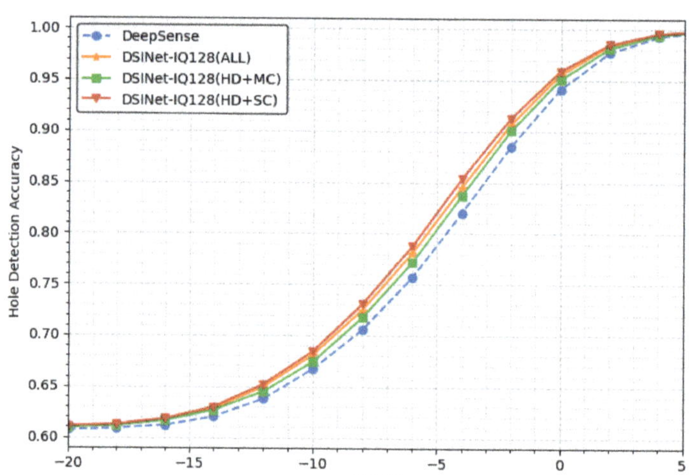

Figure 5. The hole detection accuracies achieved under different SNRs.

Interestingly, the model considering T_{hd} + T_{sc} demonstrates better performance than the model with all tasks. Upon closer inspection of Table 2, the accuracy gap between our models is approximately 1% at −4 dB. Similarly, the $F_1 score$ ranks in descending order are as follows: DSINet-IQ128(hd+sc), DSINet-IQ128(all), and DSINet-IQ128(hd+mc). Our results indicate that T_{sc} has the most significant impact on T_{hd} performance.

Table 2. Performance comparison between DSINet-IQ128 and DeepSense (SNR: −4 dB).

Methods	Accuracy	Recall	Precision	F_1 score
DeepSense	0.8202	0.8161	0.8767	0.8452
DSINet-IQ128(all)	0.8447	0.8459	0.8906	0.8675
DSINet-IQ128(hd+mc)	0.8369	0.8264	**0.8947**	0.8590
DSINet-IQ128(hd+sc)	**0.8531**	**0.8601**	0.8919	**0.8756**

5.1.2. Signal Classification Performance

It can be seen that WTC-CNNRSSI and WTC-CNNSTFT have relatively poor results compared not only to those of our models but also to those of WTC-CNNIQ, as shown in Figure 6. Such results were also shown in the single-task dataset in their original paper. WTC-STFT does not perform particularly well, and it is estimated that most information is lost in the process of resizing to a 64 × 64 image. Our three models as well as WTC-CNNIQ all achieve approximately 100% accuracy above −6 dB. The effect of MTL appears below an SNR of −8 dB, which is the starting point of the performance difference between WTC-CNNIQ and our models.

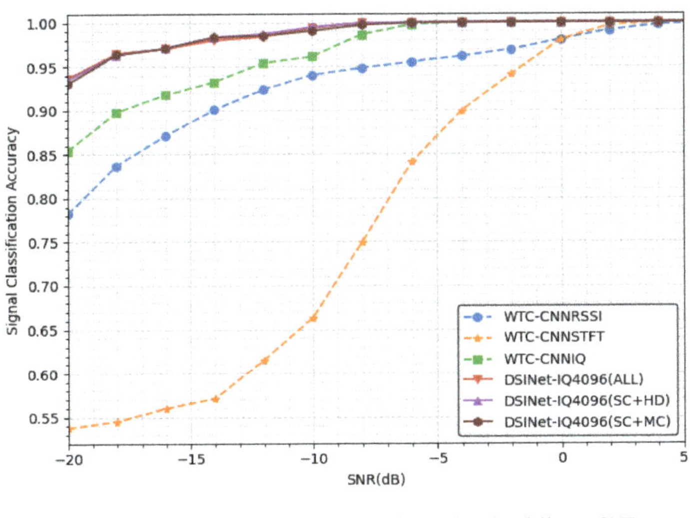

Figure 6. The signal classification accuracies achieved under different SNRs.

Table 3 shows that all our models perform signal classification with similar performance to each other. Because the task is easy, the performance differences between our models are insignificant, up to 0.4% (we do not consider the case where two or more signals exist simultaneously, so the task is easy to solve).

Table 3. Performance comparison between DSINet-IQ4096 and WTC (SNR: −6 dB).

Methods	Accuracy	Recall	Precision	F_1 score
WTC-CNNIQ	0.9539	0.9309	0.9314	0.9265
WTC-CNNRSSI	0.9235	0.8811	0.8964	0.8769
WTC-CNNSTFT	0.6149	0.4200	0.5093	0.4199
DSINet-IQ4096(all)	0.9839	0.9769	0.9741	0.9750
DSINet-IQ4096(sc+hd)	**0.9870**	0.9801	**0.9791**	0.9791
DSINet-IQ4096(sc+mc)	0.9851	**0.9832**	0.9757	**0.9793**

5.1.3. Modulation Classification Performance

The modulation classification task compares our three models with RSC models [42]. Figure 7 shows that T_{mc} is much more difficult than T_{sc}. Even at an SNR of 4 dB, only 90% accuracy is achieved. In addition, the performance of DSINet-1024, especially DSINet-1024(mc+sc), is better than that of the other models in the SNR range from −18 dB to −8 dB. DSINet-1024(mc+sc) exhibits an accuracy difference of 19% at −12 dB from RSC-VGGNet, which is the model with the lowest performance.

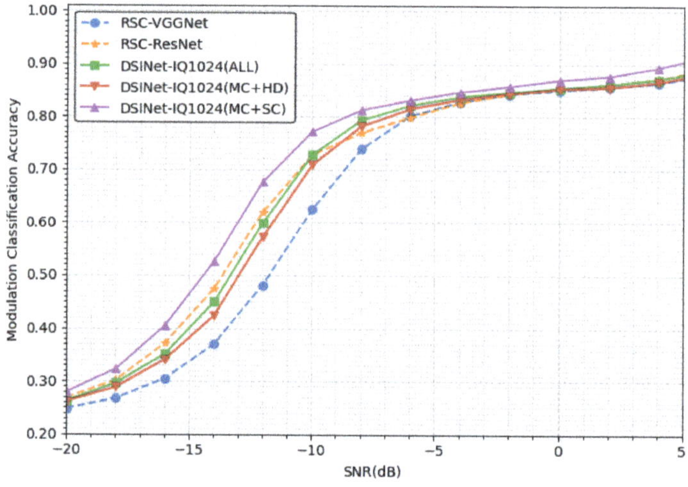

Figure 7. The modulation classification accuracies achieved under different SNRs.

What is unusual is that RSC-ResNet, a single modulation model, achieves better performance than our DSINet-1024(all). Sharing information about all tasks actually interrupts the MC performance improvement. In other words, there is competition among tasks. However, it can be said that our model is relatively stable, given that DSINet-1024(all) has lower accuracy than RSC-ResNet but high recall (the correct identification percentage attained for a certain class), as shown in Table 4. The best model is DSINet-1024(mc+sc), which only shares two hole detection tasks, and it is better than RSC-ResNet.

Table 4. Performance comparison between DSINet-1024 and RSC (SNR: −12 dB).

Methods	Accuracy	Recall	Precision	F_1 score
RSC-VGGNet	0.4818	0.3944	0.5022	0.4719
RSC-ResNet	0.6194	0.4949	0.5968	0.5149
DSINet-1024(all)	0.5986	0.5395	0.5827	0.5501
DSINet-1024(mc+hd)	0.5721	0.5074	0.5492	0.5181
DSINet-1024(mc+sc)	**0.6765**	**0.6481**	**0.6487**	**0.6454**

5.1.4. Model Performance in Terms of Execution Time

At the end of CS-1, we examine the model size and FLOPs, as described in Table 5. As emphasized in the DeepSense paper, DeepSense can perform hole detection with a very small model size and high speed. Our model also exhibits high computational performance on the order of three decimal places in GFLOPs. Although the model size is twice as large, increasing the amount of memory is not a major problem in terms of cost and system complexity. The fact that it produces results for three tasks simultaneously with almost the same size and speed can be considered a remarkable performance. Furthermore, it may be even better when only two tasks are considered, e.g., DSINet(hd+sc).

RSSI and STFT are small and fast models but require additional modules, such as a preprocessing module for calculating RSSIs or STFTs. In particular, it can be expected that the inference speed of the system equipped with WTC-STFT might be very slow because I/Q samples must be transformed with the STFT and even converted to images.

Table 5. Comparison between the model sizes and FLOPs of DSINet and DeepSense, RSCs, WTCs.

Methods	Model Size (K)	FLOPs (G)
DeepSense	65.040	0.00119
RSC-VGGNet	159.431	0.0262
RSC-ResNet	166.791	0.053
WTC-IQ	212.935	0.113
WTC-RSSI	43.747	0.00175
WTC-STFT	55.430	0.0184
DSINet-IQ128	136.796	0.00216
DSINet-IQ1024	180.284	0.0123
DSINet-IQ4096	207.228	0.0464

5.2. Analysis of Case Study 2

In CS-2, since only the path loss and AWGN channels are applied as the average power estimation task; the performance differences between the models are not clearly revealed. Therefore, we analyze the average performance of each model across the entire test dataset (all SNRs).

5.2.1. Task Ablation Study for Signal Classification

As seen in Table 6, all models classify signals well, even when the signals have mixed data samples (the *Accuracy*, *Recall*, *Precision*, and $F_1 score$ values all exceed 90%). In particular, as in CS-1, DSINet(sc+mc) achieves the highest performance. Since not much noise is contained in the data sample, there are almost no performance differences between the models, but the single model performs slightly better than DSINet(all). However, DSINet(all) estimates two additional tasks with little performance degradation. Even in this case study, the results show that to improve the performance of a task, only task information related to that task should be shared.

Table 6. Signal classification performance comparison according to a task ablation study.

Methods	Accuracy	Recall	Precision	F_1 score
SSINet(sc)	0.9430	0.9228	0.9227	0.9222
DSINet(all)	0.9422	0.9218	0.9219	0.9215
DSINet(sc+pe)	0.9446	0.9252	0.9248	0.9245
DSINet(sc+mc)	**0.9454**	**0.9260**	**0.9259**	**0.9257**

5.2.2. Task Ablation Study for Power Estimation

The average power levels estimated by the models are shown in Table 7. It can be seen that almost all models estimate the average power well, around plus or minus 2.75 (the mean absolute error). When the evaluation metric is the $R^2 score$, DSINet(pe+sc), which achieves the best performance, produces a score of 94.2%. The power estimation yielded by a model with a resolution of 3 dBm is sufficient for performing adaptive CCA. DSINet provides very useful information for developing a CTI mitigation strategy by classifying signals while estimating the power of each signal.

Table 7. Power estimation performance comparison according to a task ablation study.

Methods	Mean Absolute Error	R^2 score
SSINet(pe)	2.7512	0.9356
DSINet(all)	**2.7495**	**0.9407**
DSINet(pe+sc)	**2.7509**	**0.9419**
DSINet(pe+mc)	2.7646	0.9374

5.2.3. Task Ablation Study for Modulation Classification

Unlike CS-1, in this study, the difficulty of modulation classification is increased by mixing signals. In particular, the recall values of all models are low, as shown in Table 8, which means that the probability of being correct is low. In modulation classification, when signals are mixed simultaneously, the I/Q values are mixed and distorted in the time domain, so a low value is obtained.

Table 8. Modulation classification performance comparison according to a task ablation study.

Methods	Accuracy	Recall	Precision	F_1 score
SSINet (mc)	0.9377	0.5530	0.6994	0.5693
DSINet (all)	0.9367	0.5344	0.6620	0.5461
DSINet (mc + sc)	0.9369	0.5324	0.6616	0.5428
DSINet (mc + pe)	**0.9377**	**0.5554**	**0.7097**	**0.5711**

We can see that signal classification is not helpful for modulation classification. However, the power values that exist in different spectrum spaces are helpful to some extent, as shown in the performance of DSINet (mc + pe).

In summary, it is confirmed that DSINet achieves better performance than the single-task learning model (when specific tasks are considered) and that it solves many tasks simultaneously with just one model. This means that DSINet can provide abundant spectrum information, which is necessary for establishing a strategy corresponding to each wireless technology in a scenario where CTI frequently occurs.

5.3. *Analysis of Case Study 3*

5.3.1. Residual Unit vs. Plain Unit

DSINet's performance is significantly affected by how it utilizes the feature extraction units within a block, as with all other AI models. As described in Section 3.3, we consider two feature extraction units, a residual unit and a plain unit, and the performance achieved on each task when these units are applied to DSINet is shown in Figure 8. The scores obtained for the classification tasks, including T_{sc},T_{cc} and T_{mc}, represent accuracy, while the scores obtained for the regression task, which is T_{pe}, represent $R^2 scores$. The residual unit achieves good performance in terms of deep signal identification as well as in various fields. The residual unit solves the vanishing problem, and the network depth is relatively deep compared to the plain unit, so it achieves scores of 0.908, 0.975, 0.894, and 0.930 for each task.

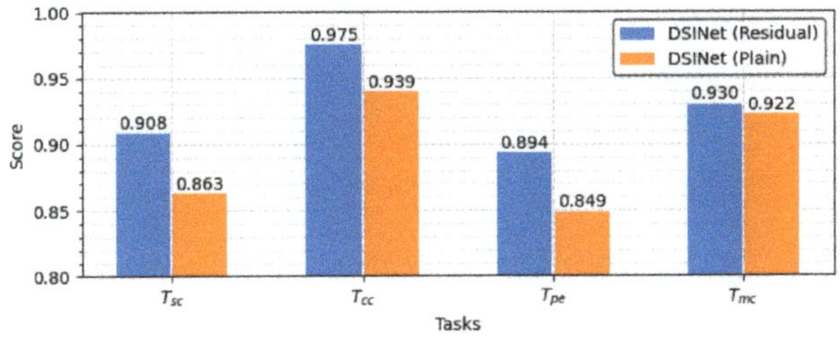

Figure 8. Comparison between the residual unit and plain unit.

5.3.2. Analysis of the Proportions of HBs and SBs

As mentioned in design principle 1, our model has a hybrid shared network structure with HBs and SBs. To investigate how the proportions of HBs and SBs affect the performance of DSINet, we evaluate DSINet-HBx:SBy + TB_{pe} variants in which the proportions of x and y are set as 0:4, 1:3, 2:2, 3:1, 4:0.

Figure 9 shows the performance of the DSINet-HBx:SBy + TB_{pe} variants, in which the y-axis indicates the score, and the x-axis is the ratio of HBs to SBs. For classification tasks including T_{sc},T_{cc}, and T_{mc}, the accuracy metric is used for the scoring function. For T_{pe}, which is the regression task, we use the $R^2 score$ as the scoring function.

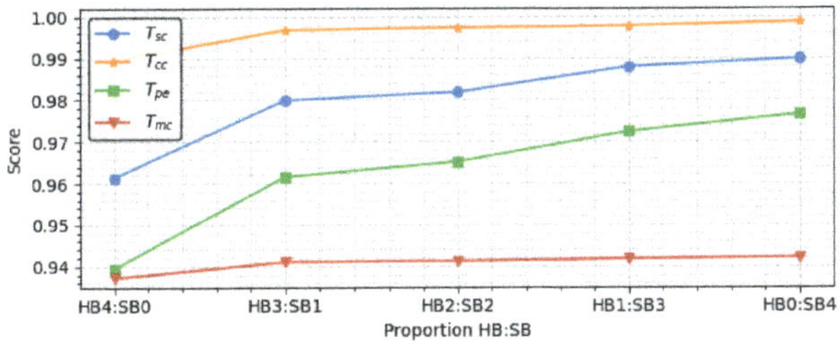

Figure 9. Score comparison according to the proportions of HBs and SBs (SNR: −4 dB).

As depicted in Figure 9, the DSINet variant with HB4:SB0 achieves the worst performance among all the variants. HB4:SB0 means that all feature extraction layers (weight parameters) are fully shared between all tasks. Therefore, each task cannot learn its own features, leading to the task competition problem, which frequently occurs in heterogeneous multitask learning settings, such as CS-3.

The overall performance of DSINet increases sharply at the HB3:SB1 point on the x-axis, and thereafter, the performance tends to increase slightly as the proportion of SBs increases. Since the SBs allow each task to learn its own features through independent weight parameters, the overall performance is enhanced. However, the use of too many SBs results in high memory usage and a slow inference speed. Figures 10 and 11 show the number of parameters (memory usage) and the number of floating point operations (FLOPs) according to the increase in the SB proportion, respectively. The number of parameters indicates a small memory usage level of approximately 199.4K in SB0, but this usage gradually increases as SB increases, and in SB4, the memory usage of 356.6K exceeds that of four single models (gray box). The number of FLOPs also increases gradually, similar to the number of parameters, and it increases sharply, especially starting from SB2. The numbers of parameters and FLOPs can increase more rapidly as the number of tasks increases. Therefore, we select the DSINet variant with HB3:SB1 as the best model, as it exhibits rapid performance improvement without significantly increasing the numbers of parameters and FLOPs.

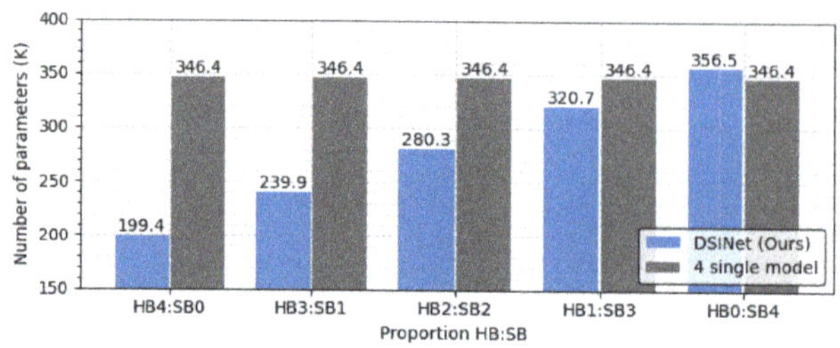

Figure 10. Memory usage comparison according to the proportions of HBs and SBs.

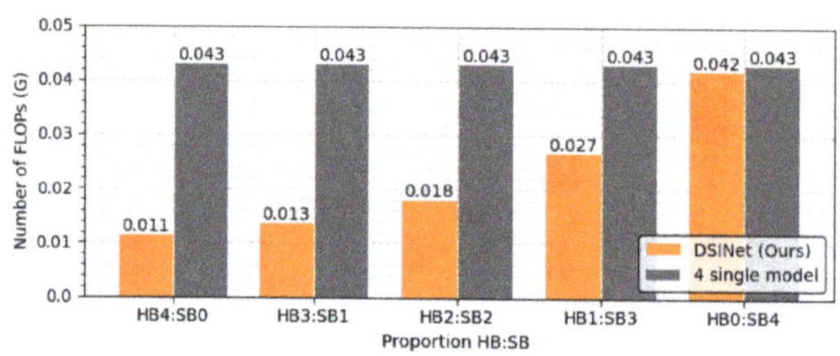

Figure 11. FLOPs comparison according to the proportions of HBs and SBs.

5.3.3. Analysis of the Effects of the TBs

According to design principle 2, we employ TBs as additional building blocks for DSINet. In this subsection, DSINet-HB3:SB1 + TB_{pe}, which is the best model selected in the previous subsection, is compared with DSINet-HB3:SB1 and DSINet-HB3:$TB_{all}1$ to evaluate the effect of the TB on DSINet. DSINet-HB3:SB1 has only HBs and SBs but no TBs. On the other hand, DSINet-HB3:$TB_{all}1$ consists only of HBs and TBs. TB_{all} means that all tasks have a specific TB, while TB_{pe} means that only task T_{pe} has a TB.

On average, as described in Table 9, DSINet-HB3:SB1 + TB_{pe} outperforms the other models with scores of 0.9084, 0.9752, 0.8935, and 0.9296. Among DSINet-HB3:SB1 and DSINet-HB3:$TB_{all}1$, DSINet-HB3:$TB_{all}1$ achieves better performance.

Table 9. Performance comparison according to the block combinations of DSINet.

Methods	T_{sc} *Acc* Score	T_{cc} *Acc* Score	T_{pe} R^2Score	T_{mc} *Acc* Score
DSINet-HB3:$TB_{all}1$	0.8804	0.9551	0.8668	0.9250
DSINet-HB3 + SB1	0.8701	0.9452	0.8445	0.9238
DSINet-HB3+SB1 + TB_{pe}	**0.9084**	**0.9752**	**0.8935**	**0.9296**

This trend appears similarly in the SNR-based performance comparison. Figure 12 shows the performance of DSINet-HB3:SB1+TB_{pe}, DSINet-HB3:SB1 and DSINet-HB3:$TB_{all}1$ by SNR and by task. In a high-SNR environment, the performances of all DSINets are similar to each other, but as the SNR decreases, the performance of each model is significantly different. Even at SNRs of −5 or lower, DSINet-HB3:SB1 + TB_{pe} achieves the highest performance by a large margin. Among the other two models, DSINet-HB3:$TB_{all}1$ exhibits slightly better performance than DSINet-HB3:SB1, at approximately 4%. This indicates that even if an independent layer is provided for each task through the use of SBs, the performance is not improved without TBs.

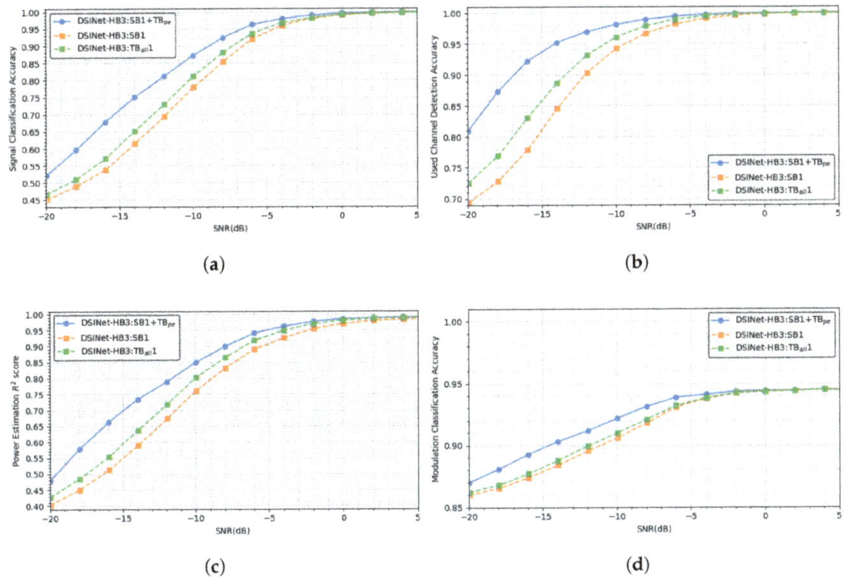

Figure 12. Performance comparison among the DSINet variants according to the block combinations in (**a**) T_{sc}, *Accuracy* of signal classification; (**b**) T_{cc}, *Accuracy* of used channel detection; (**c**) T_{pe}, R^2*score* of power estimation; and (**d**) T_{mc}, *Accuracy* of modulation classification.

Again, CS-3 contains highly heterogeneous tasks, including the classification tasks T_{sc}, T_{cc}, and T_{mc} and the regression task T_{pe}. In a classification task, the output is a discrete value of 0 or 1, whereas in a regression task, the output has a continuous value between 0 and 1. In this case, the cross-stitch unit of the SBs shares information between task layers through the linear transformation of discrete features and continuous features, so the performance between tasks may degrade. On the other hand, if a TB is used for a specific task (T_{pe} in our case), task-specific layers near the output are additionally employed, so the SB features are not significantly affected by the output characteristics. Therefore, DSINet-HB3:SB1 + TB_{pe} can accommodate the advantages of SBs without compromising performance due to the output characteristics.

5.3.4. Analysis of the Model Size and FLOPs with a Task Size Increase

To see the model size and FLOPs more clearly, we generalize the task to a task with the same one-hot labels and derive the results. In Figure 13, the optimal combination of DSINet solves four tasks with less than 30% of the parameters used by a single model set, for which the model size is 344.1K. In terms of computational performance, our model does not slow down because its FLOPs increase almost insignificantly, unlike the single model set that is slowed by n times every time the task increases, as shown in Figure 14. Although the results are not shown in the figure, our model performs well with a model size of 271.6K vs. 430.2K and FLOPs of 0.014 GFLOPs vs. 0.053 GFLOPs, even when the number of tasks is 5. We believe that it is desirable to apply multitask learning to our DSINet to derive rich spectrum state information while maintaining acceptable model size and speed levels in wireless systems.

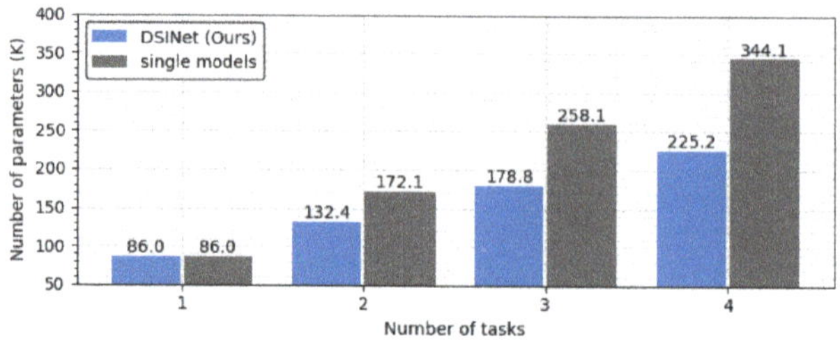

Figure 13. Memory usage comparison between DSINet and the single models.

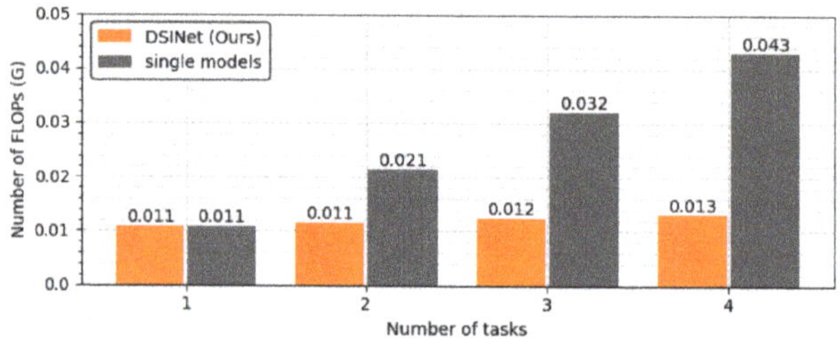

Figure 14. FLOPs comparison between DSINet and the single models.

6. Conclusions

In this paper, we proposed DSINet, a multitask learning-based deep signal identification network for advanced spectrum sensing systems, with a focus on preparing for the upcoming 5G-advanced and 6G systems. We presented three important types of blocks within DSINet, hard sharing blocks, soft sharing blocks, and task-specific blocks, that maximize the effect of sharing characteristics in multitask learning and support the deep identification of signal footprints in multidimensional spectrum spaces.

We evaluated DSINet in three challenging case studies where heterogeneous devices used the spectrum with complex dynamics. In Case Study 1, we compared DSINets with three other model types, including DeepSense, WTCs, and RSCs, in different tasks, such as hole detection, signal classification, and modulation classification. Our results showed that DSINet outperformed these models, with performance improvements of 3.3%, 3.3%, and 5.7%, respectively. In Case Study 2, we conducted task ablation studies to understand the contributions of tasks and found that DSINet provided rich information while exhibiting better performance than single models in all tasks. Through experiments, we discovered that sharing only related tasks is better than sharing all identification tasks unilaterally in terms of performance. The block-based model structure makes it easy to consider this factor in the model design stage. In the last case study, we analyzed DSINets designed with different numbers of HBs, SBs, and TBs under the model design principles to determine the performance differences yielded for each task and derived the best model. The best DSINet achieved 98%, 99%, and 94% accuracy and an $R^2 score$ of 0.96 in signal classification, channel classification, modulation classification, and power estimation, respectively, even at a relatively low SNR of –4 dB. This performance is significant in practical use, as it came from a model with 225.2 K model parameters and 0.013 GFLOPs (the individual model sets required 344.1 K model parameters and 0.043 GFLOPs).

We expect that DSINet will play a crucial role in supporting seamless spectrum sharing and resource management in the future network environments of 5G-advanced and 6G, where dynamic spectrum sharing and cross-technology interference frequently occur.

Author Contributions: Conceptualization, H.K., Y.-J.K. and W.-T.K.; methodology, H.K. and Y.-J.K.; software, H.K. and Y.-J.K.; validation, H.K. and Y.-J.K.; formal analysis, H.K. and Y.-J.K.; investigation, H.K. and Y.-J.K.; resources, W.-T.K.; data curation, H.K. and Y.-J.K.; writing—original draft preparation, H.K. and Y.-J.K.; writing—review and editing, H.K., Y.-J.K. and W.-T.K.; visualization, H.K. and Y.-J.K.; supervision, W.-T.K.; project administration, W.-T.K.; funding acquisition, W.-T.K. All authors have read and agreed to the published version of the manuscript.

Funding: This work was supported by the VRAR Content Industry Promotion Project (D0747-23-1001) by the National IT Industry Promotion Agency (NIPA), the Smart Edge Device Technology Development Project (2022-0-00454) by the Institute of Information and Communications Technology Panning and Evaluation (IITP), and the Post-Doctoral Program of KOREATECH.

Institutional Review Board Statement: Not applicable.

Informed Consent Statement: Not applicable.

Data Availability Statement: Data are contained within the article.

Conflicts of Interest: The authors declare no conflict of interest.

References

1. Yang, P.; Kong, L.; Chen, G. Spectrum sharing for 5G/6G URLLC: Research frontiers and standards. *IEEE Commun. Stand. Mag.* **2021**, *5*, 120–125. [CrossRef]
2. Qian, B.; Zhou, H.; Ma, T.; Yu, K.; Yu, Q.; Shen, X. Multi-operator spectrum sharing for massive IOT coexisting in 5G/B5G Wireless Networks. *IEEE J. Sel. Areas Commun.* **2021**, *39*, 881–895. [CrossRef]
3. Bhatti, F.A.; Khan, M.J.; Selim, A.; Paisana, F. Shared spectrum monitoring using deep learning. *IEEE Trans. Cogn. Commun. Netw.* **2021**, *7*, 1171–1185. [CrossRef]
4. Ali, A.; Hamouda, W. Advances on spectrum sensing for cognitive radio networks: Theory and applications. *IEEE Commun. Surv. Tutor.* **2016**, *19*, 1277–1304. [CrossRef]
5. Soni, B.; Patel, D.K.; López-Benítez, M. Long Short-Term Memory Based Spectrum Sensing Scheme for Cognitive Radio Using Primary Activity Statistics. *IEEE Access* **2020**, *8*, 97437–97451. [CrossRef]
6. Junmei, Y.; Lou, W.; Xie, R.; Jiao, X.; Wu, K. Mitigating Cross-Technology Interference through Fast Signal Identification. *IEEE Trans. Veh. Technol.* **2023**, *72*, 2521–2534.
7. Grimaldi, S.; Mahmood, A.; Hassan, S.A.; Gidlund, M.; Hancke, G.P. Autonomous interference mapping for industrial Internet of Things networks over unlicensed bands: Identifying cross-technology interference. *IEEE Ind. Electron. Mag.* **2021**, *15*, 67–78. [CrossRef]
8. Gorcin, A.; Arslan, H. Signal identification for adaptive spectrum hyperspace access in wireless communications systems. *IEEE Commun. Mag.* **2014**, *52*, 134–145. [CrossRef]
9. Kulin, M.; Kazaz, T.; Moerman, I.; De Poorter, E. End-to-end learning from spectrum data: A deep learning approach for wireless signal identification in spectrum monitoring applications. *IEEE Access* **2014**, *6*, 18484–18501. [CrossRef]
10. Dobre, O.A.; Oner, M.; Rajan, S.; Inkol, R. Cyclostationarity-based robust algorithms for QAM signal identification. *IEEE Commun. Lett.* **2014**, *16*, 12–15. [CrossRef]
11. Chavali, V.G.; da Silva, C.R.C.M. Maximum-likelihood classification of digital amplitude-phase modulated signals in flat fading non-Gaussian channels. *IEEE Trans. Commun.* **2014**, *59*, 2051–2056. [CrossRef]
12. Liu, W.; Kulin, M.; Kazaz, T.; Shahid, A.; Moerman, I.; De Poorter, E. Wireless technology recognition based on rssi distribution at sub-nyquist sampling rate for constrained devices. *Sensors* **2017**, *17*, 2081. [CrossRef] [PubMed]
13. Mendis, G.J.; Wei-Kocsis, J.; Madanayake, A. Deep learning based radio-signal identification with hardware design. *IEEE Trans. Aerosp. Electron. Syst.* **2019**, *55*, 2516–2531. [CrossRef]
14. Wang, L.; Chen, Z.; Zhang, Y. Wireless signal modulation identification method based on RF I/Q data distribution. *Sci. Rep.* **2021**, *11*, 21383. [CrossRef] [PubMed]
15. Tekbıyık, K.; Akbunar, Ö; Ekti, A.R.; Görçin, A.; Kurt, G.K.; Qaraqe, K.A. Spectrum Sensing and Signal Identification with Deep Learning Based on Spectral Correlation Function. *IEEE Trans. Veh. Technol.* **2021**, *70*, 10514–10527. [CrossRef]
16. Meaning of Identification in English, Cambridge Dictionary. Available online: https://dictionary.cambridge.org/dictionary/english/identification (accessed on 10 October 2023).
17. Simeone, O. A very brief introduction to machine learning with applications to communication systems. *IEEE Trans. Cogn. Commun. Netw.* **2018**, *4*, 648–664. [CrossRef]
18. Zhao, X.; Li, H.; Shen, X.; Liang, X.; Wu, Y. A modulation module for multi-task learning with applications in image retrieval. In Proceedings of the European Conference on Computer Vision, Munich, Germany, 8–14 September 2018.
19. Tang, Y.; Pino, J.; Wang, C.; Ma, X.; Genzel, D. A general multi-task learning framework to leverage text data for speech to text tasks. In Proceedings of the IEEE International Conference on Acoustics, Speech and Signal Processing, Toronto, ON, Canada, 6–11 June 2021.

20. Worsham, J.; Jugal, K. Multi-task learning for natural language processing in the 2020s: Where are we going? *Pattern Recognit. Lett.* **2020**, *136*, 120–126. [CrossRef]
21. Salahdine, F.; El Ghazi, H.; Kaabouch, N.; Fihri, W.F. Matched filter detection with dynamic threshold for cognitive radio networks. In Proceedings of the International Conference on Wireless Networks and Mobile Communications, Marrakech, Morocco, 20–23 October 2015.
22. Nguyen, T.; Mark, B.L.; Ephraim, Y. Spectrum sensing using a hidden bivariate Markov model. *IEEE Trans. Wirel. Commun.* **2013**, *12*, 4582–4591. [CrossRef]
23. Gao, J.; Yi, X.; Zhong, C.; Chen, X.; Zhang, Z. Deep learning for spectrum sensing. *IEEE Wirel. Commun. Lett.* **2019**, *8*, 1727–1730. [CrossRef]
24. Liu, C.; Wang, J.; Liu, X.; Liang, Y.-C. Deep CM-CNN for Spectrum Sensing in Cognitive Radio. *IEEE J. Sel. Areas Commun.* **2019**, *37*, 2306–2321. [CrossRef]
25. He, J.; Chen, W.; Jia, L.; Wang, T. An Effective Reconstruction Algorithm Based on Modulated Wideband Converter for Wideband Spectrum Sensing. *IEEE Access* **2020**, *8*, 152239–152247. [CrossRef]
26. Junmei, Y.; Zheng, X.; Xie, R.; Wu, K. Cross-technology communication for heterogeneous wireless devices through symbol-level energy modulation. *IEEE Trans. Mob. Comput.* **2021**, *21*, 3926–3940.
27. Yuan, W.; Linnartz, J.P.M.; Niemegeers, I.G. Adaptive CCA for IEEE 802.15. 4 wireless sensor networks to mitigate interference. In Proceedings of the IEEE Wireless Communication and Networking Conference, Sydney, Australia, 18–21 April 2010.
28. Dikmese, S.; Renfors, M. Optimized FFT and filter bank based spectrum sensing for Bluetooth signal. In Proceedings of the IEEE Wireless Communications and Networking Conference, Paris, France, 1–4 April 2012.
29. Yubo, Y.; Yang, P.; Li, X.Y.; Zhang, Y.; Lu, J.; You, L.; Xiong, Y. Wizbee: Wise zigbee coexistence via interference cancellation with single antenna. *IEEE Trans. Mob. Comput.* **2014**, *14*, 2590–2603.
30. Phillips, J.; Martinez, J.; Bârsan, I.A.; Casas, S.; Sadat, A.; Urtasun, R. Deep multi-task learning for joint localization, perception, and prediction. In Proceedings of the IEEE/CVF Conference on Computer Vision and Pattern Recognition, Nashville, TN, USA, 20–25 June 2021.
31. Guo, M.; Haque, A.; Huang, D.A.; Yeung, S.; Fei-Fei, L. Dynamic task prioritization for multitask learning. In Proceedings of the European Conference on Computer Vision, Munich, Germany, 8–14 September 2018.
32. Zhang, Y.; Yang, Q. A Survey on Multi-Task Learning. *IEEE Trans. Knowl. Data Eng.* **2022**, *34*, 5586–5609. [CrossRef]
33. Vandenhende, S.; Georgoulis, S.; Van Gansbeke, W.; Proesmans, M.; Dai, D.; Van Gool, L. Multi-task learning for dense prediction tasks: A survey. *IEEE Trans. Pattern Anal. Mach. Intell.* **2022**, *44*, 3614–3633. [CrossRef]
34. Gong, J.; Qin, X.; Xu, X. Multi-Task Based Deep Learning Approach for Open-Set Wireless Signal Identification in ISM Band. *IEEE Trans. Cogn. Commun. Netw.* **2022**, *8*, 121–135. [CrossRef]
35. Zhang, T.; Qiang, Y. An overview of multi-task learning. *Natl. Sci. Rev.* **2018**, *5*, 30–43. [CrossRef]
36. Misra, I.; Shrivastava, A.; Gupta, A.; Hebert, M. Cross-stitch networks for multi-task learning. In Proceedings of the IEEE Conference on Computer Vision and Pattern Recognition, Las Vegas, NV, USA, 26 June–1 July 2016.
37. Tai, Y.; Yang, J.; Liu, X. Image super-resolution via deep recursive residual network. In Proceedings of the IEEE Conference on Computer Vision and Pattern Recognition, Honolulu, HI, USA, 21–26 July 2017; pp. 2790–2798.
38. Cypher, D. *NIST Smart Grid Interoperability Panel Priority Action Plan 2: Guidelines for Assessing Wireless Standards for Smart Grid Applications*; National Institute of Standards and Technology: Gaithersburg, MD, USA, 2014.
39. Klambauer, G.; Unterthiner, T.; Mayr, A.; Hochreiter, S. Self-normalizing neural networks. In Proceedings of the Advances in Neural Information Processing Systems, Long Beach, CA, USA, 4–9 December 2017; Volume 30.
40. Fontaine, J.; Fonseca, E.; Shahid, A.; Kist, M.; DaSilva, L.A.; Moerman, I.; De Poorter, E. Towards low-complexity wireless technology classification across multiple environments. *Ad Hoc Netw.* **2019**, *91*, 101881. [CrossRef]
41. Uvaydov, D.; D'Oro, S.; Restuccia, F.; Melodia, T. Deepsense: Fast wideband spectrum sensing through real-time in-the-loop deep learning. In Proceedings of the IEEE INFOCOM 2021—IEEE Conference on Computer Communications, Vancouver, BC, Canada, 10–13 May 2021; pp. 1–10.
42. O'Shea, T.J. ; Roy, T.; Clancy, T.C. Over-the-air deep learning based radio signal classification. *IEEE J. Sel. Top. Signal Process.* **2018**, *12*, 168–179. [CrossRef]

sensors

MDPI

Article

Efficient Cumulant-Based Automatic Modulation Classification Using Machine Learning

Ben Dgani * and Israel Cohen *

Andrew and Erna Viterbi Faculty of Electrical & Computer Engineering, Technion-Israel Institute of Technology, Haifa 3200003, Israel

* Correspondence: bendgani@campus.technion.ac.il (B.D.); icohen@ee.technion.ac.il (I.C.)

Abstract: This paper introduces a new technique for automatic modulation classification (AMC) in Cognitive Radio (CR) networks. The method employs a straightforward classifier that utilizes high-order cumulant for training. It focuses on the statistical behavior of both analog modulation and digital schemes, which have received limited attention in previous works. The simulation results show that the proposed method performs well with different signal-to-noise ratios (SNRs) and channel conditions. The classifier's performance is superior to that of complex deep learning methods, making it suitable for deployment in CR networks' end units, especially in military and emergency service applications. The proposed method offers a cost-effective and high-quality solution for AMC that meets the strict demands of these critical applications.

Keywords: cumulants; modulation classification; machine learning

1. Introduction

The use of smart transmitters, adaptive receivers and intelligent communication networks has revolutionized radio communication in the modern era. These technologies have helped us optimize our usage of wireless channels. One crucial task in this regard is the implementation of intelligent radio modulation detection. Automatic modulation classification plays a vital role in various civilian and military applications, which are diverse and often massive. Established radio networks have been in operation for years and require fast and adaptive radio networks that can be used to combat various tasks daily and in times of need. To achieve this, the SDR platform is considered a feasible option.

Software-defined radio (SDR) is a communication platform that works through software programs rather than traditional hardware components. This means that the radio's functions can be adapted and changed dynamically based on different networks and tasks. SDR is considered the modern realization of the term "software radio", which was coined by Mitola [1]. By converting signal processing to the digital domain, SDR-based networks can incorporate complex algorithms that improve the communication rate, accuracy, and distance.

One example of the benefits of SDR is the concept of Cognitive Radio, which was also introduced by Mitola in 1999 [2]. The idea behind Cognitive Radio is to develop a radio unit that can manage its communication network dynamically and automatically using software programs. The reason for this is to solve the problem of spectral congestion in modern communication. As the number of daily users increases, spectrum resources become more expensive and less available. By using Cognitive Radio, the radio unit can adapt to its environment and establish unlicensed communication networks that share the same wireless spectrum with other users. This helps solve the problem of spectral congestion and makes it easier for unlicensed users, such as the military and emergency services, to communicate effectively in times of crisis.

Setting up the shared network involves a method called spectrum sensing, where the Cognitive Radio unit can detect and analyze the available spectrum, and then manage the communication network accordingly. By utilizing software instead of expensive and complex hardware, Cognitive Radio offers a more efficient and cost-effective solution to the problem of spectral congestion. During the spectrum sensing step, a Cognitive Radio performs a scan of all available channels to find "empty holes", or unused frequencies in the spectrum. By sensing the channel parameters, the adaptive qualities of the Cognitive Radio enable it to dynamically change the modulation scheme used for transmission in order to improve the transmission rates and quality. However, dynamic modulations require a recognition method for the received signal modulation scheme, which is known as automatic modulation classification, or modulation recognition and modulation detection in the literature [3–5]. Spectrum sensing is essential for adaptive networks that must change according to their location without causing any interruption to local users' daily communication needs, especially in times of emergency, where every message counts.

Automatic modulation classification (AMC) is a process used to detect and identify the unknown modulation scheme of a radio signal. It has been around since the 1980s [6,7] but gained more popularity after the publication of the Cognitive Radio idea. Today, the modern AMC literature is mainly focused on deep learning and neural networks [8–10]. However, the classic AMC algorithms are split between two methods [11]. The process of identifying radio signals can be challenging due to the high computational complexity that certain algorithms require. The first set of methods, called likelihood-based algorithms [12], are accurate but can be expensive to implement, which is a drawback for government-funded services that rely on cost-efficient projects.

On the other hand, feature-based (FB) methods offer a more efficient approach to signal analysis. By extracting features like high-order statistical moments (HOS) or instantaneous frequency and phase parameters, these algorithms can help understand signal characteristics. FB algorithms rely on a predetermined set of features, reducing the computational burden when compared to more complex machine learning techniques. To use FB algorithms, the process starts by extracting classification features from the signal. Once the algorithm has identified the key features, it can train the selected classification algorithm. Some examples of feature-based algorithms are support vector machines (SVMs), decision trees, and K-means. After the model is trained, it can be applied with minimal complexity and runtime, making it a useful tool for identifying radio signals.

In recent years, the field of neural networks has experienced massive growth, leading to most new studies on AMC utilizing neural networks for feature extraction, classification, or both. Neural networks have demonstrated optimal results concerning classification accuracy, achieving high detection probability, even at low SNR values. However, like likelihood-based methods, neural networks can suffer from high computational complexity and require significant system storage if a deep network model is used. This can become a challenge when dealing with a quick-setup radio network that relies on end units and needs to be power- and time-efficient. This is particularly important when considering emergency services, where there is a need for a new, easy-to-set-up network that can quickly adapt itself to the spectral environment. Therefore, the training process for the desired modulation must be efficient.

This paper focuses on the FB method for the AMC, where we utilize high-order cumulants (HOCs) as key features. Cumulants are statistical measures that are alternatives to moments such as mean and variance. The HOCs are popular in the AMC literature as a detection feature [13,14], due to their additive property that separates the desired signal from its additive noise. While most of the literature discusses digital modulation classification [15,16], our work aims to include analog modulations as well. In our study, we analyze the statistics of eleven modulation schemes, explore the best classification feature for each modulation, and discover the statistical connection between pairs of modulations.

The rest of this paper is organized as follows. Section 2 presents the modulated signal models and the theory behind HOS and cumulants. In Section 3, we evaluate the simulation results for our proposed FB method. Finally, in Section 4, we discuss the conclusions from the results.

2. Background and Algorithm

This section is divided into four subsections to provide a comprehensive explanation of our work. The first part briefly reviews the theory of high-order statistics and the cumulants, which provide an alternative approach to the more familiar moments. In the second part, we introduce the dataset that we used for our work. The third subsection explains the theoretical aspect of the classification process. Finally, the last subsection briefly explains the method we used for our work.

2.1. High-Order Statistics

In statistics, moments of a random variable provide quantitative measures related to the shape of its probability density graph and the distribution behavior. Specifically, the nth-order moment of a random variable x is defined by

$$m_n(x) = \int x^n f(x)dx, \tag{1}$$

where $f(x)$ is the probability density function of x. If the nth-order moment integral diverges, the nth-order moment of the variable does not exist. If the nth-order moment of x exists, so does the $(n-1)$th-order moment, and thus all lower-order moments. We define a central moment by $m_n(x-\mu)$, which is the mean value of x, and define the normalized moment as $m_n(x/\sigma)$, where σ is the standard deviation.

The cumulants c_n of a probability distribution are an alternative to the moments of the distribution. Any pair of probability distributions with identical moments will also share identical cumulants, and vice versa. The first, second, and third cumulants are equal to the first three central moments—known as the mean, variance, and skewness—but fourth- and higher-order cumulants diverge from the central moments. The cumulant of a random variable x is defined using the cumulant-generating function

$$C_t(x) = log(E[e^{tx}]), \tag{2}$$

and the cumulants c_n are obtained from a power series expansion of the cumulant-generating function:

$$C_t(x) = \sum c_n \frac{t^n}{n!}. \tag{3}$$

This expansion is a Maclaurin series expansion of the logarithm function, so we obtain that the nth-order cumulant can be obtained by differentiating the expansion n times at $t = 0$. The use of cumulants instead of moments as features for AMC in the literature is usually based on their cumulative property: assume X and Y are independent random variables, then

$$c_n(X+Y) = c_n(X) + c_n(Y). \tag{4}$$

Despite its benefits, the cumulant-generating function can be challenging to calculate, especially when we are not familiar with the exact distribution of the signal. However, since moments and cumulants are shared when the distribution is the same, we can use the moments to calculate the cumulants based on the work in [14,17]. We use

$$c_{4.0} = m_{4.0} - 3m_{2.0}^2 \quad (5)$$
$$c_{4.1} = m_{4.1} - 3m_{2.1}m_{2.0} \quad (6)$$
$$c_{4.2} = m_{4.2} - |m_{2.0}|^2 - 2m_{2.1}^2 \quad (7)$$
$$c_{6.0} = m_{6.0} - 15m_{4.0}m_{2.0} + 30m_{2.0}^3 \quad (8)$$
$$c_{6.3} = m_{6.3} - 9c_{4.2}m_{2.1} - 6m_{2.1}^2 \quad (9)$$
$$c_{8.0} = m_{8.0} - 28m_{6.0}m_{2.0} - 35m_{4.2}^2 + 420m_{4.0}m_{2.0}^2 - 630m_{2.0}^4 \quad (10)$$

where $m_{q,p}$ is the mixed moment of the signal, defined as $E(x^q \times \bar{x}^{p-q})$. This set of equations enables simple and fast calculation for the HOC, which will act as features for the classification algorithm. Since our work is based on digital signals, we will use the discrete approximation for the mixed moment of N samples from a signal:

$$m_{q,p} = \frac{1}{N}\sum_{n=1}^{N} x^q[n] \times \bar{x}^{p-q}[n]. \quad (11)$$

The classification method in our paper uses the cumulants above as input features, along with the matching cumulants $C_{p,q}(x_d)$ for the derivative of the signal

$$x_d[n] = x[n+1] - x[n]. \quad (12)$$

After experimenting with different features, we added the cumulants of the derivative to the features list, including separated cumulants for the real and imaginary parts of the signal and the cumulants of the amplitude and phase of each signal. Only the derivative's cumulants noticeably improved the classification accuracy.

When discussing AWGN (additive white Gaussian noise), the primary benefit of utilizing cumulants becomes evident. The noise is a random process that follows a normal distribution. Therefore, we can use its cumulant-generating function, which is calculated in [18,19] using

$$C_t(w) = log(M_t(w)) = \mu t + \sigma^2 \frac{t^2}{2}, \quad (13)$$

where μ and σ are the mean and standard deviation, respectively. The function in (13) is a second-order polynomial, so its 3rd- and higher-order derivatives are zero-valued, meaning the HOC cumulants of AWGN are equal to zero. Hence, the HOC of the sum between AWGN and another independent random process will equal the HOC of the second random process alone. Since the derivative of AWGN is distributed the same, the derivative's cumulants are still effective for our method under the above assumptions.

2.2. Modulation and Signal Model

The signals used in this paper are taken from the RadioML2018.01A dataset [20]. This dataset contains 2,555,904 frames. Each frame is 1024 IQ samples from a modulated signal, with various modulation schemes and SNR values from −20 to 30 dB. The dataset contains 24 different modulations generated using SDR under the following channel specifications:

- Selective multipath Rician fading;
- Carrier frequency offset;
- Symbol rate offset;
- Non-impulsive delay spread;
- Doppler shift;
- AWGN.

We only consider 11 modulations as the "Normal" set. This includes the following: OOK, 4ASK, BPSK, QPSK, 8PSK, 16QAM, AM-SSB-SC, AM-DSB-SC, FM, GMSK, OQPSK (We will refer to AM-SSB-SC and AM-DSB-SC as SSB and DSB, respectively). We decided to evaluate our work on the "Normal" dataset since it is considered the benchmark for classification models in the literature, in contrast to the "hard" set, which contains all 24 schemes.

2.3. Decision Tree

A decision tree is a flowchart-like model for supervised learning in which each internal node represents a decision test on a single feature, each branch represents the outcome of the test, and each leaf node represents a class label. The paths from root to leaf represent the classification rules. Decision trees have several advantages as a classification model, and we are focusing on two of them in this paper. The first is that training the tree model does not require a large dataset; this is a significant advantage over neural networks that are widely popular in the literature despite the need for massive data to yield results. The second advantage is the model's simplicity, which allows it to combine with other decision techniques. The benefit of combining techniques comes when we make a tree model, where instead of a simple decision rule at each node, we can use a more complex model.

The decision tree model we used in our work is from the Python scikit-learn (Sklearn) library, which is trained based on Gini impurity. Gini impurity is a measurement for the probability of classifying the labels incorrectly. Let A be a random variable that takes one of k labels $(a_1, a_2, \dots a_k)$; then, we define the Gini impurity of a classification model as

$$G(A) = 1 - \sum_{i=1}^{k} (P(a_i))^2 . \tag{14}$$

The training process of a decision tree contains three steps:

1. Decide the feature to split the data: for each feature, the Gini impurity is calculated, and the one for which it is minimized is selected.
2. Continue splitting the data based on 1.
3. Stop splitting the data if we reach a certain tree depth.

The decision tree model's parameters are the comparison threshold at each node, barring the leaves, meaning the total number of parameters in a tree of depth d can be calculated with

$$P_n = 2^H - 1, \tag{15}$$

which is also the number of calculations a single tree applies when classifying since there is no need for multiplications or additions.

2.4. Method

Our classification algorithm is designed for integration within SDR systems operating in Cognitive Radio networks. The workflow initiates with signal acquisition by the SDR, followed by the segmentation of acquired signals into discrete packets, typically standardized at a length of 1024 samples—although this length is variable, subject to network requirements and design. These packets serve as inputs to our classification algorithm. For each packet received, we use the set of Equations (5) through (10) to extract the features vector of the cumulants. This feature extraction process is computationally straightforward, involving a singular pass through the packet data, which aligns with the predetermined packet length of the communication network's settings.

Subsequently, these extracted features undergo the classification phase, where the decision tree model is employed. The output of this classification stage yields the identification of one among the 11 available modulation schemes within the "Normal" modulations set. The detected modulation holds significant utility within the Cognitive Radio context, facilitating dynamic functionalities, such as adaptive frequency channel selection and network usage detection. Furthermore, the system's training procedure can be executed proactively in controlled environments or dynamically during operational use, enabling real-time adaptation to the evolving network landscape. This adaptability ensures the system is efficient and adaptive within varying operational scenarios, allowing seamless integration into dynamic spectrum environments.

3. Simulation and Results

The suggested classification method requires hand-picked features for each step, so we start the simulation with a small-scale experiment. We start with an experiment to determine which of the features is optimal for the classification and detection of every single modulation from our desired set of 11 schemes. This simulation includes a simple decision tree model, where only one feature from the cumulants and derivative cumulants is used each run, where a run is defined as training and testing. Table 1 shows the top two features for each modulation scheme in terms of the detection accuracy using the simple decision tree model.

Table 1. The best two features for each modulation at SNR = 10 dB.

Modulation	1st Feature	P_d	2nd Feature	P_d
SSB	$C_{4.1}$	100%	$C_{4.0}$	100%
DSB	$C^d_{4.1}$	74.69%	$C^d_{4.0}$	73.34%
BPSK	$C_{2.0}$	99.57%	$C_{4.0}$	92.73%
FM	$C_{4.1}$	99.63%	$C_{2.0}$	99.27%
GMSK	$C_{4.2}$	96.09%	$C_{8.0}$	81.23%
OOK	$C_{2.0}$	95.82%	$C_{2.1}$	93.61%
4ASK	$C_{2.1}$	95.84%	$C_{2.0}$	93.71%
QPSK	$C^d_{4.0}$	87.95%	$C^d_{8.0}$	86.69%
OQPSK	$C_{4.2}$	76.39%	$C_{4.0}$	60.53%
16QAM	$C^d_{6.0}$	100%	$C^d_{6.3}$	99.74%
8PSK	$C^d_{4.0}$	100%	$C^d_{4.2}$	99.87%

Table 1 shows the average outcomes obtained when classifying signals specifically at an SNR of 10 dB. This SNR value is deemed sufficient for a receiver to effectively detect and demodulate signals with accuracy. Utilizing the insights derived from this table, we proceed to construct the ultimate feature vector, which forms the foundational input for our classification algorithm.

This feature vector is meticulously composed of a consolidated set of cumulants, as illustrated in the table, representing a combined and optimized selection based on their performance at the specified SNR of 10 dB. This selection process is pivotal in crafting an efficient and discriminative feature set, chosen explicitly for its effectiveness in signal classification within this specific SNR context. By amalgamating these cumulants into a unified feature vector, we aim to encapsulate the most informative and distinguishing signal characteristics essential for accurate modulation classification at this SNR threshold.

The next part of the simulation is the picking process of the training set for the model. It is widely agreed that for training machine learning models, the algorithm randomly picks the training set, out of the total database. However, in our work, we consider training our model with only a subset of the data. This subset includes all the sampled signals with SNR values higher than a certain threshold. The reasoning behind our decision comes from the fact that when we study different modulation schemes, we can see that each scheme has a minimal SNR requirement for adequate demodulation, which means that using signals with an SNR below that threshold might affect our model training in negative ways. Based on these facts, we split the data into two separate groups: high SNR signals and low SNR signals. In order to pick an SNR threshold for the splitting, we conducted our second experiment. Figure 1 shows the classification results when we train the model using a signal with SNR higher than varying thresholds; we also include a zoomed-in view of the section where the classification accuracy is the highest.

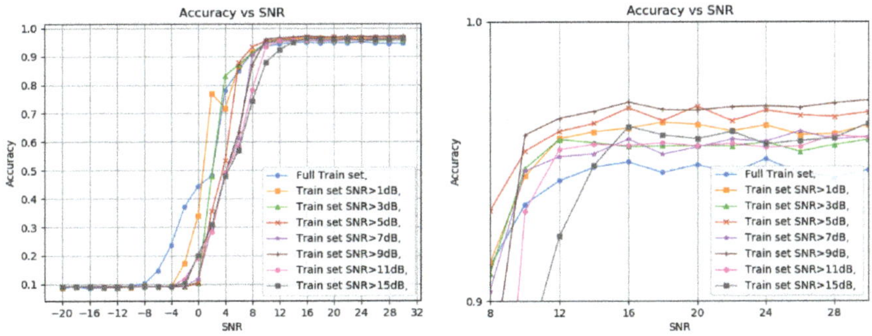

(**a**) Full view. (**b**) Zoomed in.

Figure 1. P_d based on train set SNR threshold. (**a**) Full view; (**b**) zoomed-in view.

First, picking a train set from all the SNR values gives better results at a negative SNR but deteriorates when reaching a positive value as expected. Another conclusion is that using over 5 dB as the threshold gives lesser accuracy for high and low SNR test samples. The final choice of threshold was between 1 dB and 5 dB. Although the 1 dB threshold yields better accuracy at lower SNR levels, a slight improvement in the detection accuracy is observed starting from 4 dB and higher with the 5 dB threshold. This result supports the assumption that modulation schemes have minimal SNR requirements for perfect demodulation. The depth of the decision tree was not considered for the first test. So, we want to pick a balanced tree, deep enough to view all the features we chose before but not too deep, which can result in overfitting and high complexity.

Figure 2 shows the detection probability under different decision tree depths. We start at four layers since this is the minimal depth for 11 labels, and stop at nine, as we see no significant improvement from eight. The optimal depth for each application can be chosen based on the memory and runtime requirements. We pick an 8-deep tree with a higher than 5 dB SNR train set.

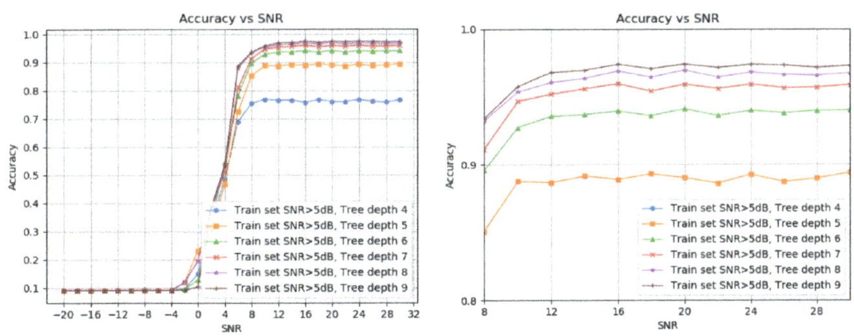

(**a**) Full view. (**b**) Zoomed in.

Figure 2. P_d based on tree depth. (**a**) Full view; (**b**) zoomed-in view.

We aim to compare our results to state-of-the-art deep learning models. First, we compare our method with three different state-of-the-art models of deep neural networks as examples. The models we examine are ResNet [21], DenseNet [22], LSTM [10], FLANs [8] and VGG [23]. We aim to compare our model to the neural networks in two manners. Figure 3 shows the first comparison of neural networks in the classification accuracy.

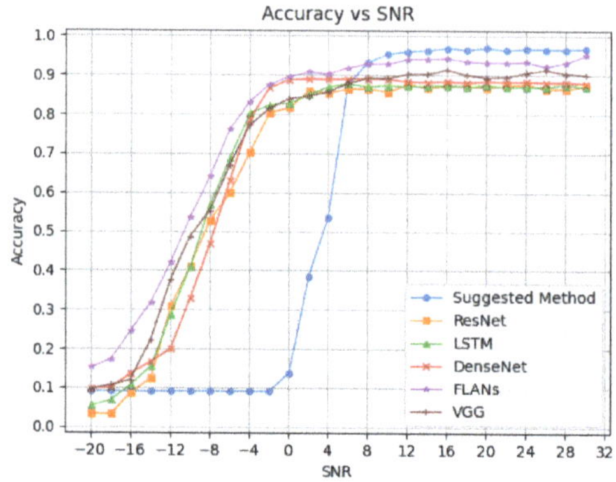

Figure 3. P_d comparison for the suggested method.

We can see the advantage of deep neural networks in the negative SNR zone, while our method fails in comparison. However, we already discussed why classification at negative SNR is usually irrelevant in terms of perfect demodulation. Starting from a splitting threshold of 6 dB SNR, our suggested method yields better detection accuracy than the popular NN methods by a margin of almost 8%.

The secondary comparison with the neural network models primarily centers on the parameters or memory requirements. We use Equation (15) to compute the total parameters necessary for diverse tree depths, subsequently contrasting these values in Table 2 alongside the proposed NN models.

Table 2. Parameter size of different models.

Model	Total Parameters
6 Deep Tree	63
7 Deep Tree	127
8 Deep Tree	255
ResNet	214,922
LSTM	770,378
DenseNet	542,850
VGG	1,621,780
FLANs	954,747

Our model, even with the inclusion of additional layers, consistently maintains a parameter count ranging in the range of a few hundreds. In contrast, the NN models register parameters in the hundreds of thousands. This substantial discrepancy highlights the sparse memory utilization inherent in our model, a characteristic that holds substantial significance. It is particularly advantageous when considering the amalgamation of multiple models to bolster accuracy, especially in scenarios necessitating low complexity and an abbreviated runtime. This streamlined memory usage serves as a pivotal advantage, especially in contexts where computational efficiency and rapid processing are imperative.

Moreover, this substantial difference in the parameter count underscores a critical limitation in NN models, notably their impracticality for deployment within SDR systems. The hefty memory requirements demanded by NN architectures, often in the realm of millions of parameters, render them impractical for many SDR implementations. In contrast, our model's ability to operate within the realm of mere hundreds or thousands of parameters renders it significantly more feasible for integration within SDR systems. This advantage positions our approach as particularly suitable for real-world deployment scenarios, where memory limitations are stringent, allowing for more practical utilization within SDR-based Cognitive Radio networks.

We examine the confusion matrix from the suggested method. Figure 4 shows the confusion matrices at 6 dB for the 7- and 8-deep decision trees under the 5 dB SNR threshold train set. We observe in the illustrated figures, as well as in the other matrices we examined, that the mislabeling of the model is consistent, meaning the labels that can become mismatched are almost always the same. Based on this observation, we can split the modulation schemes into subgroups as follows:

- Group 0—AM-SSB;
- Group 1—AM-DSB and BPSK;
- Group 2—4ASK and OOK;
- Group 3—FM and GMSK;
- Group 4—8PSK, 16QAM, QPSK and OQPSK.

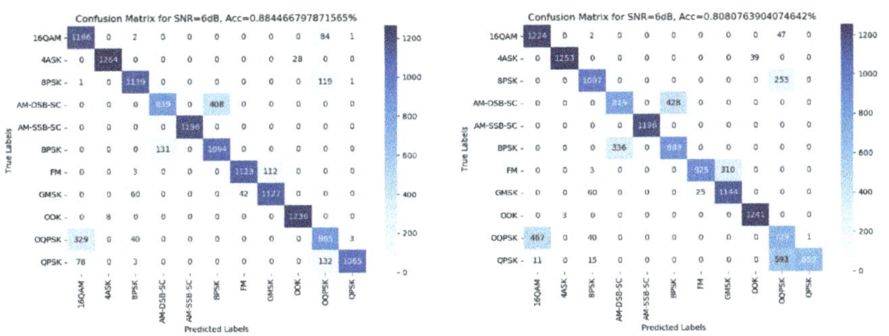

(**a**) Tree Depth = 7 (**b**) Tree Depth = 8

Figure 4. Confusion matrices for different tree depths: (**a**) depth = 7; (**b**) depth = 8.

Group 0 exclusively comprises SSB modulation due to its nearly instantaneous classification across an array of features, even at relatively low SNR values. This consistent and rapid detection is observed consistently across various models tested, establishing SSB as a distinct and easily recognizable modulation type.

On the other hand, Groups 1, 2, and 3 collectively represent pairs of modulations that exhibit a tendency to be confused or mixed up, particularly prevalent at lower SNR levels. This occurrence suggests a noteworthy statistical similarity in the signal models of these modulation schemes. This statistical resemblance often results in misclassification or overlapping characteristics, leading to the assumption that the underlying signal models within these groups share comparable statistical behaviors, thereby complicating their differentiation, especially under lower SNR conditions.

Group 4, encompassing the remaining modulations, portrays a more intricate scenario, where these modulations commonly intermingle without any particular pair exhibiting a significantly higher occurrence than the others. This lack of distinct differentiation signifies a higher level of complexity within this group, where multiple modulation schemes exhibit overlapping statistical features or behaviors, making their individual identification considerably more challenging, especially in contexts where their distinguishing characteristics are blurred due to similarities.

This categorization provides valuable insights into the behavior and detectability of various modulation groups, shedding light on the distinct characteristics observed across different SNR levels and establishing patterns of similarity or complexity among them.

4. Discussion

We conducted simulations to validate our hypothesis on the benefits of using a classic feature-based algorithm for AMC. Our initial focus was to examine the impact of individual features on detection accuracy, which led us to remove unnecessary or redundant features during the model's training phase. We then partitioned the data into high- and low-SNR groups to demonstrate how specific modulation schemes become detectable only above a certain SNR threshold. This finding allowed us to refine our model by focusing on the essential SNR levels relevant to real-world systems. The final phase of our simulation aimed to fine-tune the tree's hyperparameters to ensure the highest accuracy while maintaining the optimal complexity.

Next, we compared our finalized model against three NN models, which yielded varied results. On one hand, our proposed method exhibited superiority over NN models in terms of detection accuracy, especially beyond the specified threshold SNR values. On the other hand, our method struggled to detect signals with low SNR, where NN models excelled. However, as previously noted, many modulation schemes become irrelevant for demodulation below certain SNR thresholds. Apart from the detection accuracy, we also compared our method with NN models in terms of memory storage. While basic NN models required millions of deep parameters, our straightforward classifier required a maximum of a few hundred parameters, depending on the chosen model. This advantage significantly impacted the end units, as the variance in total parameters directly influenced memory usage and runtime during the classification process. These factors directly contribute to improved battery life, reduced costs, and enhanced speed in desired end units.

However, our suggested method does have limitations. Primarily, the algorithm may not be optimal for discussions involving low-SNR scenarios and modulations that are decodable even with an SNR lower than 2–4 dB. Additionally, we observed instances where certain modulation schemes were erroneously identified even at higher SNR. This occurrence might be attributed to similar carriers and statistics among some modulations, indicating that our method might not be suitable for those specific schemes. Despite these limitations, all outcomes affirm that the classic FB classifier for AMC can be both successful and advantageous when considering its application in real-life Cognitive Radio networks.

5. Conclusions

Our research focused on developing a simple yet effective method for automatic modulation classification in spectrum sensing. We found that our proposed method outperforms some of the most widely used neural networks in terms of both accuracy and memory efficiency. To achieve this, we chose the decision tree classifier for its simplicity, which allowed us to identify efficient statistical features like High-Order Cumulants (HOCs) that are specifically tailored for each modulation scheme. Our study went beyond the usual emphasis on digital modulations and explored the possibility of using cumulants to classify analog modulations, which has not been extensively researched before. By identifying specific cumulants for each modulation type, we discovered that there are subsets of modulations that share similar statistical behaviors. This discovery not only helped us understand modulation characteristics better but also provided a foundation for improving classification methodologies. Our approach is optimal for end units within Cognitive Radio (CR) networks, as it achieves high accuracy with minimal parameters, which is crucial for memory and speed efficiency. In our future work, we plan to leverage these identified modulation subgroups to develop a more intricate method that further improves accuracy. We also aim to expand our approach to include a wider range of modulations, making our classification framework more versatile and applicable.

Author Contributions: Conceptualization, B.D. and I.C.; methodology, B.D.; validation, B.D. and I.C.; writing—original draft preparation, B.D.; writing—review and editing, B.D. and I.C. All authors have read and agreed to the published version of the manuscript.

Funding: This research received no external funding.

Institutional Review Board Statement: Not applicable.

Informed Consent Statement: Not applicable.

Data Availability Statement: This research is based on the published dataset by Timothy James O'Shea, Tamoghna Roy and T. Charles Clancy. Available at https://www.deepsig.ai/datasets (accessed on 26 May 2023).

Conflicts of Interest: The authors declare no conflicts of interest.

Abbreviations

The following abbreviations are used in this manuscript:

AMC	Automatic modulation classification
CR	Cognitive radio
SDR	Software-defined radio
SNR	Signal-to-noise ratio
FB	Feature based
HOS	High-order statistical moments
HOC	High-order cumulants
AWGN	Additive white Gaussian noise

References

1. Mitola, J. The software radio architecture. *IEEE Commun. Mag.* **1995**, *33*, 26–38. [CrossRef]
2. Mitola, J.; Maguire, G.Q. Cognitive radio: Making software radios more personal. *IEEE Pers. Commun.* **1999**, *6*, 13–18. [CrossRef]
3. Haque, M.; Sugiura, Y.; Shimamura, T. Spectrum Sensing Based on Higher Order Statistics for OFDM Systems over Multipath Fading Channels in Cognitive Radio. *J. Signal Process.* **2019**, *23*, 257–266 [CrossRef]
4. Bozovic, R.; Simic, M. Spectrum Sensing Based on Higher Order Cumulants and Kurtosis Statistics Tests in Cognitive Radio. *Radioengineering* **2019**, *29*, 464–472. [CrossRef]
5. Salahdine, F. Spectrum sensing techniques for cognitive radio networks. *arXiv* **2017**, arXiv:1710.02668.
6. Callaghan, T.G.; Perry, J.L.; Tjho, J.K. Sampling and algorithms aid modulation recognition. *Microwaves* **1985**, *24*, 117–119.
7. Nebabin, V.G.; Sergeev, V.V. *Methods and Techniques of Radar Recognition*; Moscow Izdatel Radio Sviaz: Moscow, Russia, 1984
8. Zhang, D.; Lu, Y.; Li, Y.; Ding, W.; Zhang, B.; Xiao, J. Frequency learning attention networks based on deep learning for automatic modulation classification in wireless communication. *Pattern Recognit.* **2023**, *137*, 109345 [CrossRef]
9. Ma, J.; Hu, M.; Wang, T.; Yang, Z.; Wan, L.; Qiu, T. Automatic Modulation Classification in Impulsive Noise: Hyperbolic-Tangent Cyclic Spectrum and Multibranch Attention Shuffle Network. *IEEE Trans. Instrum. Meas.* **2023**, *72*, 5501613. [CrossRef]
10. Elsagheer, M.M.; Ramzy, S.M. A hybrid model for automatic modulation classification based on residual neural networks and long short term memory. *Alex. Eng. J.* **2023**, *67*, 117–128. [CrossRef]
11. Dobre, O.A.; Abdi, A.; Bar-Ness, Y.; Su, W. Survey of automatic modulation classification techniques: Classical approaches and new trends. *IET Commun.* **2007**, *1*, 137–156. [CrossRef]
12. Maroto, J.; Bovet, G.; Frossard, P. Maximum Likelihood Distillation for Robust Modulation Classification. In Proceedings of the ICASSP 2023-2023 IEEE International Conference on Acoustics, Speech and Signal Processing (ICASSP), Rhodes Island, Greece, 4–10 June 2023; pp. 1–5.
13. Sun, X.; Su, S.; Zuo, Z.; Guo, X.; Tan, X. Modulation classification using compressed sensing and decision tree–support vector machine in cognitive radio systems. *Sensors* **2020**, *20*, 1438. [CrossRef] [PubMed]
14. Shih, P.; Chang, D.C. An automatic modulation classification technique using high-order statistics for multipath fading channels. In Proceedings of the 2011 11th international conference on ITS Telecommunications, St. Petersburg, Russia, 23–25 August 2011; pp. 691–695.
15. Hazza, A.; Shoaib, M.; Alshebeili, S.A.; Fahad, A. An overview of feature-based methods for digital modulation classification. In Proceedings of the IEEE 2013 1st International Conference on Communications, Signal Processing, and Their Applications (ICCSPA), Sharjah, United Arab Emirates, 12–14 February 2013; pp. 1–6.
16. Swami, A.; Sadler, B.M. Hierarchical digital modulation classification using cumulants. *IEEE Trans. Commun.* **2000**, *48*, 416–429. [CrossRef]
17. Jdid, B.; Lim, W.H.; Dayoub, I.; Hassan, K.; Juhari, M.R.B.M. Robust automatic modulation recognition through joint contribution of hand-crafted and contextual features. *IEEE Access* **2021**, *9*, 104530–104546. [CrossRef]
18. Fryntov, A.E. Characterization of a Gaussian distribution by gaps in its sequence of cumulants. *Theory Probab. Appl.* **1989**, *33*, 638–644. [CrossRef]
19. Volkmer, H. A characterization of the normal distribution. *J. Stat. Theory Appl.* **2014**, *13*, 83–85. [CrossRef]
20. O'Shea, T.J.; Roy, T.; Clancy, T.C. Over-the-air deep learning based radio signal classification. *IEEE J. Sel. Top. Signal Process.* **2018**, *12*, 168–179. [CrossRef]
21. Zhang, K.; Zuo, W.; Chen, Y.; Meng, D.; Zhang, L. Beyond a gaussian denoiser: Residual learning of deep cnn for image denoising. *IEEE Trans. Image Process.* **2017**, *26*, 3142–3155. [CrossRef] [PubMed]
22. Huang, G.; Liu, Z.; Van Der Maaten, L.; Weinberger, K.Q. Densely connected convolutional networks. In Proceedings of the IEEE Conference on Computer Vision and Pattern Recognition, Honolulu, HI, USA, 21–26 July 2017; pp. 4700–4708.

23. Ghanem, H.S.; Al-Makhlasawy, R.M.; El-Shafai, W.; Elsabrouty, M.; Hamed, H.F.; Salama, G.M.; El-Samie, F.E.A. Wireless modulation classification based on Radon transform and convolutional neural networks. *J. Ambient. Intell. Humaniz. Comput.* **2023**, *14*, 6263–6272. [CrossRef]

sensors

MDPI

Article

Decision-Making Algorithm with Geographic Mobility for Cognitive Radio

Gabriel B. Cervantes-Junco, Enrique Rodriguez-Colina, Leonardo Palacios-Luengas, Michael Pascoe-Chalke *, Pedro Lara-Velázquez and Ricardo Marcelín-Jiménez

Department of Electrical Engineering, Autonomous Metropolitan University, Iztapalapa, Mexico City 09310, Mexico; brakeencj@gmail.com (G.B.C.-J.); erod@xanum.uam.mx (E.R.-C.); lpalacios@izt.uam.mx (L.P.-L.); plara@xanum.uam.mx (P.L.-V.); rmarcelin@izt.uam.mx (R.M.-J.)
* Correspondence: mpascoe@xanum.uam.mx

Abstract: The proposed novel algorithm named decision-making algorithm with geographic mobility (DMAGM) includes detailed analysis of decision-making for cognitive radio (CR) that considers a multivariable algorithm with geographic mobility (GM). Scarce research work considers the analysis of GM in depth, even though it plays a crucial role to improve communication performance. The DMAGM considerably reduces latency in order to accurately determine the best communication channels and includes GM analysis, which is not addressed in other algorithms found in the literature. The DMAGM was evaluated and validated by simulating a cognitive radio network that comprises a base station (BS), primary users (PUs), and CRs considering random arrivals and disappearance of mobile devices. The proposed algorithm exhibits better performance, through the reduction in latency and computational complexity, than other algorithms used for comparison using 200 channel tests per simulation. The DMAGM significantly reduces the decision-making process from 12.77% to 94.27% compared with ATDDiM, FAHP, AHP, and Dijkstra algorithms in terms of latency reduction. An improved version of the DMAGM is also proposed where feedback of the output is incorporated. This version is named feedback-decision-making algorithm with geographic mobility (FDMAGM), and it shows that a feedback system has the advantage of being able to continually adjust and adapt based on the feedback received. In addition, the feedback version helps to identify and correct problems, which can be beneficial in situations where the quality of communication is critical. Despite the fact that the FDMAGM may take longer than the DMAGM to calculate the best communication channel, constant feedback improves efficiency and effectiveness over time. Both the DMAGM and the FDMAGM improve performance in practical scenarios, the former in terms of latency and the latter in terms of accuracy and stability.

Keywords: cognitive radio; decision-making; geographic mobility in cognitive radio; location; handoff management

1. Introduction

Geographic mobility (GM) models describe the movement of nodes in a specific region during a given time, including changes in speed, direction, and acceleration. GM can thus adapt to a solution's specific needs, providing improved performance and flexibility. The use of wireless networks comprises two important tasks: location and handoff management. Location management ensures that the location of network nodes can be tracked, while handoff management is responsible for maintaining connections while a node moves from one network to another [1]. This GM management needs to be considered for cognitive radio networks (CRNs), given that the available radio spectrum can considerably change with location and handoff. Consequently, user GM represents a major challenge in CRNs [2] since it can significantly affect performance by interrupting the services, thus reducing service quality.

According to the current state of the art, existing mobility models can be divided into four groups: random, group, route-planned, and time-dependent models [3]. Random mobility models in wireless networks consider the random distribution of nodes within a previously defined simulation area. Each node remains in that location for a randomly selected time within a specified interval [4,5]. Similarly, the group mobility model considers that a group of nodes in a network revolve together around a common point [6,7]. Route-planned models seek to avoid unforeseen changes in speed and direction. The planning of the movements is defined by mathematical equations, in which the nodes are forced to follow these movement patterns [8]. Similarly, time-dependent mobility models are based on mathematical equations in which the nodes depend on an initial time with respect to a previous time, thus avoiding sudden changes in speed and direction [9]. Different prediction schemes can be used to control GM. For an implementation in CRNs, it is a question of finding the best GM prediction technique in order to select the most stable communication path, thus improving general performance and performance reliability.

As a consequence of the relevance of GM for a better performance of the CRN, the proposed novel algorithm, named decision-making algorithm with geographic mobility (DMAGM), includes a detailed analysis of decision-making that considers and involves GM.

In addition, an improved version of the DMAGM is also proposed where feedback of the output is incorporated. This version is named feedback-decision-making algorithm with geographic mobility (FDMAGM). This algorithm shows that a feedback system has the advantage of being able to continually adjust and adapt based on the feedback received. Furthermore, the feedback version helps to identify and correct problems, which can be beneficial in situations where the quality of communication is critical.

Grounded in this research's particular focus on decision-making, different decision-making mechanisms that include GM were studied making the following contributions:

- Proposing a detailed analysis of decision-making that considers geographic mobility (GM), a parameter that most CRN proposals have not yet explored in depth
- Developing a robust process that reduces latency to find a better communication channel; and
- Providing a feedback function that increases the precision in the selection of a better communication backup channel, based on historical data regarding network behavior through a feedback process that considers information from the evaluations of previously used channels. The value assigned to each channel thus corresponds to a relationship between current information and previous evaluations.

The evaluation process of the proposed algorithm is based on determining the attributes characterizing a communication channel through the Delphi method [2]. The decision-making attributes were proposed based on the criteria reported in the CR literature and compared with [2]. Such criteria include the signal-to-interference plus noise ratio (SINR), the bandwidth (BW), the channel availability probability (AP), the estimated channel time availability (ETA), and the random way-point mobility model (RWPM). These criteria were evaluated for two types of services, i.e., real time (RT) and best effort (BE). Test execution was conducted with an NS-3 simulator, and, in order to contrast the DMAGM and the FDMAGM, some comparisons were made with the algorithms Dijkstra [10], analytic hierarchy process (AHP) [11], fuzzy analytic hierarchy process (FAHP) [12], and modified Dijkstra decision-making algorithm (ATDDiM) [13], in which the results indicate a considerable reduction in the processing time of the proposed algorithms. In addition, there was greater precision in the selection of a communication channel since the feedback process proposed in the FDMAGM contained information regarding the evaluations of the current and previous channels. According to the analysis of the different mobility models that can be considered with the interaction among CRs and PUs, it was possible to determine which mobility model would be appropriate for this scenario. This is based on sets of given data that can be compared with wireless networks in operations such as a cellular network. The characteristics that were considered to determine user movements are speed, acceleration, pause time, and time in a location.

The rest of this paper is organized in the following sections. Section 2 describes the related works. Section 3 describes the proposed decision-making algorithm. Section 4 shows the conducted tests that compare latency with similar algorithms and mobility tests. Lastly, Section 5 presents the conclusions. At the end of this document, a table with the list of acronyms used throughout this paper is shown.

2. Related Work

To the best of the authors' knowledge, none of the studies propose research related to spectrum handoff in CRNs and solving problems associated with geographic mobility (GM). Trigui et al. [14] propose a multi-agent system (MAS) with a mobility management scheme for CRNs that considers user mobility and radio resources in order to provide optimal access and make resource allocation decisions. Obaid et al. [15] propose a cluster-based MAC protocol for CR wireless sensor networks with radio frequency (RF) energy harvesting. The protocol is based on the mobility for spectrum access decision-making. The proposed protocol requires a localization system to determine the users' current location. Priya and Kannan [16] present a frequency-band selection routing protocol with enhanced spectrum aggregation for cognitive radio ad hoc networks. The protocol considers user mobility, spectrum availability, and quality of service to make optimal routing decisions. Zhao et al. [17] propose a new approach to spectrum management in cognitive radio networks based on predicting spectrum availability to make more efficient spectrum allocation decisions. Hanif et al. [18] propose a new approach to spectrum handoff management in CRNs. This approach is based on analyzing the user traffic pattern to make more efficient spectrum handoff decisions and requires the implementation of a traffic analysis system, a spectrum handoff system, and a coordination system. The study by Jaffar et al. [19] proposes a new approach to spectrum management in heterogeneous CRNs based on user location to make proactive channel selection decisions.

The work proposed by Omer et al. [20] introduces a new approach to channel assignment in CRNs for scalable video streaming. This approach is based on adaptively assigning channels to users based on their bandwidth and quality-of-service requirements. On the other hand, Tlouyamma and Velempini [21] propose a channel selection algorithm (CSA) optimized for improved performance in CRNs. This algorithm considers multiple factors, including channel availability, interference, channel quality, user distance, and channel usage history. Thakur et al. [22] consider a new approach to spectrum mobility in CRNs using spectrum prediction and monitoring techniques. This approach is based on predicting the occurrence of primary users (PUs) and continuously monitoring the spectrum in order to detect the presence of PUs. Meanwhile, Yawada and Dong [23] show an approach to spectrum handoff/mobility (SHMA) in CRNs based on an intelligent decision-making process that takes into account multiple factors, including channel availability, interface, channel quality, user distance, and channel usage history.

Li et al. [24] propose a method based on deep learning for the prediction of user mobility using a preferential exploration and return model as a deep-learning technique to predict the future locations of the nodes. According to the proposal made by Iftikhar et al. [25], a decision-making algorithm using game theory is presented to model spectral mobility, which serves as a switching game that considers whether to change or remain in the channel. Alozie et al.'s scheme [26] presents a strategy to minimize the delay that occurs during spectrum handoff from a backup channel selection mechanism based on fuzzy data. Basically, the proposed scheme considers gathering backup channels in advance and using fuzzy logic for the selection of the best available backup channel.

Other studies have addressed the issue of user mobility in solving CR problems. For example, Sivasundarapandian et al. [27] present an approach based on a secondary user (SU) mobility model that considers SU movement and distribution. Dey and Saha propose a scheme that considers channel availability, interference, channel quality, and user distance to define the parameter MOS [28]. Al-Dulaimi et al. [29] propose a congestion control scheme, and Zhang et al. [30] propose an opportunistic programming scheme in the spectrum. However, none of these schemes solve the problem of effective handoff, nor do they consider a decision-making process that can help CR communication processes when there is GM.

The literature thus offers a wide range of proposals that consider mobility strategies. However, none of these proposals do not posit a decision-making process to determine the best communication channels, which results in ineffective handoff. In addition, the attributes that characterize a communication channel are not specified through the Delphi method [2]. To gain insights into the latency time of the proposed algorithm, other algorithms were included for comparison purposes, such as Dijkstra, AHP, FAHP, and ATDDiM.

Table 1 presents a summary that allows the comparison of common research methods, main features, limitations, and other characteristics of the algorithms found in the literature.

Table 1. Main characteristics of different decision-making algorithms for cognitive radio, where n represents the number of nodes and m represents the number of channels. The acronym NC means not considered, and NA means not available.

Algorithm	Geographic Mobility (GM)	Computational Complexity	Min./Average/Max. Decision-Making Latency (ms)	Main Features	Limitations
Dijkstra [10]	NC	$O(n^2)$	0.2/8.6/20	This approach can find the optimal path between two nodes in a graph. This approach can be used to make decisions in a variety of scenarios. This approach can be scaled to large networks.	This approach is sensitive to the accuracy of the data on the structure of the graph.
AHP [11]	NC	$O(n^2)$	0.04/2.2/6	Identifies the most cost-effective and latency-optimal routes between two points. Simplifies the route selection process for delivery services. Allows the user to specify the weights of the decision factors.	It depends on the accuracy of the specified weights by the user. It might be unreliable in networks with changing conditions, like mobility scenarios.
FAHP [12]	NC	$O(n^2)$	0.02/0.47/1	Improves the performance of spectral handoff by more efficiently selecting channels. It can be adaptable to different cognitive radio network scenarios.	It requires a large amount of data about the cognitive radio network. It is more complex than traditional spectral handoff algorithms. It depends on the accuracy of the specified weights by the user.
ATDDiM [13]	NC	$O(nlog(n))$	0.02/0.46/1	Enhances route planning performance by reducing computation time compared with Dijkstra. Reduced computation complexity. The improved algorithm implements a configurable set of numerical parameters.	It requires more preprocessing power than the original Dijkstra's algorithm.

Table 1. *Cont.*

Algorithm	Geographic Mobility (GM)	Computational Complexity	Min./Average/Max. Decision-Making Latency (ms)	Main Features	Limitations
MAS [14]	NC	$O(n^2)$	NA	Spectrum efficiency using a chaotic channel selection algorithm. The scheme is robust to changes in the radio environment. Easy to implement.	The mobility scheme requires data on the structure of the cognitive radio network. The mobility scheme is sensitive to the accuracy of the data on the structure of the cognitive radio network.
Cluster-based MAC protocol [15]	NC	$O(nlog(n) + m)$	NA	Energy efficiency. Robustness because of the mobile clustering scheme. Easy to implement.	The protocol does not use a mobility prediction model to estimate the future location of nodes. The protocol does not take into account the impact of mobility on network performance.
CSA [21]	NC	$O(nlog(n) + m)$	NA	Improves the performance of cognitive radio networks in terms of transmission success rate and signal quality. The algorithm is robust to interference.	The model does not consider user mobility to estimate the probability of interference-free channels at a given time. The model does not consider interference from other users to estimate the probability of occupied channels at a given time.
SHMA [23]	NC	$O(nlog(n) + m)$	NA	The proposed algorithm improves spectrum efficiency by selecting the best channel for each mobile user. The proposed algorithm improves QoS by considering the reception gain, interference, and other factors when selecting channels. The proposed algorithm is adaptive to mobility by considering the mobility of the user when selecting channels.	The algorithm assumes that the user's spectral mobility is known. The algorithm does not consider the impact of spectral mobility on interference. The algorithm does not consider the impact of spectral mobility on network congestion.
MOS [28]	NC	$O(mlog(n) + n))$	NA	The scheme is robust to interference because it considers the interference from primary users when making spectrum handoff switching decisions. The scheme is adaptive to mobility because it considers the mobility of secondary users when making spectrum handoff switching decisions.	The scheme assumes that primary users behave predictably. They are the only considered parameter for spectrum handoff without geographic mobility.

3. Decision-Making Algorithm with Geographic Mobility (DMAGM)

The proposed algorithm is based on the weighted sum shown in (1), which is part of a conventional technique for solving multi-objective optimization problems [30]. To find a better channel in a CR environment, with the efficiency and simplicity of using a linear combination of weights, some attributes such as the signal-to-interference plus noise ratio

(SINR), the bandwidth (BW), the channel availability probability (AP), and the estimated channel time availability (ETA) are evaluated using the random way-point mobility model (RWPM). These criteria are considered for two types of services, i.e., real time (RT) and best effort (BE). In addition, the attributes are compared with those considered in [2] with a calculation of the weights to evaluate the channels that provide the best performance.

Equation (1) shows the objective function and attributes used for analyzing communication channels.

$$Y = \sum_{j=1}^{n} w_j f_j(x), \quad (1)$$

where $w_j \in [0, 1]$, and $f_j(x)$ is the jth objective function.

Considering that it is necessary to determine the channel with the best 'fitness'/'aptitude', applying Equation (2), i.e.,

$$C_i = w_{bw}^{ts} BW + w_{sinr}^{ts} SINR + w_{ap}^{ts} AP + w_{eta}^{ts} ETA, \quad (2)$$

where

i: id-number of one of the N channels to be compared
ts: type of service (RT or BE)
BW: normalized values of BW detected by the CR for channel i
$SINR$: normalized values of SINR detected by the CR for channel i
AP: normalized values of AP estimated by the CR for channel i
ETA: normalized values of ETA estimated by the CR for channel i
w_{bw}^{ts} : assigned weight to BW depending on the selected ts
w_{sinr}^{ts} : assigned weight to SINR depending on the selected ts
w_{ap}^{ts} : assigned weight to AP depending on the selected ts
w_{eta}^{ts} : assigned weight to ETA depending on the selected ts

The values of BW, SINR, AP, and ETA, detected by the CRs, must be normalized in order to perform the weighted sum of multiple objectives. This sum is referred to as the 'objective function'. All the objective function's 'aptitudes' corresponding to the 'N' channels to be compared are thus obtained, computing a 'global maximum' among the set of channels and determining which channel has optimal communication characteristics. Figure 1 shows the flowchart of the DMAGM operation. Note that the input values are the 'service type' (ST) and the number of channels to be compared (N). The 'type of service' used consists of RT or BE.

Table 2 shows the selected values, which are values like [2] in order to compare performance. Initially, for the 'best channel's' value to be optimized, a channel is placed outside the range of channels to be analyzed, and a channel within the range of channels is determined. If, at the end of the proposed algorithm, the value of the 'best channel' is outside the range of channels, the algorithm will not be able to determine a channel with the best 'fitness'. The 'global maximum' value starts with a value of 0, which is the 'fitness' and represents that it will never have a channel analyzed, ensuring that the 'global maximum' value may be modified at least once. The 'objective function' for each channel is determined within the cycle shown in Figure 1, a value called the 'local maximum', which is compared with the value that demonstrates the 'global maximum'. If the value is greater, then the 'global maximum' takes the value of the 'local maximum', and the 'best channel' assumes the value of the channel that is currently being analyzed. If the value is lower, the analysis of the next channel continues without modifying the values of the 'best channel' and the 'global maximum'. The flowchart shows that this repetitive cycle continues until there are no more channel 'fits' to calculate, which is controlled by the 'channel' variable. Once the channel comparison cycle is complete, the output values with the 'best channel' and the 'global maximum' determined by the DMAGM are displayed.

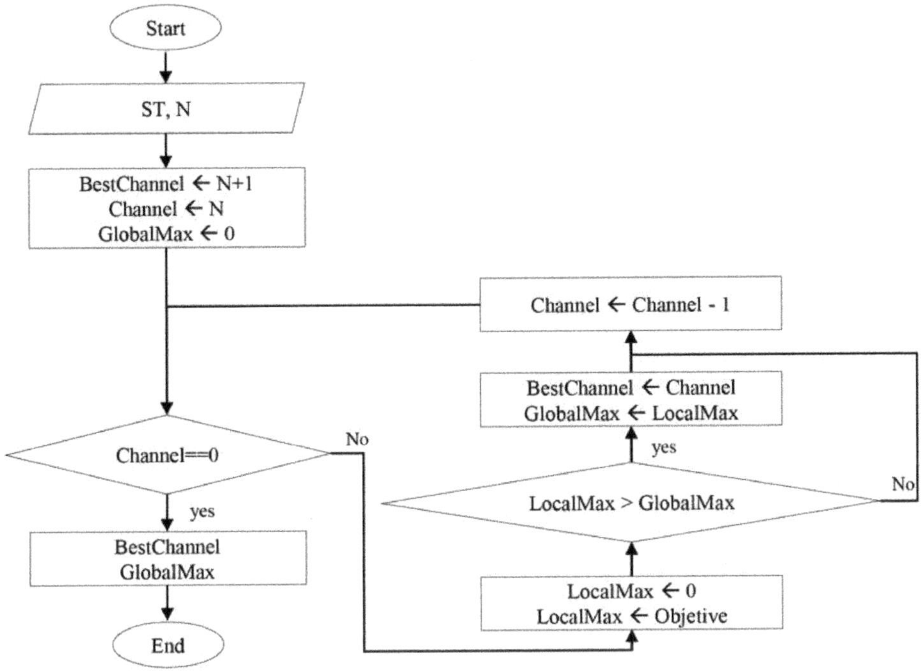

Figure 1. Proposed algorithm (DMAGM) diagram.

Table 2. Weights of attributes according to the type of service used.

Criteria	RT	BE
BW	0.1471	0.2921
SINR	0.1970	0.3949
AP	0.3593	0.1607
ETA	0.2966	0.1523

At this stage, a feedback mechanism is implemented in order to improve the accuracy with which a channel with the best communication characteristics can be determined. This procedure allows consulting the history of previous results. The process to improve the precision in selecting a channel is described below.

Feedback-Decision-Making Algorithm with Geographic Mobility (FDMAGM)

The feedback-decision-making algorithm with geographic mobility (FDMAGM) considers increasing the accuracy in selecting the best channel in a similar way to [12]. The feedback process obtains information from previous evaluations of the channels. The value allocated to each channel corresponds to a relationship between current information and past evaluations. As can be seen in Figure 2, the FDMAGM is based on the original DMAGM but includes a feedback process.

A process to determine the new 'best channel' is carried out that considers the current value, the last generated value, and the average value of a certain time interval. The final value for each channel is determined using (3).

$$SF_i = \alpha \bullet SA + \beta \bullet SPa + (1 - \alpha - \beta) \bullet SPr, \quad (3)$$

where

i: id-number of one of the N channels to be compared
SF: final value
SA: current value
SPa: previous value
SPr: mean value
$\alpha, \beta \in [0, 1]$

Once the final value is obtained for each channel, a comparison is made with the decision-making process outlined above, thus determining the best channel among the set of channels analyzed. The values of α and β are obtained in a similar way as in [12] by performing an experimental auto-regressive analysis with different combinations of α and β for a set of predetermined data. The values of α and β were considered so that the precision in the selection of the best channel was higher. These values are $\alpha = 0.60$ and $\beta = 0.35$, with an experimental precision of 87%. Both algorithms were analyzed with the two types of services: RT and BE, as well as with normalized weights.

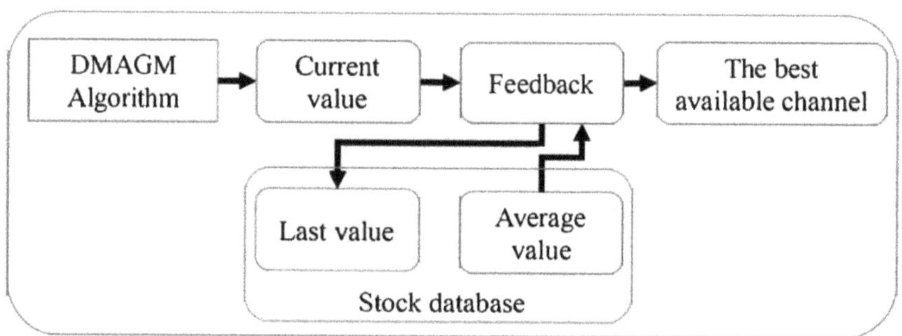

Figure 2. The feedback-decision-making algorithm scheme (FDMAGM).

4. Tests and Result Analysis

The algorithm proposed in this study for decision-making (DMAGM) has the characteristic of improving the response time (latency). The test scenario for the evaluation and validation of the DMAGM and the FDMAGM was configured with the services of RT and BE using Network Simulator 3 (NS-3). Table 3 shows the network characteristics in which it is considered that a base station (BS) with cognitive radio characteristics receives all the information detected by the CRs, i.e., the main characteristics and attributes of the channels observed and selected during previous analyses, which are: BW, SINR, AP, and ETA found within the range of channels that can be detected by the system. Initially, as shown in Figure 3, the behavior of the proposed algorithms, the DMAGM and the FDMAGM, is considered with and without a feedback mechanism, in a scenario with GM in which there is interaction of cognitive radios (CRs) and primary users (PUs). Subsequently, the latency in relation to the DMAGM is obtained and compared with the Dijkstra [10], AHP [11], FAHP [12], and ATDDiM [13] algorithms. Finally, the results obtained for the DMAGM are shown and compared with the FDMAGM (applying the feedback mechanism).

4.1. Simulation Scenario with Geographic Mobility

Figure 3 shows an approximation of the BS's geographical positions, primary users (PUs), and cognitive radios (CRs). In addition, considering that the decision-making algorithm is developed by the BS, it therefore assesses the selection of the best communication channel and establishes which channels can be occupied by the CRs. The mobility pattern followed by the CRs and PUs is defined by the random way-point mobility model (RWPM), and mobility simulation time has a duration of 3600 s. Consequently, the behavior of CRs and PUs is modeled and simulated using 124 randomly distributed nodes on a circular surface with a radius ranging from 2167 m to 4334 m, in which each node is allocated with a transmission channel.

Table 3. Parameters used in the simulation scenario with NS-3.

Parameters	Value
Frequency band	824–849 [MHz]
Communication system	Mobile
Communication technology	GSM
Number of channels	124
BW per channel	200 [kHz]
Power Tx BS	30 [dBm]
BS coverage area	2167 [m]
BS height	25 [m]
Mobile user Rx power	−80 [dBm]
Mobile user height	1 [m]

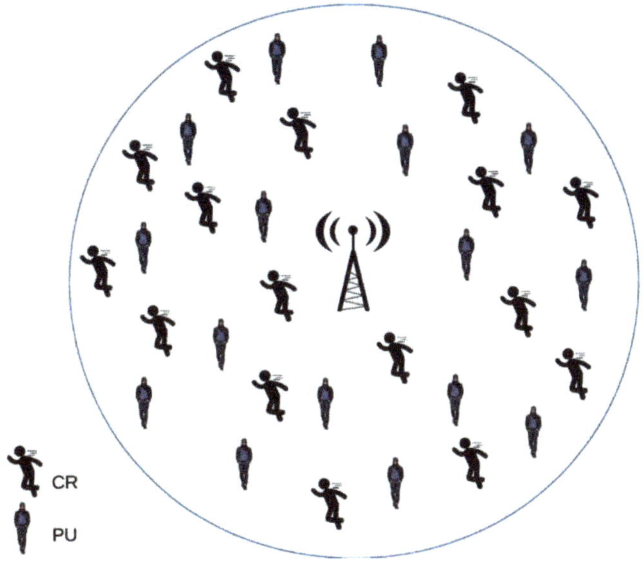

Figure 3. Geographic mobility scenario of a CRN with one BS.

In this evaluation, three GM scenarios with different occupancy percentages were generated for each channel, considering values between 25% and 75% in 25% intervals. For each of these three scenarios, three variations in PU mobility were proposed by modifying the PU mobility radius in relation to the BS coverage radius. That said, the following distances were considered:

- Same as the BS coverage radius, i.e., 2167 m
- 50% greater than the BS coverage radius, i.e., 3250.5 m
- 100% greater than the BS coverage radius, i.e., 4334 m

The simulation results yielded the average time that a channel is available. For the proposed scenarios, a simulation time of approximately 1.48 s corresponding to 2400 samples per scenario was tested. Table 4 shows the results obtained from both algorithms, with and without the feedback mechanism. Note that in the selection of the best channel, there is a similarity above 75% for the different scenarios. Therefore, algorithm application and implementation will depend on the scenario in which they are tested.

Table 4. Match rate between FDMAGM and DMAGM, with and without the feedback mechanism, for the selection of the communication channel.

Characteristics of the GM Scenario		Number of Matches over 2400 Samples	Match Rate
Mobility Radius	CPU Occupancy		
2167 m	25%	1819	75.79%
	50%	2062	85.91%
	75%	2128	88.66%
3250.5 m	25%	1839	76.62%
	50%	2090	87.08%
	75%	2162	90.08%
4334 m	25%	1821	75.87%
	50%	2075	86.45%
	75%	2190	91.25%

4.2. Comparison of the Decision-Making Algorithm with Geographic Mobility

In addition, for the purpose of comparison with existing algorithms, similar attributes as the ones presented in [2] were considered. In the first part of the tests, the algorithms' computational latency was obtained: Dijkstra [10], AHP [11], FAHP [12], ATDDiM [13], and the DMAGM using pseudo-randomly generated channels with a triangular distribution with normalized values from 3 to 38 for the ETA, 33 to 98 for the AP, 0 to 10 for the SINR, and a single value of 200 for the BW. Triangular distribution is widely used as a general approximation of any central tendency distribution (average value) and when its measure of dispersion is loosely known.

Figure 4 shows that the DMAGM has a low latency with respect to the other algorithms for 200 channels, which reveals that the time to indicate the best channel is the lowest with respect to the other algorithms.

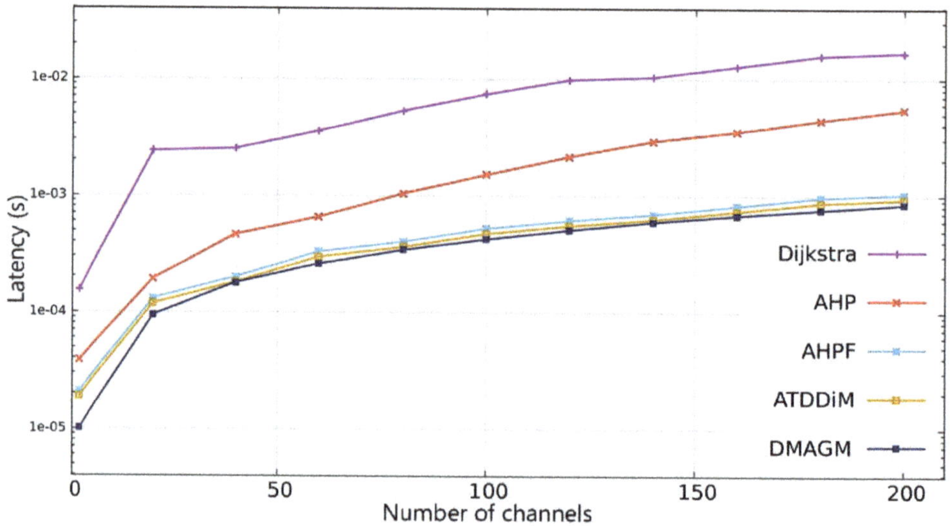

Figure 4. Latency comparison among several decision-making algorithms considering 200 channels.

It should be noted that each of the points shown in Figure 4 was obtained with an average of 100 repetitions of algorithms using different inputs. Then, the latency for each set of channels to be compared is calculated.

Figure 5 shows the percentage of time of the proposed algorithm DMAGM with respect to the other algorithms evaluated. Note that decision-making latency for the DMAGM is significantly reduced by 12.77% compared with ATDDiM, which is the second most effective algorithm and followed by FAHP with 21.42%, AHP with 71.84%, and Dijkstra with 94.27% in terms of latency reduction.

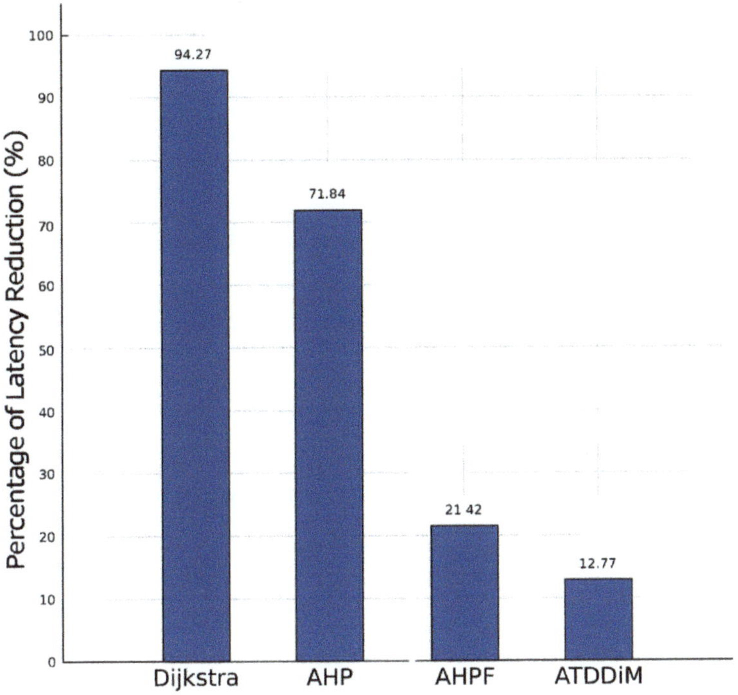

Figure 5. Percentage of latency reduction comparison of DMAGM with respect to reference algorithms.

4.3. Analysis of the Feedback-Decision-Making Algorithm with Geographic Mobility

According to the results obtained above, and in order to validate the performance of the proposed algorithm, a comparison between the FDMAGM and the DMAGM was developed, i.e., with and without considering the feedback mechanism.

Figure 6 shows the number of channels and their respective latency, which initially considers 2 channels, then 20 channels, and subsequently increases by 20 channels until reaching 200 channels.

Figure 6 also shows the confidence intervals with a level of 95% for 200 analyzed channels. Channels were randomly generated using a triangular distribution with normalized values from 3 to 38 for the ETA, 33 to 98 for the AP, 0 to 10 for the SINR, and a single value of 200 for the BW. Subsequently, to obtain the confidence interval and determine which channel is the best with respect to the proposed algorithm with and without a feedback process, the averages of 100 repetitions were computed for each of the proposed algorithms with different inputs.

In addition, Figure 6 shows that the latency of the proposed algorithm with feedback (FDMAGM), on average, increases nine times with respect to the latency of the proposed algorithm without feedback (DMAGM). This is due to the mechanism used in the feedback to check the previous state of the current channel. Consequently, according to the results, it can be suggested that the FDMAGM (with a feedback mechanism) is useful in cases in which the processing time is not an important factor when determining a channel with the best communication characteristics and with significant results in the new determination of a better communication channel.

Table 5 shows the results of the compared channels, which indicates that there is a 95% confidence level in the latency it takes to select the communication channel.

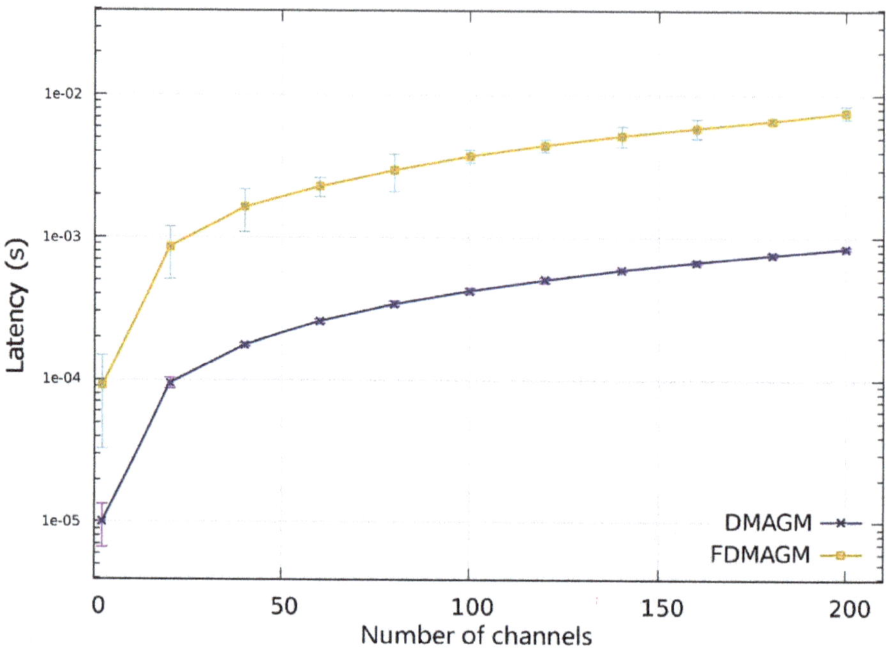

Figure 6. Latency comparison between FDMAGM and DMAGM (with and without a feedback mechanism) with 95% confidence intervals for each number of channels tested.

Table 5. Average latency values for FDMAGM and DMAGM (with and without a feedback mechanism) with a confidence interval of 95% in the selection of the communication channel.

Algorithm	Analyzed Channels	Average Latency Value $\bar{x}$ (s)	Confidence Interval (s)	
			$\bar{x}-1.96\cdot\frac{\sigma}{\sqrt{n}}$	$\bar{x}+1.96\cdot\frac{\sigma}{\sqrt{n}}$
DMAGM	2	1.0018×10^{-5}	6.6341×10^{-6}	1.3402×10^{-5}
	20	9.4471×10^{-5}	8.6513×10^{-5}	1.0243×10^{-4}
	40	1.7717×10^{-4}	1.7216×10^{-4}	1.8217×10^{-4}
	60	2.5918×10^{-4}	2.5102×10^{-4}	2.6733×10^{-4}
	80	3.4220×10^{-4}	3.2805×10^{-4}	3.5634×10^{-4}
	100	4.2337×10^{-4}	4.0347×10^{-4}	4.4327×10^{-4}
	120	5.0618×10^{-4}	4.8179×10^{-4}	5.3058×10^{-4}
	140	5.9363×10^{-4}	5.8207×10^{-4}	6.0519×10^{-4}
	160	6.7631×10^{-4}	6.6335×10^{-4}	6.8927×10^{-4}
	180	7.5883×10^{-4}	7.3926×10^{-4}	7.7841×10^{-4}
	200	8.4375×10^{-4}	8.2255×10^{-4}	8.6496×10^{-4}
FDMAGM	2	9.0613×10^{-5}	3.2976×10^{-5}	1.4825×10^{-4}
	20	8.5449×10^{-4}	5.1269×10^{-4}	1.1963×10^{-3}
	40	1.6415×10^{-3}	1.0943×10^{-3}	2.1886×10^{-3}
	60	2.2854×10^{-3}	1.9420×10^{-3}	2.6288×10^{-3}
	80	2.9847×10^{-3}	2.0981×10^{-3}	3.8714×10^{-3}
	100	3.7482×10^{-3}	3.3557×10^{-3}	4.1408×10^{-3}
	120	4.4709×10^{-3}	3.9891×10^{-3}	4.9526×10^{-3}
	140	5.2138×10^{-3}	4.3504×10^{-3}	6.0772×10^{-3}
	160	5.9145×10^{-3}	4.9912×10^{-3}	6.8378×10^{-3}
	180	6.6679×10^{-3}	6.4288×10^{-3}	6.9069×10^{-3}
	200	7.6456×10^{-3}	6.8606×10^{-3}	8.4306×10^{-3}

5. Conclusions

This work provides a novel proposal that considers multivariable decision-making for cognitive radio (CR) with geographic mobility (GM). The consulted literature shows scarce alternatives for channel allocation times when assessing the GM for the cognitive radio network (CRN), which requires in-depth analysis.

It is very helpful to use GM information in a CRN for several reasons such as reducing the number of secondary users that are connected to the network and where the quality of service can be affected. The literature reveals that decision-making is one of the main problems presented by CRNs since it is a fundamental operation for spectrum allocation, as well as spectrum sensing, spectral mobility, and sharing and cooperation. The decision-making algorithm with geographic mobility (DMAGM) is crucial for improving the channel selection and spectrum mobility.

Simulations demonstrate that the DMAGM reduces the processing time between 12% and 94% compared with other algorithms. In addition, the analysis considers geographic mobility (GM), which is, to the best of the authors' knowledge, not found in the literature for CRNs. In addition to the DMAGM, a modification to the algorithm was developed and tested using a feedback mechanism, named in this research as the feedback-decision-making algorithm with geographic mobility (FDMAGM). The FDMAGM was designed with the purpose of not only improving latency in decision-making but also to explore other options to improve the quality of the selection for the communication channel. The parameters considered for the analysis are based on a customized Delphi method, which are the bandwidth (BW), the signal-to-interference plus noise ratio (SINR), the channel availability probability (AP), and the estimated channel time availability (ETA).

Additionally, the history of previous results in the FDMAGM regarding the best selected communication channel is considered. As shown in the results, the processing time of the FDMAGM increased on average compared with the DMAGM. However, the FDMAGM can efficiently be used in the proposed scenarios without significant changes in the network. The feedback system has the advantage of being able to continually adjust and adapt based on the feedback received. In addition, the feedback version helps to identify and correct problems, which can be beneficial in situations where the quality of communication is critical. Despite the fact that the FDMAGM may take longer than the DMAGM to calculate the best communication channel, constant feedback improves efficiency and effectiveness over time. Both the DMAGM and the FDMAGM improve performance in practical scenarios, the former in terms of latency and the latter in terms of accuracy and stability.

Both the DMAGM and the FDMAGM have proven to be useful depending on the applications tested, such as best effort (BE) and real time (RT), and the different environments shown varying the radius of operation. This demonstrates that the match rate of these algorithms is similar in the worst-case scenario, about 75% on the selection of the best communication channel considering 2400 samples per scenario with the random presence of primary users (PUs).

Author Contributions: Conceptualization, E.R.-C. and P.L.-V.; methodology, results analysis, G.B.C.-J., P.L.-V. and E.R.-C.; writing—original draft preparation, L.P.-L., R.M.-J. and M.P.-C.; writing—review and editing, M.P.-C., P.L.-V., R.M.-J., L.P.-L. and E.R.-C. All authors have read and agreed to the published version of the manuscript.

Funding: This project has been supported by grants awarded by the National Council for Humanities, Science and Technology (CONAHCyT) and for the Research program RICCHUS UAM of the Autonomous Metropolitan University.

Institutional Review Board Statement: Not applicable.

Informed Consent Statement: Not applicable.

Data Availability Statement: Data are contained within the article.

Conflicts of Interest: The authors declare no conflict of interest.

Abbreviations

CR	Cognitive Radio
CRN	Cognitive Radio Network
GM	Geographic Mobility
PU	Primary User
SU	Secondary User
BS	Base Station
BW	Bandwidth
SINR	Signal-to-Interference plus Noise Ratio
AP	Channel Availability Probability
ETA	Estimated Channel Time Availability
DMAGM	Decision-Making Algorithm with Geographic Mobility
FDMAGM	Feedback-Decision-Making Algorithm with Geographic Mobility
ATDDiM	Modified Dijkstra Decision-Making Algorithm
FAHP	Fuzzy Analytic Hierarchy Process
AHP	Analytic Hierarchy Process
RWPM	Random Way-Point Mobility Model
RT	Real Time
BE	Best Effort
GSM	Global System for Mobile Communications
NS-3	Network Simulator-3

References

1. Zubair, S.; Yusoff, S.S.; Fisal, N. Mobility-Enhanced Reliable Geographical Forwarding in Cognitive Radio Sensor Networks. *Sensors* **2016**, *16*, 172. [CrossRef]
2. Hernandez, C.; Salgado, C.; López, H.; Rodriguez-Colina, E. Multivariable algorithm for dynamic channel selection in cognitive radio networks. *EURASIP J. Wirel. Commun. Netw.* **2015**, *2015*, 216. [CrossRef]
3. Wheeb, A.H.; Nordin, R.; Samah, A.A.; Alsharif, M.H.; Khan, M.A. Topology-Based Routing Protocols and Mobility Models for Flying Ad Hoc Networks: A Contemporary Review and Future Research Directions. *Drones* **2021**, *6*, 9. [CrossRef]
4. Chiang, K.-H.; Shenoy, N. A 2-D Random-Walk Mobility Model for Location-Management Studies in Wireless Networks. *IEEE Trans. Veh. Technol.* **2004**, *53*, 413–424. [CrossRef]
5. Filho, J.A.; Rosário, D.; Rosário, D.; Santos, A.; Gerla, M. Satisfactory video dissemination on FANETs based on an enhanced UAV relay placement service. *Ann. Telecommun.* **2018**, *73*, 601–612. [CrossRef]
6. Kumar, S.; Agrawal, G.S.; Sharma, S.K. Impact of Mobility on MANETs Routing Protocols Using Group Mobility Model. *Int. J. Wirel. Microw. Technol.* **2017**, *7*, 1–12. [CrossRef]
7. Abdullah, A.M.; Ozen, E.; Bayramoglu, H. Investigating the Impact of Mobility Models on MANET Routing Protocols. *Int. J. Adv. Comput. Sci. Appl.* **2019**, *10*, 25–35. [CrossRef]
8. Liu, X.; Du, X.; Zhang, X.; Zhu, Q.; Guizani, M. Evolution-algorithm-based unmanned aerial vehicles path planning in complex environment. *Comput. Electr. Eng.* **2019**, *80*, 106493. [CrossRef]
9. Nagaraju, M.; Rao, K.R.; Vijayalakshmi, K. Efficient way of implementing the random and GM (Gauss-Markov) mobility model in MANET. *Int. J. Eng. Technol.* **2018**, *7*, 270. [CrossRef]
10. Fan, D.; Shi, P. Improvement of Dijkstra's algorithm and its application in route planning. In Proceedings of the 2010 Seventh International Conference on Fuzzy Systems and Knowledge Discovery, Yantai, China, 10–12 August 2010; IEEE: New York City, NY, USA, 2010; pp. 1901–1904. [CrossRef]
11. Uchida, N.; Takahata, K.; Zhang, X.; Takahata, K.; Shibata, Y. Min-Max Based AHP Method for Route Selection in Cognitive Wireless Network. In Proceedings of the 2010 13th International Conference on Network-Based Information Systems, Takayama, Japan, 14–16 September 2010; IEEE: New York City, NY, USA, 2010; pp. 22–27. [CrossRef]
12. Hernández-Suárez, C.A.; Pedraza-Martínez, L.F.; Rodríguez-Colina, E. Fuzzy feedback algorithm for the spectral handoff in cognitive radio networks. *Rev. Fac. Ing. Univ. Antioq.* **2016**, *81*, 47–62. [CrossRef]
13. Méndez-Martínez, L.; Rodriguez-Colina, E.; Medina-Ramírez, C. Toma de Decisiones Basadas en el Algoritmo de Dijkstra. *Redes Ing.* **2014**, *4*, 35. [CrossRef]
14. Trigui, E.; Esseghir, M.; Merghem-Boulahia, L. A mobility scheme for cognitive radio networks. In Proceedings of the 2013 12th Annual Mediterranean Ad Hoc Networking Workshop (MED-HOC-NET), Ajaccio, France, 24–26 June 2013; IEEE: New York City, NY, USA, 2013; pp. 97–102. [CrossRef]
15. Obaid, A.; Fernando, X.; Jaseemuddin, M. A mobility-aware cluster-based MAC protocol for radio-frequency energy harvesting cognitive wireless sensor networks. *IET Wirel. Sens. Syst.* **2021**, *11*, 206–218. [CrossRef]

16. Deva priya, S.; Kannan, N. Enhanced spectrum aggregation based frequency-band selection routing protocol for cognitive radio ad-hoc networks. *Concurr. Comput.* **2019**, *31*, e4911. [CrossRef]
17. Zhao, Y.; Hong, Z.; Luo, Y.; Wang, G.; Pu, L. Prediction-Based Spectrum Management in Cognitive Radio Networks. *IEEE Syst. J.* **2018**, *12*, 3303–3314. [CrossRef]
18. Hanif, I.; Zeeshan, M.; Ahmed, A. Traffic Pattern Based Adaptive Spectrum Handoff Strategy for Cognitive Radio Networks. In Proceedings of the 2016 10th International Conference on Next Generation Mobile Applications, Security and Technologies (NGMAST), Cardiff, UK, 24–26 August 2016; IEEE: New York City, NY, USA, 2016; pp. 18–23. [CrossRef]
19. Jaffar, J.; Ksyusof, S.; Ahmad, N.; Mustapha, J. Location assisted proactive channel in heterogeneous cognitive radio network. *J. Telecommun. Electron. Comput. Eng.* **2016**, *8*, 49–53.
20. Omer, A.E.; Hassan, M.S.; El-Tarhuni, M. An Adaptive Channel Assignment Approach for Streaming of Scalable Video over Cognitive Radio Networks. In Proceedings of the 2016 UKSim-AMSS 18th International Conference on Computer Modelling and Simulation (UKSim), Cambridge, UK, 6–8 April 2016; IEEE: New York City, NY, USA, 2016; pp. 305–310. [CrossRef]
21. Tlouyamma, J.; Velempini, M. Channel Selection Algorithm Optimized for Improved Performance in Cognitive Radio Networks. *Wirel. Pers. Commun.* **2021**, *119*, 3161–3178. [CrossRef]
22. Thakur, P.; Singh, G. Spectrum Mobility in Cognitive Radio Networks Using Spectrum Prediction and Monitoring Techniques. In *Spectrum Sharing in Cognitive Radio Networks: Towards Highly Connected Environments*; Wiley: Hoboken, NY, USA, 2021; pp. 147–166. [CrossRef]
23. Yawada, P.S.; Dong, M.T. Intelligent Process of Spectrum Handoff/Mobility in Cognitive Radio Networks. *J. Electr. Comput. Eng.* **2019**, *2019*, 7692630. [CrossRef]
24. Li, F.; Sun, Z.; Lam, K.-Y.; Sun, L.; Shen, B.; Peng, B. Dynamic Spectrum Optimization for Internet-of-Vehicles with Deep-Learning-Based Mobility Prediction. *Res. Sq.* **2022**, 1–20. [CrossRef]
25. Iftikhar, A.; Rauf, Z.; Khan, F.A.; Ali, M.S.; Kakar, M. Bayesian game-based user behavior analysis for spectrum mobility in cognitive radios. *Phys. Commun.* **2019**, *32*, 200–208. [CrossRef]
26. Alozie, E.; Faruk, N. Intelligent Process of Spectrum Handoff in Cognitive Radio Network. *SLU J. Sci. Technol.* **2022**, *4*, 205–218. [CrossRef]
27. Sivasundarapandian, S.; Sakthiprabha, R.; Vedanarayanan, V.; Aranganathan, A.; Gomathi, T.; Rajinikanth, E. Spectrum Reallocation Algorithm in Cognitive Radio Networks Based on Secondary User Mobility Model. In Proceedings of the 2023 International Conference on Networking and Communications (ICNWC), Chennai, India, 5–6 April 2023; IEEE: New York City, NY, USA, 2023; pp. 1–5. [CrossRef]
28. Dey, S.; Misra, I.S. A novel spectrum handoff switching decision scheme for improved performance of secondary users in cognitive radio networks. *Int. J. Commun. Syst.* **2023**, *36*, e5413. [CrossRef]
29. Al-Dulaimi, A.; Al-Rubaye, S.; Cosmas, J. Adaptive congestion control for mobility in cognitive radio networks. In Proceedings of the 2011 Wireless Advanced, London, UK, 20–22 June 2011; IEEE: New York City, NY, USA, 2011; pp. 273–277. [CrossRef]
30. Zhang, L.; Zeng, K.; Mohapatra, P. Opportunistic spectrum scheduling for mobile cognitive radio networks in white space. In Proceedings of the 2011 IEEE Wireless Communications and Networking Conference, Cancun, Mexico, 28–31 March 2011; IEEE: New York City, NY, USA, 2011; pp. 844–849. [CrossRef]

sensors

MDPI

Article

Environment-Aware Rate Adaptation Based on Occasional Request and Robust Adjustment in 802.11 Networks

Weijie Yu [1], Li Wang [1,*], Jin Song [1], Lijun He [1,2] and Yanting Wang [1]

[1] School of Software, Northwestern Polytechnical University, Xi'an 710072, China; 17710904549@mail.nwpu.edu.cn (W.Y.); 2025180890@mail.nwpu.edu.cn (J.S.); lijunhe@nwpu.edu.cn (L.H.); yantingwang@nwpu.edu.cn (Y.W.)

[2] Yangtze River Delta Research Institute, Northwestern Polytechnical University, Taicang 215400, China

* Correspondence: wang.li@nwpu.edu.cn

Abstract: The IEEE 802.11 standard provides multi-rate support for different versions. As there is no specification on the dynamic strategy to adjust the rate, different rate adaptation algorithms are applied according to different manufacturers. Therefore, it is often hard to interpret the performance discrepancy of various devices. Moreover, the ever-changing channels always challenge the rate adaptation, especially in the scenario with scarce spectrum and low SNR. As a result, it is important to sense the radio environment cognitively and reduce the unnecessary oscillation of the transmission rate. In this paper, we propose an environment-aware robust (EAR) algorithm. This algorithm employs an occasional small packet, designs a rate scheme adaptive to the environment, and enhances the robustness. We verify the throughput of EAR using network simulator NS-3 in terms of station number, motion speed and node distance. We also compare the proposed algorithm with three benchmark methods: AARF, RBAR and CHARM. Simulation results demonstrate that EAR outperforms other algorithms in several wireless environments, greatly improving the system robustness and throughput.

Keywords: IEEE 802.11; rate adaptation; robustness; NS-3

1. Introduction

1.1. Background

The devices in a wireless local area network transmit and receive data through wireless electromagnetic waves, where the spectrum is extremely limited. This determines that the devices are required to enhance performance cognitively by interaction with the radio environment. In 802.11 networks, the wireless medium is shared among multiple nodes, and channel conditions can be unpredictable due to factors like node mobility, channel fading and interference [1]. As a result, the quality of wireless links can vary over time, leading to potential disruptions in network performance. Therefore, it is crucial for network devices to actively engage with the environment in order to mitigate these influential factors. Although higher rates can result in higher throughout and lower the channel occupancy time, the difficulty of parsing at the receiving end is greater, and messages are not easily received correctly [2]; on the contrary, although lower rates are easier to parse, increasing transmission time affects efficiency and increases the probability of interference. Interference-causing reception errors may result in packet retransmission or even discarding, significantly reducing data transmission efficiency. Any transmission rate that is too high or too low can potentially affect transmission efficiency. Rate selection plays a pivotal role in the implementation of IEEE 802.11 standard. A well-designed rate adaptation algorithm can not only improve data transmission efficiency, but also balance the stability of data transmission [3]. On the contrary, using inappropriate rate adaptation algorithm can ultimately lead to poor internet experience and even inability to communicate properly. Therefore, adaptively selecting an appropriate transmission rate for the given channel quality becomes an important task in improving the performance of each wireless link. This is the task of rate control at the MAC layer.

In the original IEEE 802.11 standard, all frames were transmitted at inherent rates. Due to the continuous changes in channel environment, it is not possible to switch to higher rates to fully utilize channel resources when the channel conditions improve; when channel conditions deteriorate, excessively high rates result in the continuous retransmission of data packets [4]. From this perspective, a single transmission rate will seriously affect the utilization of channel resources and network performance, resulting in a significant reduction in system throughput.

In the subsequent IEEE 802.11 standard, support for different rates was provided by changing the modulation and encoding methods of the physical layer. For example, in 802.11b, four transmission rates are provided: 1 Mbps, 2 Mbps, 5.5 Mbps and 11 Mbps. In 802.11a, eight transmission rates are provided: 6 Mbps, 9 Mbps, 12 Mbps, 18 Mbps, 24 Mbps, 36 Mbps, 48 Mbps and 54 Mbps. Although the IEEE 802.11 physical layer provides support for multiple rates, the standard only specifies the allowed rate set and does not design relevant schemes to achieve rate switching. Because rate selection is crucial for improving the performance of wireless network, the research on this topic has great practical significance for engineering guidance.

In Section 3, we propose an environment-aware robust (EAR) algorithm based on the analysis of various rate adaptation algorithms. The four key contributions of this paper are as follows:

1. EAR uses the method of comparing the packet loss rates of neighboring windows to perceive changes in the channel environment and adjusts the transmission rate accordingly;
2. By using the frequency-based RTS/CTS mechanism, it effectively distinguishes the cause of packet loss whether it is due to low SNR or a hidden terminal;
3. The dynamic window and threshold adjustment mechanism make the real-time and accuracy of rate adjustment much higher than some classic rate adaptation algorithms currently available;
4. The algorithm's oscillation detection mechanism can effectively suppress or utilize abnormal rate fluctuations, further improving system throughput.

1.2. Starting Point

One of the important factors affecting wireless network performance is link layer retransmission. According to the IEEE 802.11 standard, all unicast frames sent require ACK frames for confirmation. If there is a conflict or loss during data transmission, the CRC (Cyclic Redundancy Check) at the receiving end will fail, making it impossible to reply to the sending end with an ACK frame. The transmitter did not receive the ACK frame corresponding to the packet within the SIFS (Short Interframe Space), and will attempt to resend the data frame. In addition, even if the receiving end replies to the ACK frame normally, encountering a loss of the ACK frame can lead to the occurrence of retransmission.

Retransmission will have two levels of impact on the performance of a wireless network. From a micro level perspective, retransmission will increase the overhead of the data link, leading to worsening channel congestion and thereby reducing the throughput of the entire wireless network coverage range of the site. From a macro perspective, for upper layer applications, retransmission can cause message jitter and delay. Real-time application services, such as voice and video calls, heavily rely on continuous message transmission. For these applications, real-time performance is higher than data accuracy. Because even if a short segment is missing from the sound or image, users may not have a clear perception, but due to the high delay caused by retransmission, the discontinuity or lag of the sound or image will greatly affect the normal operation of the application.

There are many explanations as to why retransmission occurs, such as multipath effects [5], radio frequency interference, low SNR, hidden terminal, near/far effects [6], power mismatch and adjacent-channel interference [7], which can all lead to retransmission. Although the multipath phenomenon is a serious problem in traditional 802.11a/b/g devices, with the emergence of 802.11n technology and the application of MIMO (Multiple-Input Multiple-Output) and MRC (Maximum-Ratio Combining) [8] signal processing technology, multipath phenomenon has actually had a positive impact on system throughput. The method of using CSMA/CA media access can effectively avoid near/far effects. In addition, issues such as radio frequency interference, power mismatch, and adjacent frequency interference are usually difficult to solve at the software level. Therefore, perceivable and controllable factors, such as SNR and hidden terminal, have become the key to planning most of rate adaptation algorithms.

In the presence of a hidden terminal, the channel may experience significant bit error and packet loss [9]. The traditional algorithm believes that as long as the bit error and packet loss rate increases, the sender needs to reduce the transmission rate. Actually, doing so does not solve the collision problem [10]. Reducing the transmission rate can lead to more intense channel conflicts due to the prolonged transmission time of each packet, thereby exacerbating the conflicts. This reduction can also cause a further decrease in transmit rate, which could have a detrimental effect on the system's throughput performance, creating a vicious cycle. Consequently, EAR's key challenge is to identify the cause of packet loss, whether it is collision or channel noise, and adopt adaptive strategies to regulate the transmission rate in varying environments. Therefore, it is imperative that the EAR be designed in accordance with three fundamental principles:

1. When the wireless network environment is favorable, try to utilize high-rate to transmit data.
2. In situations where the SNR is low, the likelihood of successfully receiving data packets decreases as the transmission rate increases. Therefore, opting for a lower transmission rate becomes imperative.
3. In collision scenarios caused by a hidden terminal, a decrease in transmission rate may actually result in channel deterioration and a subsequent reduction in overall system throughput. As a result, it is imperative to select a more suitable transmission rate.

The remainder of this paper is as follows. Section 2 introduces some research achievements in rate adaptation proposed in recent years. Section 3 explains our proposed EAR scheme. Section 4 explains the configuration of the simulation and its results. Finally, the conclusion is given in Section 5.

2. Related Works

In recent years, rate adaptation algorithms have been widely studied and applied in the field of wireless communication. As shown in Table 1, based on different dimensions, the most common classification methods are as follows: based on the sending end or based on the receiving end; and based on historical statistics or based on current measurements [11]. The above classification methods are not completely independent, and in practical applications, it is often necessary to cross the two or combine other classification methods to design appropriate rate adaptation algorithms.

Table 1. Classic algorithm classification.

Algorithm	Sender or Receiver	History or Current
ARF [12]	Sender	History
AARF [13]	Sender	History
Sample Rate [14]	Sender	History

Table 1. *Cont.*

Algorithm	Sender or Receiver	History or Current
RBAR [15]	Receiver	Current
OAR [16]	Receiver	Current
CHARM [17]	Sender	Current
CARA [18]	Sender	History
SGRA [19]	Sender	Current
BARA [20]	Sender	Current
Minstrel [21]	Sender	History

The algorithm based on the sender determines the rate of transmission based on the conditions of the sender itself [22]. The receiver-based algorithm will feedback the appropriate rate to the sender by evaluating the current channel conditions, and adjust the next transmission rate after the sender receives the feedback information [23]. The algorithms based on the sending end include ARF, AARF, ONOE, etc. These algorithms have strong compatibility with various wireless network products and do not need to modify the existing IEEE 802.11 frame format, making them easy to implement. Rate adaptation algorithms based on the receiving end include OAR, RBAR, RARA, and so on. Although these algorithms can theoretically obtain more accurate reference data, additional feedback mechanisms are needed to transmit this information to the sender [24]. Usually, CTS frames or ACK frames are used to transmit feedback information, and it is generally necessary to modify the IEEE 802.11 standard frame structure. Therefore, many algorithms cannot be implemented in existing IEEE 802.11 devices and can only be used for theoretical research.

According to algorithms based on historical statistics or current measurements, channel information can be obtained in principle, but the difference between the two lies in the real-time nature of the information. Historical statistics algorithm using the data transmission situation in the previous period to make predictions about the transmission situation in the next period, thereby making rate adjustment decisions in advance. Therefore, it is generally suitable for environments with slow channel changes, and its performance is not very good for scenarios with drastic changes in channel environment. Typical algorithms based on historical statistics include ARF and the improved AARF algorithm based on ARF. These two algorithms record the success and failure status of each transmission frame to determine whether to adjust the rate.

ARF takes data packets as the statistical unit and starts a timer at the beginning of the algorithm. When the data packet fails to be sent twice in a row, the sender will directly reduce the transmit rate. Whenever 10 consecutive packets are successfully sent or the timer reaches the specified time, a higher rate packet will be sent for detection. If the detection packet is successfully sent, that is, the sending end successfully receives the ACK from the receiving end, then it will be sent at this detection rate next time. If the detection frame fails to be sent, the transmission will continue at the previous rate thereafter. After each rate adjustment, the timer will be reset. To stabilize the rate and reduce the occurrence of rate oscillation, when AARF detects a detection packet transmission failure, it will increase the threshold for the rate increase to twice the original value. For example, if the initial threshold is 10 and the detection packet fails to transmit, the threshold will be updated to 20 (the maximum value does not exceed 50), which means that the next time we want to increase the rate again, we need to successfully transmit 20 consecutive times. If the rate decreases due to two consecutive transmission failures during normal transmission, the threshold for the rate increase will be reset to the initial value of 10. AARF increases the time for rate improvement in a stable channel environment, resulting in lower-rate oscillation. However, the relatively fixed adjustment modes of these two algorithms are difficult to adapt to flexible and ever-changing channel environments. Overall, algorithms based on historical statistics infer the next channel state based on channel states in the past.

In channel-measurement-based methods, it can be further divided into measurement RSS (Received Signal Strength), SNR (Signal-to-Noise Ratio), BER (Bit Error Rate), PLR (Packet Loss Rate) and other indicators [25]. For example, in algorithms such as CHARM and SGRA, the sender determines the channel status by the signal strength RSS value in ACK. CHARM utilizes trial and error to find better transmission rates, and therefore, this mechanism is not well suited for mobile high dynamic environments. Before sending the data packet to a specific destination, CHARM calls the path-loss prediction algorithm to estimate the current path loss to the destination. The sender uses its own transmission power and the noise level at the receiver to obtain the SINR estimation of the receiver. Finally, a set of transmission rates is determined by looking up the SINR threshold table. CHARM utilizes channel reciprocity to enable the sender to obtain relevant receiver information, without incurring RTS/CTS overhead during the algorithm process. Although the combination of these technologies enables CHARM to quickly respond to changes in dynamic channels, CHARM inevitably encounters three problems: (1) The transmission of power and noise level information relies on introducing an additional 802.11 information element into the beacon, probe request and probe response, which violates the existing 802.11 frame format standards. (2) If the sending end is using a power control algorithm, it will have a certain degree of impact on the accuracy of CHARM, depending on the speed of power change. (3) CHARM assumes a symmetric link between the sender and receiver, which is not true in wireless channels due to mobility, fading and interference [26].

In RBAR, the format of the RTS frame is modified to no longer carry the reserved channel time, but rather the rate to be used and the size of the data frame. Neighbor nodes calculate the reserved channel time based on these two values. After receiving the RTS frame, the receiving end selects an appropriate rate based on the SNR value of the RTS frame and carries the rate in CTS to inform the sender. The receiving end defines two SNR thresholds for each rate. If the measured SNR falls precisely between these two thresholds, the rate is selected. The disadvantage of RBAR is that it requires enabling the RTS/CTS mechanism for a prolonged period of time to assist in rate control, which is relatively expensive. In addition, RBAR requires modifications to the frame format of RTS/CTS, which brings compatibility issues. The rapid changes in channel conditions require the responsive adaptation of transmission rates. Ideally, RBAR selects an optimal rate for every frame transmission. However, it requires low-latency hardware that may be dedicated to processing MAC-layer frames, such as FPGA, that is capable of selecting the rate on a per-frame basis. From this perspective, RBAR is also difficult to apply to general network devices.

3. Framework and Design

3.1. Initialization

In the 802.11 network, each node possesses cognitive abilities to enhance its interaction with the environment. During algorithm initialization, it is necessary to configure several fundamental parameters. Table 2 below provides a list of these parameters along with their respective setting values.

Table 2. Initialization parameters.

Parameter	Symbol	Value
Initial rate	R_{org}	Protocol maximum rate
Initial window size	$Wind_{org}$	30
Maximum window size	$Wind_{max}$	40
Minimum window size	$Wind_{min}$	20
Original decrease rate threshold	P_{down_org}	0.4
Original increase rate threshold	P_{up_org}	0.2
Threshold update coefficient	α	0.2

The initial rate R_{org} is set to the highest supported rate under the current protocol, for example, the initial rate of 802.11b is 11 Mbps, and the initial rate of 802.11a/g is 54 Mbps. The window value is the number of statistical packets, which is the foundation for calculating the packet loss rate. The default value of the initial window is set at 30. In addition, during the algorithmic operation, the window size is subject to dynamic updates based on the current packet loss rate, threshold range and rate. The setting of the window range is particularly important for the algorithm. On the one hand, it can prevent the lag of environment perception caused by a window that is too large, and on the other hand, it can reduce the impact of computing overhead and accidental packet loss caused by a window that is too small. In order to ensure that the window size changes within a reasonable range, we combine the above related papers based on window algorithm and our own simulation experience. When the maximum and minimum values of the window are set to 40 and 20, respectively, the algorithm performs well.

After extensive experiments, we have made the following limitations on the range of threshold values: the increase rate threshold range is set at 0.1–0.3; The decrease rate threshold range is set at 0.3–0.5. Initialize P_{up_org} in the table is the initialization threshold for increasing the rate, which defaults to 0.2. P_{down_org} is the initialization threshold for reducing the rate, which defaults to 0.4. Set coefficient α to 0.2. It has been demonstrated through simulation that the range of threshold values and the coefficients for updating them can enable the algorithm to perform optimally under this setting.

3.2. Packet Loss Analysis

The sending end will compare the number of ACKs received with the total number of packets sent to determine the packet loss rate of this window, The calculation formula for loss is as follows Equation (1).

$$P_{loss} = \frac{Transmitted\_frames - ACK\_frames}{Transmitted\_frames} \tag{1}$$

Among them, $Transmitted\_frames$ represents the total number of packets sent during the window period. ACK_frames represents the number of ACK frames received, and subtracting the two represents the number of packet losses during this window. This is in accordance with P_{loss}, P_{up} and P_{down}, which are used to determine the update strategy for the next window.

3.3. Window Update

The setting of window size will greatly affect the performance of the algorithm in different environments. If the window size setting is too small, the packet loss rate statistics will be too contingent, which will result in the wrong rate selection; on the contrary, if the window set is too large, the algorithm will react slowly and cannot adapt to changes in the channel in a timely manner.

Different window sizes can affect transmission times for multiple rates. In high-speed states, increasing the window size results in more data packets being transmitted during the window period. This adjustment is ideal when subsequent channel conditions are well predicted and facilitates the transmission of high-speed data packets. However, to prevent rate increases from causing mismatches with channel conditions, reducing the window size enables timely callback at high rates, avoiding higher packet loss rates. In low-speed states, reducing the window size results in fewer transmission opportunities at that rate, allowing the system to adjust to high-speed rates faster. Nonetheless, when the SNR is too low, the packet loss rate at the lowest rate remains too high. To mitigate this issue, increasing the window size helps to reduce the use of the RTS mechanism and further minimizes system losses. Therefore, EAR proposes an idea of dynamically adjusting the window size, which adopts the following Algorithm 1 based on the current packet loss rate and rate:

Algorithm 1: Window Update Algorithm

```
if (Current window is RTS window || Next window is RTS window) {
    Wind_next = Wind_curr;
    break;
}
else {
    if (P_loss ≤ P_up) {
        if (R_curr < R_max) {
            Wind_next = Wind_min;
        }
        else {
            Wind_next = Wind_curr + ceil(Wind_curr * (P_up − P_loss));
            if (Wind_next ≥ Wind_max) {
                Wind_next = Wind_max;
            }
        }
    }
    else if(P_loss ≥ P_down) {
        if (R_curr > R_min) {
            Wind_next = Wind_curr − ceil(Wind_curr * (P_loss − P_down));
            if (Wind_next ≤ Wind_min) {
                Wind_next = Wind_min;
            }
        }
        else {
            Wind_next = Wind_max;
        }
    }
    else {
        Wind_next = Wind_curr;
    }
}
```

- When the current or next window is an RTS window, the size of the next window equals the size of the current window.
- When the packet loss rate P_{loss} is less than or equal to P_{up}, trigger the rate increase strategy (binary search), and if the rate at this point does not reach the maximum rate supported by the protocol, the size of the next window should be set to the minimum window value defined at initialization. Because if the increased rate is not suitable for the current channel environment, the algorithm can quickly callback the rate and stop the loss as soon as possible when the rate mismatch leads to continuous packet loss.
- Unlike the above situation, the current rate has reached the maximum rate supported by the protocol, and the packet loss rate is low. In order to maximize system throughput, the algorithm allows the high-speed rate to have more transmission opportunities, and increases the window size while maintaining the maximum rate in the next window.
- When the packet loss rate P_{loss} is more than or equal to P_{down}, and the rate of the current window is not the lowest rate, the size of next window needs to be reduced. In order to speed up the window size update frequency and increase the probability of sending high-speed data packets. Similar to the logic of window enlargement, it guides the setting of subsequent window sizes based on the current window size and packet loss rate.

- Undertake the above content, and if the current window is the lowest rate, the next window size should be adjusted to the maximum to minimize the use of RTS and reduce resource consumption.
- The threshold of window packet loss rate is between P_{up} and P_{down}, and the size of the next window remains unchanged.

3.4. Threshold Update

In addition to elastically scaling the window size according to the current channel environment, the threshold of the window also needs to be adjusted in real time. The dynamic threshold can effectively predict the channel environment, and the rate within the most suitable range can be stabilized. The lower the threshold for increasing the rate, the greater the difficulty in increasing the rate; on the contrary, the higher the threshold for increasing the rate, the simpler it is to increase the rate. The same principle applies to reducing the threshold of the rate. The setting of the threshold needs to be determined before the window executes the sending task. This paper adopts a method similar to EWMA (Exponentially Weighted Moving Average) [27], which guides the subsequent threshold setting through short-term historical thresholds. The threshold update is shown in Figure 1 below.

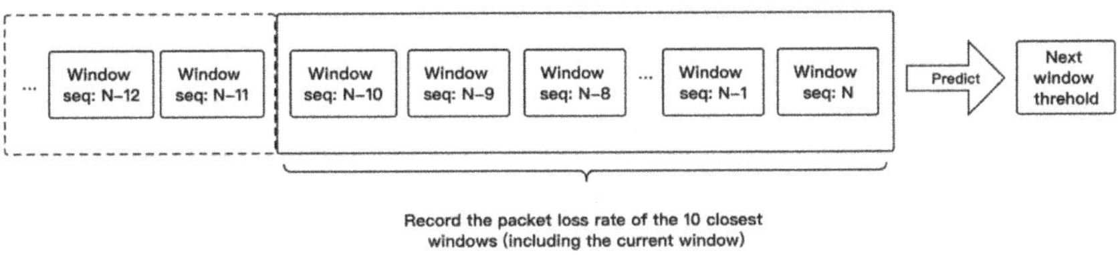

Figure 1. The packet loss rate information of the last ten windows is recorded to calculate the threshold for the following windows.

Inferring the packet loss rate W for the next window based on the recent performance of the packet loss rate, we obtain Equations (2) and (3).

$$W = \alpha * \frac{\sum_{i=1}^{N-1} P_i}{N-1} + (1-\alpha) * P_N, \qquad (1 < N < 10) \tag{2}$$

$$W = \alpha * \frac{\sum_{i=N-9}^{N-1} P_i}{N-1} + (1-\alpha) * P_N, \qquad (N \geq 10) \tag{3}$$

where P_i is the packet loss rate of the $N-i$ history window closest to the current window, and $W \leq \left(P_{up\_org} + P_{down\_org}\right)/2$ and $P_N \leq P_{up\_N}$ indicate that packet loss rate is relatively low in the near future, and the next window has met the requirements for increase rate. P_N and P_{up_N} represent the packet loss rate and increase rate threshold of the current window. The P_{up} increase in the next window makes it easier to trigger the mechanism of rate increase to achieve higher throughput, where $W > \left(P_{up\_org} + P_{down\_org}\right)/2$ and $P_N \geq P_{down\_N}$ indicates that after inference, not only the current window, but also the next window may have a higher packet loss rate. Then, P_{down} will be decrease to prevent the sustained occurrence of excessive packet loss rate. The pseudocode for threshold update is shown in Algorithm 2 below.

Algorithm 2: Threshold Update Algorithm

```
if (Current window is RTS window || Next window is RTS window){
    P_up_next = P_up_curr;
    P_down_next = P_down_curr;
}
else {
    W = 0.2 * (sum(history_loss)/history_loss.length) + 0.8 * P_loss;
    Avg = (P_up_org + P_down_org)/2;
    if (W ≤ Avg && P_loss < P_up_curr) {
        P_up_next = (1 − W) * P_up_org + P_up_curr;
        P_down_next = P_down_curr + W * P_down_org;
        if (P_up_next > 0.3) { P_up_next = 0.3; }
        if (P_down_next > 0.5) { P_down_next = 0.5; }
        break;
    }
    if (W > Avg && P_loss > P_down_curr) {
        P_down_next = P_down_curr − W * P_down_org;
        P_up_next = P_up_curr − (1 − W) * P_up_org;
        if (P_down_next < 0.3) { P_down_next = 0.3; }
        if (P_up_next < 0.1) { P_up_next = 0.1; }
        break;
    }
    P_up_next = P_up_curr;
    P_down_next = P_down_curr;
}
history_loss.append(P_loss);
if (history_loss.length ≥ 10) {
    history.pop(0);
}
```

3.5. Environment-Aware Robust (EAR) Algorithm

After the counter is cleared, calculate the packet loss rate P_{loss} of the current window, and if $P_{loss} \leq P_{up}$, then the sending rate of the next window will increase. In order to find the most suitable rate faster, a binary search method is used to select the rate of the next window; If $P_{down} > P_{loss} > P_{up}$, then the rate within the next window remain unchanged; If $P_{loss} \geq P_{down}$, the rate will not immediately decrease. Instead, add a window $Wind_{rts}$ to enable the RTS mechanism later, during this window period, before sending each data packet, an RTS frame is sent to make a reservation for the channel. According to the packet loss rate P_{rts_loss} during the $Wind_{rts}$ period, the following four strategies will be adopted to update the rate:

- If $P_{rts\_loss} \geq P_{loss}$: The packet loss rate not only does not decrease, but also worsens the channel conditions and increases the packet loss rate. This indicates that the reason for the high packet loss rate in the current environment is due to the low SNR. Therefore, the next window should decrease the rate (stepped type) and turn off the RTS mechanism.
- If $P_{down} < P_{rts\_loss} < P_{loss}$: Although the packet loss rate has slightly improved, it still does not fall below the threshold of decrease the rate. In this case, the benefits of enabling RTS are not enough to offset the costs of RTS. Therefore, the next window should maintain the original rate and turn off the RTS mechanism.
- If $P_{up} < P_{rts\_loss} \leq P_{down}$: The packet loss rate has reduced to a certain extent, and the throughput benefits brought by RTS are slightly greater than the inherent costs brought by RTS. The factors leading to a higher packet loss rate in the upper window are mostly caused by hidden station issues. Therefore, the next window maintains

the original rate of transmission, and the RTS mechanism is still enabled in the next window.

- If $P_{rts\_loss} \leq P_{up}$: After enabling the RTS mechanism, the packet loss rate has been greatly reduced, indicating that the previous high packet loss rate was entirely caused by hidden terminal issues. At this point, the opening of RTS brings great benefits and eliminates the impact of SNR factors. Therefore, in order to maximize the benefits, the next window increases the rate of transmission and still open RTS mechanism.

In the case of hidden terminal, reducing the rate will make the conflict more intense, as it prolongs the transmission time of each data packet, exacerbating channel conflicts and affecting system throughput. The execution flowchart of the EAR algorithm is shown in Figure 2 below.

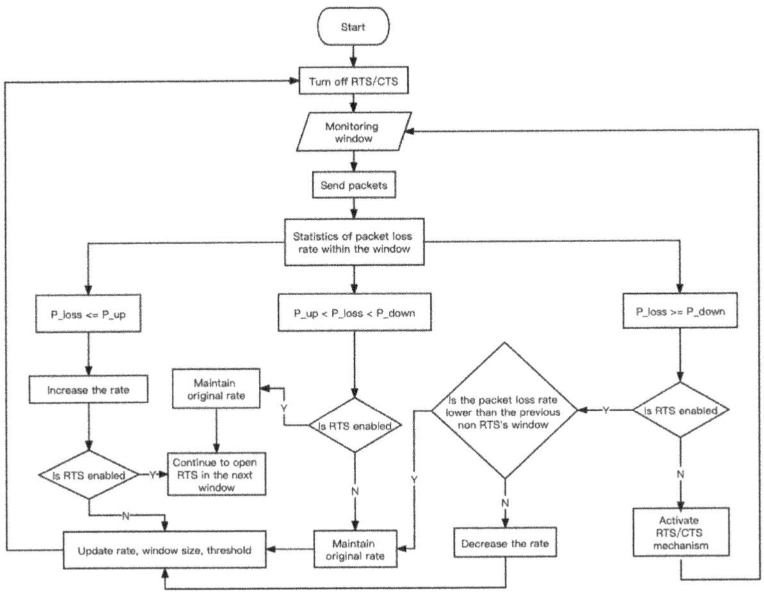

Figure 2. EAR execution flowchart.

3.6. Rate Oscillation Detection

In a relatively stable channel environment, if the threshold selection is not appropriate, it may cause the rate selection mechanism to continuously alternate between adjacent high-speed rates, thereby forming rate jitter, which may affect the network throughput [28]. For example, the current speed is 12 Mbps, and the window packet loss rate is 10%, which meets the requirement of increasing the speed to 18 Mbps. Afterwards, the window packet loss rate increased to 50%, and due to the high packet loss rate, it returned to 12 Mbps, repeating this cycle. In theory, if the channel is stable, the corresponding packet loss rate at each rate should be very stable. Therefore, according to calculations, the throughput at 12 Mbps is actually higher than the throughput provided by 18 Mbps. In order to avoid such problems, we added an oscillation detection mechanism to the algorithm: let R be the current rate and R^+ be the higher-order rate. If the history of rate selection shows a sequence like $R \rightarrow R^+ \rightarrow R \rightarrow R^+ \rightarrow R \rightarrow R^+$, it is determined that there is an oscillation reaction at that rate.

When oscillation occurs, in order to maximize throughput, it is necessary to combine the oscillation sequence and compare the average throughput of R^+ and R. If the average throughput provided by R^+ is higher than R, the sequence can be ignored and subsequent oscillations can be continuously detected until the oscillation sequence is broken or R's average throughput is higher than R^+. If the average throughput provided by R is higher than R^+, it indicates that the threshold for the rise rate is not set properly. In order to stabilize the current rate, it is necessary to lower the threshold for the rise rate, making the requirements for the rise more stringent.

In order to implement the rate oscillation mechanism, the algorithm needs to continuously record the relevant information of the six windows closest to the current window during execution. Set the window rate sequence as R_{ol} $[R_1, R_2 \ldots R_6]$, the window start timestamp sequence as T_{begin} $[T_{b_1}, T_{b2} \ldots T_{b6}]$, the corresponding window end timestamp sequence as T_{end} $[T_{e1}, T_{e2} \ldots T_{e6}]$, and the number of packets successfully sent by the window sequence as S_{suc} $[S_1, S_2 \ldots S_6]$. If R_{ol} satisfies the law of $R \to R^+ \to R \to R^+ \to R \to R^+$, then the following Equations (4) and (5) needs to be used to make a detection.

$$Throughput_R = \frac{(S_1 + S_3 + S_5) * L_{data}}{(T_{e1} - T_{b1}) + (T_{e3} - T_{b3}) + (T_{e5} - T_{b5})} \quad (4)$$

$$Throughput_{R+} = \frac{(S_2 + S_4 + S_6) * L_{data}}{(T_{e2} - T_{b2}) + (T_{e4} - T_{b4}) + (T_{e6} - T_{b6})} \quad (5)$$

Among them, L_{data} represents the payload size of each packet, $Throuthput_R$ represents the average throughput of R within R_{ol}, $Throuthput_{R+}$ represents the throughput corresponding to R^+ within R_{ol}. If $Throuthput_R \leq Throuthput_{R+}$, it indicates that the oscillation leads to an increase in throughput, and the threshold does not need to be updated to fully utilize the advantages brought by oscillation. If $Throuthput_R > Throuthput_{R+}$, it indicates that the oscillation brings about a decrease in throughput, and it is necessary to reduce the threshold for increasing the rate to half of the original value to impose punishment. This makes the requirements for increasing the rate more stringent to avoid the occurrence of sustained-rate oscillations in the future.

Figure 3 below shows a sequence of rate oscillations that we discovered during the 802.11b simulation process. It can be seen that after the addition of rate oscillation detection mechanism, the fluctuation of the rate has been significantly suppressed and that the packet loss rate has been reduced.

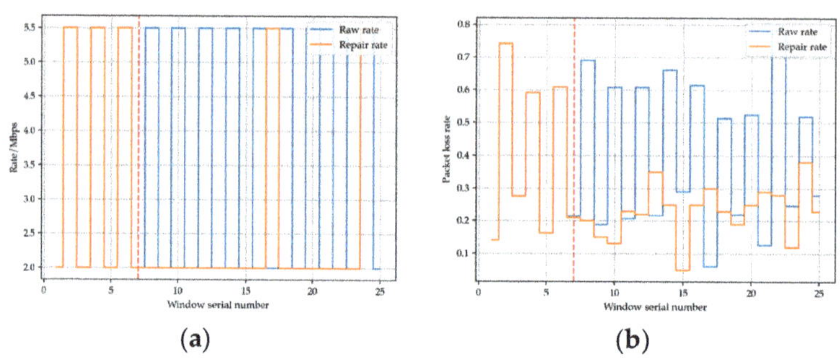

Figure 3. (**a**) Transmission rate comparison before and after adding rate oscillation detection mechanism. (**b**) Packet loss rate comparison before and after adding rate oscillation detection mechanism.

In the first six windows, the rate switches back and forth between 2 Mbps and 5.5 Mbps. After adding the oscillation detection mechanism, in window number 7 (red dotted line), an average throughput evaluation of the oscillation sequence is performed by combining the historical packet length and transmission time. The evaluation results indicate that in this historical sequence, the throughput provided by using the lower rate of 2 Mbps is higher than that provided by 5.5 Mbps. Therefore, a rate of 2 Mbps will be used for transmission in window number 7, and the threshold for increasing the rate will be lowered, making the conditions for increasing the rate more stringent. It can be observed that in the subsequent windows, the rate stabilizes, and the packet loss rate remains low.

4. Simulation Results

In this section, we use NS-3 to verify the performance of EAR in different scenarios. First, we designed a series of experiments to validate the necessity of dynamic parameters in the EAR algorithm. Secondly, by constructing a hidden terminal model, we demonstrated the positive impact of the dynamic RTS/CTS mechanism on the performance improvement of the EAR algorithm. Finally, we investigated the influence of node quantity and node movement speed on different algorithms. The final simulation results indicate that the throughput performance of the EAR algorithm is superior to other comparative algorithms in different scenarios.

4.1. Case Study—Adjustment of Window and Threshold

To demonstrate the necessity of parameters updates, we designed a set of parameter validation experiments. In this experiment, fixed threshold range and window size were used for throughput testing. This experimental scenario consists of two nodes: an Access Point (AP) node and an STA (Station) node. The UDP traffic of 802.11b 11 Mbps CBR generated by the STA node to the AP node is sent at different distances using the LogDistancePropagationLossModel [29] as the propagation loss model for NS-3. The path loss varies depending on distance, as shown in Equation (6) below.

$$L = L_0 + 10nlog(\frac{d}{d_0}) \tag{6}$$

where n is the path loss distance exponent, d_0 is the reference distance (m), L_0 is the path loss at reference distance (dB); and d and L represent the distance and path loss (dB).

In the scenario, the movement step of STA is set to 1 m, and the movement time interval is 1 s, which means that the STA node moves away from the AP node at a distance of 1 m per second. STA node starts at the coordinates (5,0,0), and AP node remains at the coordinates (0,0,0) throughout the simulation process. The distance scene model is shown in the following Figure 4.

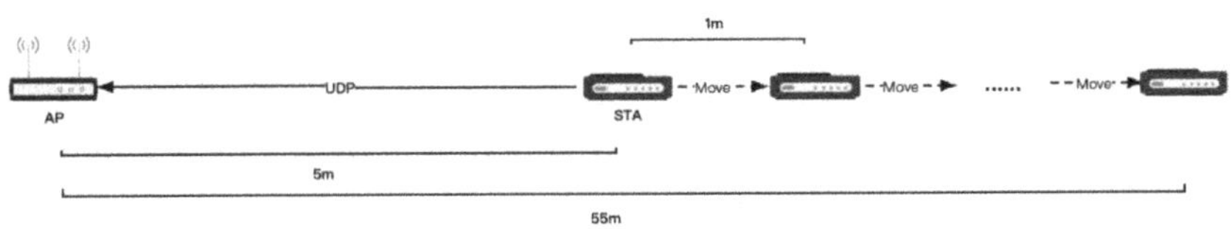

Figure 4. Pull distance scene model.

We repeated the experiment 200 times, and the average throughput under each threshold and window during the entire process is shown in Table 3 below.

Table 3. Comparison of throughput between fixed and dynamic parameters in high dynamic channel environments.

Threshold \ Window	20	30	40	Dynamic
0.1–0.2	2.33 Mbps	2.43 Mbps	2.12 Mbps	2.54 Mbps
0.1–0.3	2.54 Mbps	2.68 Mbps	2.55 Mbps	2.74 Mbps
0.1–0.4	2.43 Mbps	2.83 Mbps	2.63 Mbps	3.12 Mbps
0.2–0.3	2.21 Mbps	2.44 Mbps	2.36 Mbps	3.01 Mbps
0.2–0.4	2.12 Mbps	2.88 Mbps	2.34 Mbps	2.97 Mbps
0.3–0.4	1.74 Mbps	2.13 Mbps	2.29 Mbps	2.53 Mbps
Dynamic	2.64 Mbps	3.26 Mbps	3.03 Mbps	3.45 Mbps

It can be seen that the average throughput varies under different threshold ranges and window sizes. When the threshold range is fixed, dynamic windows provide more throughput. When the window size range is fixed, dynamic thresholds also provide more throughput. Finally, when the threshold range and window size can be dynamically adjusted, the throughput provided is the highest. The above experimental results demonstrate that dynamic parameters have better environmental adaptability and provide higher throughput compared to fixed parameters when the channel environment changes significantly.

4.2. Case Study—Effect of Hidden Terminal

In order to explore the impact of a hidden terminal on rate selection, this part used the NS-3 platform to build a WLAN environment based on 802.11g. AARF, RBAR, CHARM and EAR algorithms were used for simulation experiments in the hidden terminal environment, and the experimental results were compared and analyzed. During the simulation experiment, to ensure the fairness of the experiment, the sender uses UDP one-way transmission protocol and sends UDP data streams at a CBR (Constant Bit Rate) of 60 Mbps. The specific parameter settings for each module of NS-3 are shown in Table 4.

Table 4. Parameter settings for each module of NS-3.

Parameter	Value
Protocol standard	IEEE 802.11g
Frequency	2.4 GHz
Physical layer model	YansWifiPhy
Transmission method	CBR
Data flow type	UDP
Packet length	1400 Bytes
Data flow rate	60 Mbps
Channel model	YansWifiChannel
Channel fading model	MatrixPropagationLossModel

The model for the hidden terminal model is shown in the following Figure 5.

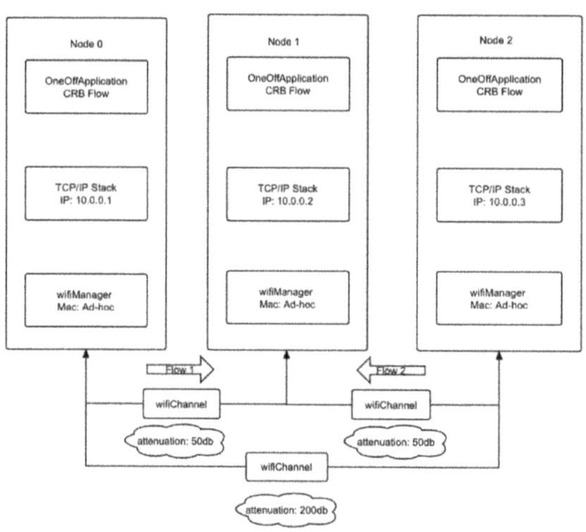

Figure 5. Hidden terminal model.

The channel propagation loss model selection for this experiment is the MatrixPropagationLossModel of NS-3. The propagation loss of each pair of nodes in this model is fixed and does not depend on their actual positions. And by default, the propagation loss is symmetric. When the propagation loss is 50 dB, two nodes can send and receive information to each other. When the propagation loss reaches 200 dB or above, the signals of the two nodes are weak and cannot be detected. This design can effectively construct a hidden terminal experimental model.

The sender uses the OnOffApplication class to create a CBR source, which can generate traffic for a single destination. However, its sending and stopping states are alternated by default. Therefore, setting its OnTime and OffTime properties to 1 and 0, respectively can keep the source in the sending state. The receiving end uses the PacketSink class in NS-3 to install application services for the node, which are responsible for receiving and consuming the traffic generated to the IP address and port. Finally, we install the EarWifiManager algorithm for flow 1 and select ConstantRateWifiManager as the rate algorithm for flow 2.

We conducted 200 simulation experiments, each lasting for 5 s. To facilitate the display of the EAR algorithm operation process, Table 5 below shows a representative set of data.

Table 5. Performance of the EAR on each data stream.

Indicators	Flow 1	Flow 2
Tx Packets	26,785	26,825
Tx Bytes	38,248,980	38,306,100
Tx Offered	61.1984 Mbps	61.2898 Mbps
Rx Packets	2286	1565
Rx Bytes	3,264,408	2,234,820
Average Throughput	5.22305 Mbps	3.57571 Mbps

In Flow 1, Node 0 sent a total of 26,785 packets, and 2286 packets were successfully received by Node 1. The average throughput was calculated to be 5.22305 Mbps. In Flow 2, Node 2 sent a total of 26,825 packets, with 1565 successfully reaching Node 1 and an average throughput of 3.57571 Mbps. Meanwhile, this section utilized the tracking system provided by NS-3 to conduct a statistical analysis of the data transmission changes during the Flow 1 experiment, as shown in Figure 6.

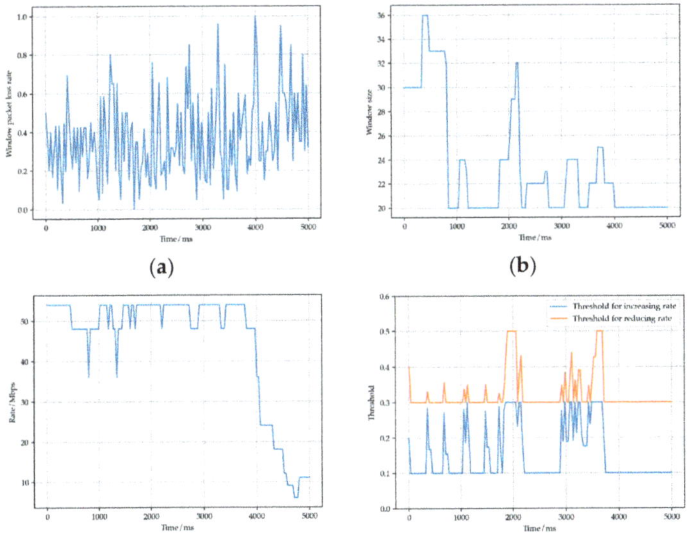

Figure 6. (**a**) Changes in packet loss rate. (**b**) Changes in window size. (**c**) Changes in transmit rate. (**d**) Changes in threshold.

In addition, the model experiment also incorporates AARF, RBAR and CHARM algorithms for simulation, and the comparative performance is shown in the following Figure 7. It can be seen that it is difficult for algorithms that do not have the addition of RTS mechanism to overcome the packet loss problem in the case of a hidden terminal scenario. And the throughput performance of EAR's adaptive RTS mechanism in this scenario is also better than the RBAR using full RTS mode. The high collision rate caused by hidden terminal can cause AARF and CHARM errors to reduce transmission rates, thereby exacerbating channel degradation. The performance of both on Flow 1 is similar, while on Flow 2, CHARM performs better than AARF. But overall, the throughput provided by both in this scenario is not ideal.

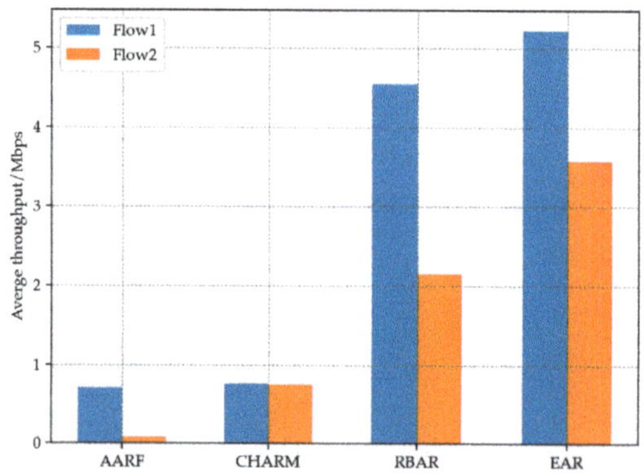

Figure 7. Throughput performance of different algorithms in hidden terminal scenario.

4.3. Comprehensive Scenario

In the comprehensive testing scenario, we constructed a basic BSS (Basic Service Set) network consisting of one AP and 20 STA. In this network, STA cannot communicate directly with each other and must transmit data through AP. The distribution range of STA and AP is in a rectangular area of 300 m × 300 m. The experimental topology is shown in the following Figure 8.

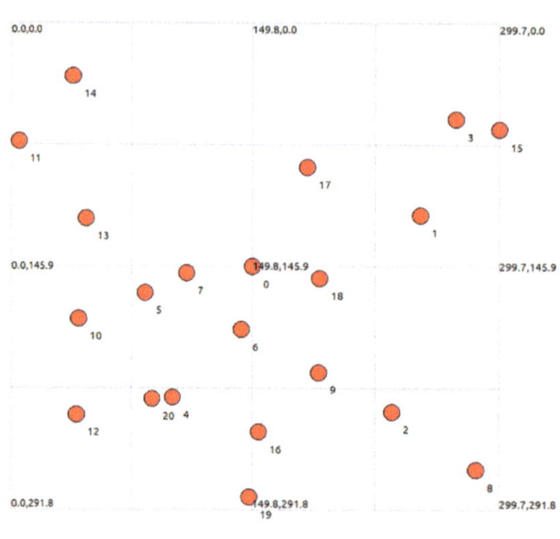

Figure 8. Topological structure of the experimental area.

AP is always at a fixed position in the center of the region (150, 150, 0), and STA is randomly distributed within the region. Each STA continuously generates a 100 Mbps CBR data stream. At the same time, in order to be closer to the real scene, the LogDistancePropagationLossModel is introduced in the environment to perform signal fading on STA nodes at different distances.

4.3.1. Different STA Quantities

This group of experiments verified the impact of different STA numbers on WLAN system throughput by setting different numbers of STA access AP in the above areas. Figure 9 shows the performance of four rate adaptive algorithms, EAR, AARF, CHARM and RBAR, as the number of STA nodes increases. It can be seen that as the number of STA sites increases, the average throughput shows a downward trend. This is because as the number of STA increases, the likelihood of collisions between STAs also increases. At the same time, the average time for each STA to access the AP will also decrease, which will lead to a decrease in system average throughput. Among them, the AARF algorithm is the most sensitive to changes in the number of STAs, and its throughput decreases the fastest. The EAR algorithm, due to the addition of the RTS adaptive mechanism, exhibits relatively good stability and maintains relatively high throughput even when the number of STAs is high. Overall, the RBAR algorithm exhibits the most stable throughput performance during the process of node increase, and when the number of nodes exceeds 13, the average throughput exceeds that of the EAR algorithm. This is due to the RBAR algorithm using RTS for channel reservation before each packet is sent. As the number of STAs in WLAN continues to increase, the probability of packet collisions increases, and the benefits brought by RTS become more apparent. When the number of STAs is less than 10, the throughput performance of CHARM and RBAR is extremely similar. However, as the number of STAs increases, the fading trend of CHARM is more pronounced than that of EAR and RBAR due to the lack of collision detection mechanism.

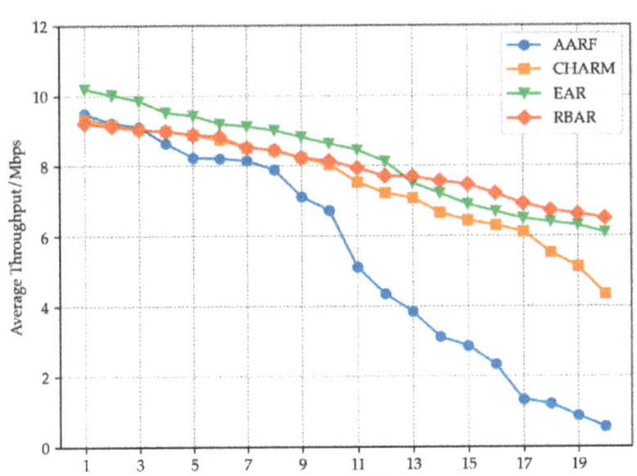

Figure 9. Adding nodes in the WLAN, the average throughput changes of different algorithms.

4.3.2. Different Mobile States

The movement of nodes can cause multipath effects, leading to changes in the channel environment [30]. Figure 10 shows the SNR curves corresponding to two different node movement speeds in the simulation experiment. It can be seen that as the speed of node movement increases, the fluctuation of SNR becomes more severe, channel fading also becomes more severe, and the overall channel environment becomes worse. This is because the movement of nodes can cause changes in frequency and phase, leading to signals traveling through different channel propagation paths. Due to different path lengths and scattering characteristics, signal synthesis or interference may occur, thus affecting the effectiveness of the channel. When the node moves faster, the time interval between adjacent signal transmissions is shorter, and the overlap and interference between signals become more significant, thereby more severely affecting channel quality.

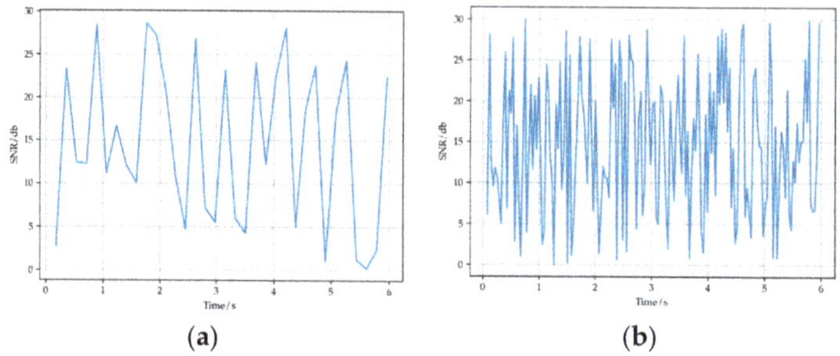

Figure 10. (a) Changes in SNR over time when the movement speed is slow. (b) Changes in SNR over time when the movement speed is fast.

By configuring the RandomWalk2dMobility model of NS-3, STA nodes were randomly moved within the aforementioned 300 × 300 m area and simulated two different motion states: When the motion speed is 0.5 m/s, the state of slow node movement is simulated; and when the motion speed is 5 m/s, the state of rapid node movement is simulated.

Figure 11 shows the average throughput variation of STA at different speeds and distances. It can be seen that when the STA mobility speed is slow, the EAR algorithm has higher throughput than AARF, RBAR and CHARM algorithms at different distances, and as the distance increases, the advantage is most obvious when reaching a distance of 30 m. This is because when the channel state of STA is relatively stable, the EAR algorithm tends to adopt a high transmission rate and maintain a relatively stable rate. The dynamic window and threshold update strategy ensure that the EAR algorithm can accurately capture changes in the channel. In high-speed mobile mode, the throughput of all four algorithms decreases sharply, with an average throughput only reaching about half of that at low speeds. At 10 m, 20 m and 30 m, the throughput performance of EAR algorithm and CHARM algorithm is similar thanks to CHARM's channel evaluation and rate adaptation mechanism. AARF and RBAR perform poorly in this scenario, only reaching about half of the throughput of EAR algorithm.

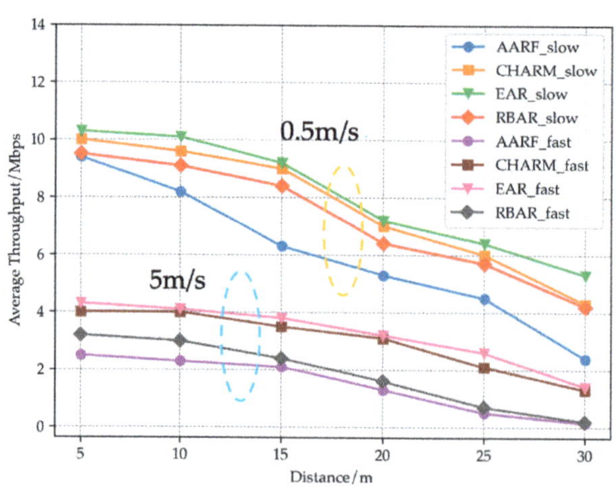

Figure 11. Comparison of throughput of four algorithms at different speeds and distances.

Based on mobile scenarios, compared to the CHARM, RBAR and AARF algorithms, EAR has increased UDP's throughput by 7%, 25% and 51%, respectively, improving the throughput performance of wireless networks in mobile environments.

5. Conclusions

The proposed EAR rate adaptation algorithm, as outlined in this article, effectively identifies the source of channel degradation and differentiates whether the degradation is due to hidden terminal or low SNR. Diverse rate adjustment measures can be implemented based on the various channel environments. The dynamic window and threshold adjustment mechanism significantly enhance the real-time and accuracy of rate adjustment compared to certain conventional rate adaptive algorithms that are currently available. However, there are still shortcomings in the current work. The design concept of EAR is based on the characteristics of traditional standards, such as 802.11a/b/g. There has been no targeted adaptation and optimization of relevant standards, such as 802.11n/ac/ax. This is also a common limitation of many rate adaptation algorithms at present. In future work, EAR will make corresponding improvements based on the characteristics of different standards. For example, the 802.11n standard has added new features, such as aggregate frames, MIMO (Multiple Input Multiple Output), channel bonding and ShortGI (Short Guard Interval). Additionally, 802.11 n allows for many more rate options than 802.11a/b/g mode, ranging from 6.5 Mbps to 600 Mbps. As the rate increases, loss does not monotonically grow with rates in different modes. Therefore, future work has to take these additional factors into account. In addition, 802.11e enhances its support for QoS (Quality of Service) by setting priorities, which can effectively ensure the quality of service for different priority services. In 802.11e devices, the competition unit of the channel in EDCA (Enhanced Distributed Channel Access) mode changes from a set of data to a transmission opportunity. Therefore, the unit of the statistical window in EAR also changes from the number of data packets to a period of time. Additionally, threshold initialization and threshold update strategies are implemented under four different AC (Access Category) conditions to meet the requirements for QoS.

Author Contributions: Conceptualization, W.Y. and L.W.; methodology, W.Y. and L.W.; software, W.Y.; validation, W.Y. and J.S.; formal analysis, W.Y., L.W. and L.H.; investigation, W.Y.; resources, W.Y., L.W. and Y.W.; data curation, W.Y. and J.S.; writing—original draft preparation, W.Y.; writing—review and editing, L.W., L.H. and Y.W.; visualization, W.Y.; supervision, L.W.; project administration, W.Y.; funding acquisition, L.W., L.H. and Y.W. All authors have read and agreed to the published version of the manuscript.

Funding: This research was supported in part by the National Natural Science Foundation of China (61901388, 62002293), in part by the Fundamental Research Funds for the Central Universities, in part by the China Postdoctoral Science Foundation (BX20200280), in part by the Basic Research Programs of Taicang (TC2021JC25), and in part by the Natural Science Basic Research Program of Shaanxi Province (2022JQ-615).

Institutional Review Board Statement: Not applicable.

Informed Consent Statement: Not applicable.

Data Availability Statement: Not applicable.

Conflicts of Interest: The authors declare no conflict of interest.

References

1. Ngo, T. Why Wi-Fi stinks—And how to fix it. *IEEE Spectr.* **2016**, *53*, 44–50. [CrossRef]
2. Cuzanauskas, T. Hybrid rate control mechanisms in IEEE 802.11 networks. In Proceedings of the 2017 Open Conference of Electrical, Electronic and Information Sciences (eStream), Vilnius, Lithuania, 27 April 2017.
3. Sharma, S.; Singh, A.; Tripathi, P.N. Performance of Wireless Ad Hoc Network with All the Data Rate of IEEE 802.11b Using NS-3. In *VLSI, Microwave and Wireless Technologies: Select Proceedings of ICVMWT 2021*; Springer Nature Singapore: Singapore, 2023; pp. 421–430.

4. Mahesh, M.; Pavan, B.S.; Harigovindan, V.P. Data Rate-Based Grouping to Resolve Performance Anomaly of Multi-Rate IEEE 802.11ah IoT Networks. *IEEE Netw. Lett.* **2020**, *2*, 166–170. [CrossRef]
5. Ma, Y.H.; So, P.L.; Gunawan, E. Performance Analysis of OFDM Systems for Broadband Power Line Communications Under Impulsive Noise and Multipath Effects. *IEEE Trans. Power Deliv.* **2005**, *20*, 674–682. [CrossRef]
6. Goodman, D.J.; Saleh, A. The near/far effect in local ALOHA radio communications. *IEEE Trans. Veh. Technol.* **1987**, *36*, 19–27. [CrossRef]
7. Valenti, M.C.; Torrieri, D.; Talarico, S. Adjacent-Channel Interference in Frequency-Hopping Ad Hoc Networks. In Proceedings of the IEEE International Conference on Communications, Budapest, Hungary, 9–13 June 2013.
8. Wang, Z.; Cao, Y.; Zhang, D.; Hua, X.; Gao, P.; Jiang, T. User Selection for MIMO Downlink With Digital and Hybrid Maximum Ratio Transmission. *IEEE Trans. Veh. Technol.* **2021**, *70*, 11101–11105. [CrossRef]
9. Gu, C.; Xu, H.; Yao, N.; Jiang, S.; Zheng, Z.; Xu, Y. Enabling A MAC Protocol with Self-Localization Function to Solve Hidden & Exposed Terminal Problems in Wireless Ad Hoc Networks. *IEICE Trans. Fundam. Electron. Commun. Comput. Sci.* **2022**, *E105/A*, 613–621.
10. Yin, W.; Hu, P.; Indulska, J.; Portmann, M.; Mao, Y. MAC-Layer Rate Control for 802.11 Networks: Lesson Learned and Looking Forward. *arXiv* **2018**, arXiv:1807.02827.
11. Albar, R.; Arif, T.Y.; Munadi, R. Modified Rate Control for Collision-Aware in Minstrel-HT Rate Adaptation Algorithm. In Proceedings of the 2018 International Conference on Electrical Engineering and Informatics (ICELTICs), Banda Aceh, Indonesia, 19–20 September 2018.
12. Kamerman, A.; Monteban, L. WaveLAN-II: A high-performance wireless LAN for the unlicensed band. *Bell Labs Tech. J.* **1997**, *2*, 118–133. [CrossRef]
13. Lacage, M.; Manshaei, M.H.; Turletti, T. IEEE 802.11 rate adaptation: A practical approach. In Proceedings of the 7th ACM international symposium on Modeling, analysis and simulation of wireless and mobile systems, Venice, Italy, 4–6 October 2004; pp. 126–134.
14. Bicket, J. Bit-Rate Selection in Wireless Networks. Ph.D. Thesis, Massachusetts Institute of Technology, Cambridge, MA, USA, 2006.
15. Holland, G.; Vaidya, N.; Bahl, P. A rate-adaptive MAC protocol for multi-Hop wireless networks. *Proc. Acm/IEEE Mobicom.* **2001**, *41*, 236–251.
16. Sadeghi, B.; Kanodia, V.; Sabharwal, A.; Knightly, E. Opportunistic media access for multirate ad hoc networks. In Proceedings of the 8th annual international conference on Mobile computing and networking, Atlanta, GA, USA, 23–28 September 2002; pp. 24–35.
17. Judd, G.; Wang, X.; Steenkiste, P. Efficient channel-aware rate adaptation in dynamic environments. In Proceedings of the International Conference on Mobile Systems, Breckenridge, CO, USA, 17–20 June 2008; pp. 118–131.
18. Kim, J.; Kim, S.; Choi, S.; Qiao, D. CARA: Collision-Aware Rate Adaptation for IEEE 802.11 WLANs. In Proceedings of the Infocom IEEE International Conference on Computer Communications, Barcelona, Spain, 23–29 April 2006.
19. Zhang, J.; Tan, K.; Zhao, J.; Wu, H.; Zhang, Y. A Practical SNR-Guided Rate Adaptation. In Proceedings of the IEEE INFOCOM 2008-The 27th Conference on Computer Communications, Phoenix, AZ, USA, 13–18 April 2016; pp. 2083–2091.
20. Holl, G.; Vaidya, N.; Bahl, P. *A Rate-Adaptive MAC Protocol for Wireless Networks*; Texas A & M University: College Station, TX, USA, 2000.
21. Arif, T.Y.; Munadi, R. Evaluation of the minstrel-HT rate adaptation algorithm in IEEE 802.11n WLANs. *Int. J. Simul. Syst. Sci. Technol.* **2017**, *18*, 11.11–11.17. [CrossRef]
22. Zhao, Y.; Kang, K.; Qian, H.; Luo, X.; Jin, Y. Greedy Hybrid Rate Adaptation in Dynamic Wireless Communication Environment. In Proceedings of the ICASSP 2020–2020 IEEE International Conference on Acoustics, Speech and Signal Processing (ICASSP), Barcelona, Spain, 4–8 May 2020.
23. Khan, S.; Mahmud, S.A.; Noureddine, H.; Al-Raweshidy, H.S. Rate-adaptation for multi-rate IEEE 802.11 WLANs using mutual feedback between transmitter and receiver. In Proceedings of the IEEE International Symposium on Personal Indoor & Mobile Radio Communications, Istanbul, Turkey, 26–30 September 2010.
24. Park, C.S.; Park, S. Implementation of a Fast Link Rate Adaptation Algorithm for WLAN Systems. *Electronics* **2021**, *10*, 91. [CrossRef]
25. Huang, T.; Li, S.; Lu, X.; Gao, S. An Interference-Aware Rate and Channel Adaptation Scheme for Dense IEEE 802.11n Networks. *Wirel. Commun. Mob. Comput.* **2019**, *2019*, 1902463. [CrossRef]
26. Petermann, M.; Wübben, D.; Kammeyer, K.-D. Evaluation of encoded MU-MISO-OFDM systems in TDD mode with non-ideal channel reciprocity. In Proceedings of the 2010 International ITG Conference on Source and Channel Coding (SCC), Siegen, Germany, 18–21 January 2010; pp. 1–6.
27. Rathod, V.; Tahiliani, M.P. Geometric Series based effective RTO estimation Technique for CoCoA. *Ad Hoc Netw.* **2022**, *130*, 102801. [CrossRef]
28. Yin, W.; Hu, P.; Indulska, J.; Bialkowski, K. Performance of mac80211 rate control mechanisms. In Proceedings of the International Symposium on Modeling Analysis & Simulation of Wireless & Mobile Systems, Miami, FL, USA, 31 October–4 November 2011; pp. 427–436.

29. Khaled, Z.E.; Ajib, W.; Mcheick, H. Log Distance Path Loss Model: Application and Improvement for Sub 5 GHz Rural Fixed Wireless Networks. *IEEE Access* **2022**, *10*, 52020–52029. [CrossRef]
30. Zhang, Z.; Chen, B.; Yang, M. Moving Target Detection Based on Time Reversal in a Multipath Environment. *IEEE Trans. Aerosp. Electron. Syst.* **2021**, *57*, 3221–3236. [CrossRef]

sensors

MDPI

Article

Study of Interference Detection of Rail Transit Wireless Communication System Based on Fourth-Order Cyclic Cumulant

Peng Wang [1], Junliang Yao [2,*], Yong Pu [2], Shuyuan Zhang [2] and Lu Wen [3]

[1] China Railway First Survey and Design Institute Group Co., Ltd., Xi'an 710043, China; th_wangpeng6.yy@crcc.cn

[2] School of Automation and Information Engineering, Xi'an University of Technology, Xi'an 710048, China; 17726694754@163.com (Y.P.); zsyxaut@163.com (S.Z.)

[3] State Key Laboratory of Rail Transit Engineering Informatization (FSDI), Xi'an 710043, China; wenlu.yy@crcc.cn

* Correspondence: yaojunliang@xaut.edu.cn

Abstract: The wireless communication system is used to provide dispatching, control, communication and other services for rail transit operations. In practice, interference from other wireless communication systems will affect the normal operation of trains, so it is an urgent problem to study the interference detection algorithms for rail transit applications. In this paper, the fourth-order cyclic cumulant (FOCC) of signals with different modulation modes is analyzed for the narrow-band wireless communications system of rail transit. Based on the analysis results, an adjacent-frequency interference detection algorithm is proposed according to the FOCC of the received signal within the predetermined cyclic frequency range. To detect interference with the same carrier frequency, a same-frequency interference detection algorithm using the relationship between the FOCC and the received power is proposed. The performance of the proposed detection algorithms in terms of correct rate and computational complexity is analyzed and compared with the traditional second-order statistical methods. Simulation results show that when an interference signal coexists with the expected signal, the correct rates of the adjacent-frequency and the same-frequency interference detection algorithms are greater than 90% when the signal-to-noise ratio (SNR) is higher than 2 dB and −4 dB, respectively. Under the practical rail transit wireless channel with multipath propagation and the Doppler effect, the correct rates of the adjacent-frequency and the same-frequency interference detection algorithms are greater than 90% when the SNR is higher than 3 dB and 7 dB, respectively. Compared with the existing second-order statistical methods, the proposed method can detect both the adjacent-frequency and the same-frequency interference when the interference signals coexist with the expected signal. Although the computational complexity of the proposed method is increased, it is acceptable in the application of rail transit wireless communication interference detection.

Keywords: rail transit wireless communications; interference detection; cyclic cumulant; multipath propagation; Doppler effect

1. Introduction

Urban rail transit is a high-capacity and high-efficiency mode of public transportation. In recent years, China's rail transit industry has achieved rapid development [1]. By the end of 2021, a total of 9192.62 km of urban rail transportation routes were in operation in more than 50 cities, 78.9% of which are metro lines. As an important component of the urban rail transit system, wireless communication not only connects various types of equipment but also provides effective control to ensure the normal operation of the train. Currently, the main standards and wireless frequency spectrum used in urban rail transit mainly include 800 MHz TETRA (Terrestrial Trunked Radio) [2] used for vehicle-to-ground wireless communication trunk dispatching systems; 2.4 GHz and 5.8 GHz WiFi used for passenger information systems [3]; and 1.8 GHz LTE-M (LTE-Machine-to-Machine) [4] used for communication-based train control (CBTC) systems [5]. Among them, 2.4 GHz and 5.8 GHz are public frequency bands, which are inevitably interfered with by external user's equipment. The 1.8 GHz LTE-M is a specially used system. In addition to rail transit, it is also used in industries such as electric power, airports, petroleum, etc. Therefore, it could be interfered with by the same-frequency signal. What is more, the 1.8 GHz signal will also be interfered with by the telecom operators' adjacent-frequency systems, such as the China Telecom FDD-LTE and China Mobile DCS1800 systems. If the interference power exceeds a certain threshold, the communication service will be interrupted, and the train will be emergency braked. Therefore, interference signal detection is important to ensure the safe and stable operation of trains [6,7].

Classical interference detection algorithms include time-domain energy detection [8–10], frequency-domain detection [11–13], time–frequency-domain detection [14,15], array-based signal processing [16], etc. Guo proposed a blind interference detection algorithm based on an energy detector [10], which is very simple and widely used. However, the performance of the algorithm is greatly affected by the estimation of the noise power, and the estimation error will cause the algorithm's accuracy to decrease. In addition, the time-domain method lacks information on the interference frequency and cannot locate the interference source, so it is not effective to eliminate the interference of the rail transit wireless communication system. The continuous mean elimination (CME) algorithm for interference detection based on frequency-domain analysis has been proposed [12], which sets the detection threshold based on minimizing the false-alarm probability. The frequency-domain method can not only judge the presence of interference signals but also obtain the location information of interference frequency points, which makes up for the deficiency of the time-domain method. However, the CME algorithm requires a large carrier frequency interval between the interference signal and the expected signal, so it is unable to detect the same-frequency interference that often occurs in the rail transit wireless communication system.

In fact, it is not enough to analyze the characteristics of the time domain or frequency domain independently, and the relationship between them should also be explored. With the help of Wigner–Ville distribution [17] and short-time Fourier transform, a time–frequency-domain-based interference detection method is proposed [18], but the algorithm complexity is too high for online use. In addition, array signal processing methods can also be used for interference detection in multi-antenna systems [19,20]. Due to the limited number of antenna elements in railway communication terminals, the traditional array signal processing method cannot be applied.

To address these problems, the single-channel source number estimation and signal separation algorithms based on a higher-order cyclic cumulant are presented in [21,22]. With only one antenna at the receiver, the observation matrix is represented by a single parameter, and an evaluation function is designed using the properties of the high-order cumulant to obtain the optimal value of this parameter [21]. After recognizing the observation matrix, four effective feature parameters of the high-order cumulant were extracted, and single-channel time-frequency overlapping signals were separated through signal reconstruction [22]. The method has good anti-noise performance and does not require estimation of the prior information such as noise power, so it is suitable for interference detection in rail transit wireless communications. In this paper, a fourth-order cyclic cumulant (FOCC)-based interference detection method is proposed. First, the single-channel time-frequency overlapping signal model for rail transit downlink communication is described. Second, the detection algorithms for both the adjacent-frequency and the same-frequency interference are implemented. By calculating the FOCC of the received signal and judging the amplitude of the cyclic cumulant at different cyclic frequencies, the interference signals can be detected. Finally, simulation tests under typical rail transit wireless channels, such as two paths and large Doppler frequencies, are performed to verify the effectiveness of the proposed methods. We highlight our main contributions in the following summary of the paper:

- Two FOCC-based interference detection algorithms are proposed for application in the train transit wireless communication scene, which is susceptible to both adjacent-frequency interference and same-frequency interference.
- The correct rate and computational complexity of the proposed algorithms are analyzed and compared with the traditional second-order statistical methods.
- The impact of multipath and Doppler on algorithm performance is analyzed through simulation tests.
- The advantages of the proposed algorithms include the following:
  (1) The FOCC of the received time-frequency overlapping signal is used as a feature for interference detection. Compared with the traditional second-order statistical properties, the anti-noise performance is better because it is 0 for the cyclic cumulant of white Gaussian noise to be larger than the second order. This is very useful in rail transit wireless communication applications because the background noise varies with the arrival and departure of trains.
  (2) Traditional time-domain energy detectors are only effective during quiet periods, i.e., they can only detect interference when expected signals are not present, and the detection performance largely depends on the accuracy of the noise estimation. However, in trail transit wireless communication scenarios, the interference signals usually coexist with the expected signal, and there are errors in estimating the noise variance. In this case, the energy detectors would no longer be applicable, and the proposed method in this paper has significant advantages over the existing one.
  (3) The traditional frequency-domain method can only detect adjacent-frequency interference. The proposed methods can detect both adjacent-frequency interference and same-frequency interference because the spectral power of the expected signal is considered to detect same-frequency interference.
  (4) In a rail transit wireless communication scenario, the multipath propagation and Doppler effect will lead to the performance degradation of interference detection algorithms. The proposed FOCC-based method can overcome the Doppler effect because its influence on the cyclic frequencies can be ignored. Although the proposed method is also affected by multipath propagation, it still works well at medium SNRs.

2. Signal Model

In this section, we define the signal received by the downlink communication as a single-channel time-frequency overlapping signal, since the terminal is usually equipped with a single antenna in the rail transit wireless communication. In practice, rail traffic has a dedicated frequency band for communication, so the probability of multiple interference sources occurring simultaneously is low. In this paper, it is assumed that the maximum number of source signals will not be more than three; then, the received signal can be expressed as follows:

$$x(t) = \sum_{i=1}^{r} s_i(t) + n(t) \tag{1}$$

where $r \leq 3$ is the number of source signals, and $s_i(t) = A_i(t)e^{j(2\pi f_{ci}t+\theta_i(t)+\phi_i)}$ represents the ith source signal, where $A_i(t)$, f_{ci}, $\theta_i(t)$ and ϕ_i are the signal amplitude, carrier frequency, modulation phase and initial phase, respectively. $n(t)$ is the additive white Gaussian noise.

Due to the complexity of the railway wireless communications environment, the influence of multipath and Doppler on interference detection should also be considered. In the presence of the multipath, the received signal contains multiple signals at the same frequency and at different power levels, making it more difficult to detect interfering signals.

Multipath signals have different carrier phases and delays corresponding to the main path signal, and the received signal through a multipath fading channel can be modeled as follows [23]:

$$x(t) = \sum_{i=1}^{r}\sum_{l=1}^{L} h_l s_i(t-\tau_l)e^{j2\pi f_d t} + n(t) \tag{2}$$

where L is the number of paths, and h_l and τ_l are the attenuation coefficient and the time delay corresponding to path l from the transmitter to the receiver. $f_d = f_c \cdot v/c$ is the Doppler shift, where v is the speed at which the train is traveling with an average speed of 30 m/s, and c is the speed of light with a magnitude of 3×10^8 m/s. For simplicity, this paper assumes that the carrier synchronization of the received signal $x(t)$ is perfect, and only the strongest path information is considered, ignoring other weaker path signals; that is, the following research considers only the two-path situation for the detection of interfering signals.

3. Algorithm Principle and Implementation

3.1. Fourth-Order Cyclic Cumulant

In this section, the FOCC of various modulated signals will be derived. According to the definition, the kth-order cyclic moment of the cyclic smooth signal $x(i), i = 0, 1, \ldots, N-1$ is calculated as follows [24]:

$$\begin{aligned} &M_{kx}^{\alpha}(\tau_1, \tau_2, \ldots, \tau_{k-1}) \\ &= \frac{1}{T}\sum_{i=0}^{N-1} x(i)x(i+\tau_1)\cdots x(i+\tau_{k-1})e^{-j2k\pi\alpha t} \\ &= \left\langle x(i)x(i+\tau_1)\cdots x(i+\tau_{k-1})e^{-j2k\pi\alpha t}\right\rangle_t \end{aligned} \tag{3}$$

where k is the order of the cyclic moment, $\tau_1, \tau_2, \ldots, \tau_{k-1}$ is the fixed time delay, and α is the cyclic frequency. N is the number of sample points during the time period T, and $\langle\cdot\rangle_t$ represents the time averaging operation. According to the transfer relation between the cyclic moment and the cyclic cumulant, the FOCC $C_{40,x}^{\alpha}(\tau_1, \tau_2, \tau_3)$ of the signal $x(t)$ can be expressed as [24]

$$\begin{aligned} C_{40,x}^{\alpha}(\tau_1, \tau_2, \tau_3) &= M_{4x}^{\alpha}(\tau_1, \tau_2, \tau_3) \\ &-M_{2x}^{\alpha}(\tau_1)M_{2x}^{\alpha}(\tau_3-\tau_2) \\ &-M_{2x}^{\alpha}(\tau_2)M_{2x}^{\alpha}(\tau_1-\tau_3) \\ &-M_{2x}^{\alpha}(\tau_3)M_{2x}^{\alpha}(\tau_2-\tau_1) \end{aligned} \tag{4}$$

when $\tau_1 = \tau_2 = \tau_3 = 0$,

$$\begin{aligned} C_{40,x}^{\alpha}(0,0,0) &= M_{4x}^{\alpha}(0,0,0) - 3M_{2x}^{\alpha}(0)M_{2x}^{\alpha}(0) \\ &= \langle x^4(t)e^{-j8\pi\alpha t}\rangle_t - 3\langle x^2(t)e^{-j4\pi\alpha t}\rangle_t^{\,2} \end{aligned} \tag{5}$$

Since it is 0 for the cyclic cumulant of white Gaussian noise to be larger than the second order, the FOCC of the noise in Equation (1) is 0. Therefore, the method has better anti-noise performance. Then, the FOCC of the time-frequency overlapping signal in Equation (1) is

$$\begin{aligned} C_{40,x}^{\alpha}(0,0,0) &= \sum_{i=1}^{r} C_{40,s_i}^{\alpha}(0,0,0) + C_{40,n}^{\alpha}(0,0,0) = \sum_{i=1}^{r} C_{40,s_i}^{\alpha}(0,0,0) \\ &= \sum_{i=1}^{r}\langle s_i{}^4(t)e^{-j8\pi\alpha t}\rangle_t - \sum_{i=1}^{r} 3\langle s_i{}^2(t)e^{-j4\pi\alpha t}\rangle_t^{\,2} \\ &= \sum_{i=1}^{r}\left\langle A_i{}^4(t)e^{j4(\pi(2f_{ci}-2\alpha)t+\theta_i+\phi_i)}\right\rangle_t - \sum_{i=1}^{r} 3\left\langle A_i{}^2(t)e^{j2(\pi(2f_{ci}-2\alpha)t+\theta_i+\phi_i)}\right\rangle_t^{\,2} \end{aligned} \tag{6}$$

When the cyclic frequency α is equal to the carrier frequency f_{ci} of the ith signal component, the signal modulation is BPSK, QPSK, 8 QAM or 16 QAM, and the initial phase ϕ_i obeys the uniform distribution in the range of $[0, 2\pi]$, Equation (6) can be derived as

$$\begin{aligned} C_{40,x}^{\alpha=f_{ci}}(0,0,0) &= \left\langle A^4(t)e^{j4(\pi(2f_{ci}-2\alpha)t+\theta(t)+\phi)}\right\rangle_t - 3\left\langle A^2(t)e^{j2(\pi(2f_{ci}-2\alpha)t+\theta(t)+\phi)}\right\rangle_t^2 \\ &= A^4e^{j4\phi}\left\langle e^{j4\theta(t)}\right\rangle_t - 3A^2e^{j4\phi}\left\langle e^{j2\theta(t)}\right\rangle_t^2 \\ &= \begin{cases} -2A^4e^{j4\phi}, & \text{BPSK} \\ A^4e^{j4\phi}, & \text{QPSK} \\ 0.34A^4e^{j4\phi}, & \text{8QAM} \\ 0.68A^4e^{j4\phi}, & \text{16QAM} \end{cases} \end{aligned} \tag{7}$$

For the cyclic frequency not equal to the carrier frequency, Equation (6) can be derived as

$$\begin{aligned} C_{40,x}^{\alpha\neq f_{ci}}(0,0,0) &= \left\langle A_i^4(t)e^{j4(\pi(2f_{ci}-2\alpha)t+\theta_i(t)+\phi_i)}\right\rangle_t - 3\left\langle A_i^2(t)e^{j2(\pi(2f_{ci}-2\alpha)t+\theta_i(t)+\phi_i)}\right\rangle_t^2 \\ &= \left\langle A_i^4(t)e^{j4\phi_i}\cdot e^{j8\pi(f_{ci}-\alpha)t+4\theta_i(t)}\right\rangle_t - 3\left\langle A_i^2(t)e^{j2\phi_i}\cdot e^{j4\pi(f_{ci}-\alpha)t+2\theta_i(t)}\right\rangle_t^2 \\ &= 0 \end{aligned} \tag{8}$$

Figure 1a–d show the amplitude spectrum $\left|C_{40,x}^{\alpha}(0,0,0)\right|$ with respect to the cyclic frequency α for BPSK, QPSK, 8 QAM and 16 QAM modulated signals, respectively. The carrier frequencies of the modulated signals are 5 MHz, the symbol rate is 0.5 Mbps, and the SNR is 4 dB.

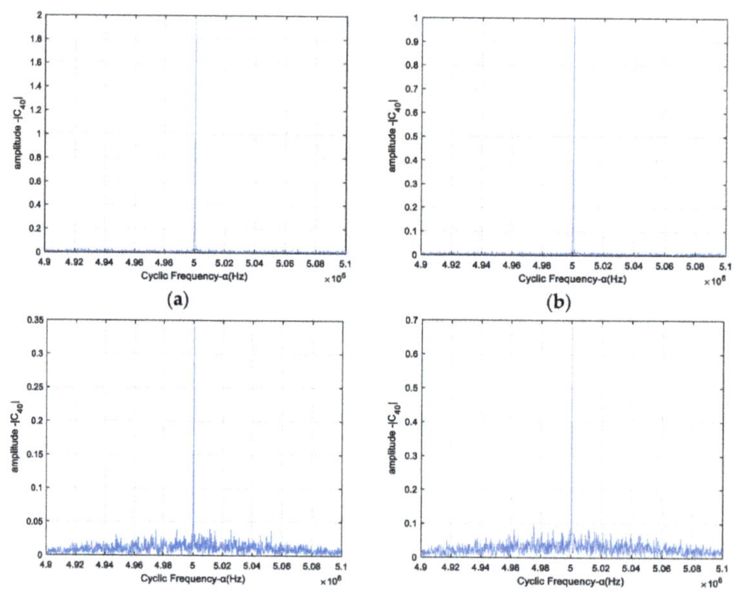

Figure 1. Amplitude spectrum of the FOCC of a single modulated signal. (**a**) BPSK, (**b**) QPSK, (**c**) 8 QAM, (**d**) 16 QAM.

We can see from Figure 1 that the amplitude is not zero when the cyclic frequency is equal to the carrier frequency, and the amplitude values are different for different modulation modes. The amplitudes of the FOCC of a single BPSK, QPSK, 8 QAM and 16 QAM modulated signal are 2, 1, 0.34 and 0.68, respectively. When the cyclic frequency is not equal to the carrier frequency, the amplitude of FOCC is close to zero. These results are consistent with the theoretical analysis of Equation (7), which provides a basis for judging the number of mixed signals.

3.2. Detection Algorithm for the Adjacent-Frequency Interference

To detect the adjacent-frequency interference, the source signals with different carrier frequencies are mixed, and the FOCC of the mixed signal is analyzed.

In the case of $r = 2$, Figure 2a shows the FOCC amplitude of the time-frequency overlapping signals with respect to the cyclic frequency. The modulation modes of two source signals are BPSK and QPSK, and the carrier frequencies are 5 MHz and 5.01 MHz, respectively. The symbol rates of both source signals are 0.5 Mbaud, and the SNR is 4 dB. We can see that, when the carrier frequencies are different, the FOCCs of the mixed signals have discrete spectral lines at the carrier frequencies of each source. The amplitudes for the BPSK and QPSK signals are similar to the results in Figure 1. The amplitudes at non-carrier frequencies are slightly higher than the results in Figure 1 due to the mutual interference between two source signals. In the case of $r = 3$, the modulation modes of three source signals are BPSK, QPSK and 16 QAM, and the carrier frequencies are 5 MHz, 5.01 MHz and 5.02 MHz, respectively. From Figure 2b, we can see that the amplitudes at carrier frequencies are similar to the results in Figure 2a. The amplitudes at non-carrier frequencies are higher than those in Figure 2a because the mutual interference increases. The FOCCs of two sources with the same modulation order 16 QAM and different carrier frequencies are shown in Figure 2c, and the FOCCs of three sources with the same modulation order QPSK and different carrier frequencies are shown in Figure 2d. We can see that, when the carrier frequencies are different, the FOCCs of the mixed signals have discrete spectral lines at the carrier frequencies of each source. The amplitudes for each signal are similar to the results in Figure 1.

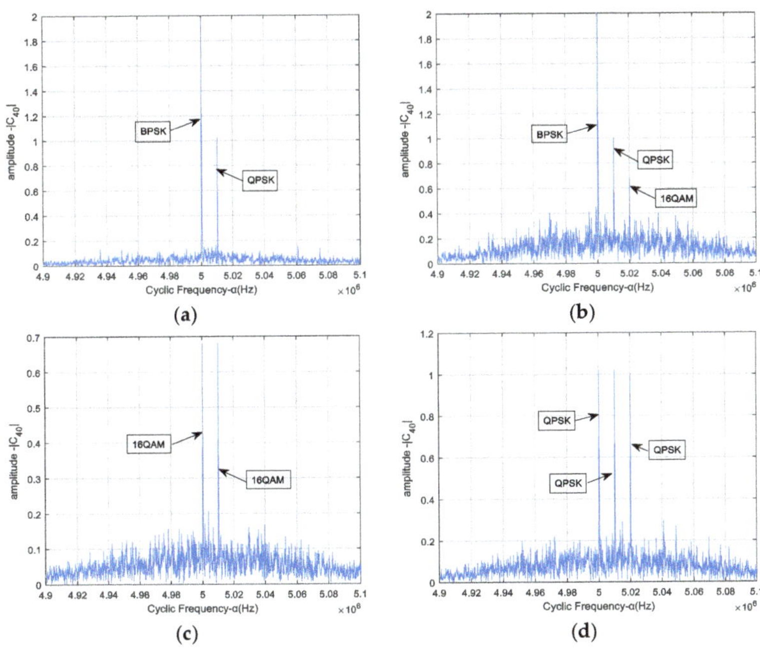

Figure 2. Amplitude spectrum of FOCC of mixed signals with different carrier frequencies. (**a**) BPSK&QPSK, (**b**) BPSK&QPSK&16 QAM, (**c**) 16 QAM&16 QAM, (**d**) QPSK&QPSK&QPSK.

Based on the above analysis results, the adjacent-interference detection algorithm is proposed by determining the number of discrete spectral lines in the amplitude spectrum. The adjacent-frequency interference exists when the number of discrete spectral lines is not equal to the number of the expected signal, which is known in advance. The steps of the algorithm are described as follows (Algorithm 1).

Algorithm 1 FOCC-based Adjacent-Frequency Interference Detection Algorithm

(Step 1) The power of the received mixed signal $x(t)$ is normalized, and the power spectrum is calculated to determine the search range of the cyclic frequency α.
(Step 2) The FOCC of $x(t)$ is calculated, and its amplitude spectrum for cyclic frequencies $\alpha - \left|C_{40,x}^{\alpha}(0,0,0)\right|$ is obtained.
(Step 3) A suitable threshold is set to calculate the number of discrete spectral lines in the amplitude spectrum. In this paper, the threshold is set to 0.5 according to the analysis of Equation (7).
(Step 4) The adjacent-frequency interference occurs when the number of discrete spectral lines is greater than the number of expected signals; otherwise, there is no adjacent-frequency interference.

3.3. Detection Algorithm for the Same-Frequency Interference

For adjacent-frequency interference, the FOCCs of different carrier frequencies are separated. The above-proposed algorithm is not applicable to the detection of same-frequency interference because the carrier frequencies of the expected and the interference signals are the same and there is only one discrete spectral line in the amplitude spectrum. In this subsection, the relationship between the FOCC and the received power is analyzed, which is used to design the detection algorithm for the same-frequency interference.

In practice, the power of the received signal can be obtained, and the case of one interference signal is considered in the analysis. The case of more interference signals is similar. We denote the received power as $k = k_1 + k_2$, where k_1 is the power of the expected signal, and k_2 is the power of the interference signal. Assuming that the modulation of the expected signal is known in advance, the FOCCs $C_{40,x}^{\alpha}(0,0,0)$ of the received signal $x(t)$ at the cyclic frequency $\alpha = f_{ci}$ are shown in Table 1.

Table 1. Relationship between the FOCC and the received power of the received signal.

	Without Interference	With One BPSK Interference	With One QPSK Interference
Expected signal is BPSK	$C_{40,x}^{\alpha=f_{ci}} = 2k_1^2 = 2k^2$	$C_{40,x}^{\alpha=f_{ci}} = 2(k_1^2 + k_2^2)$	$C_{40,x}^{\alpha=f_{ci}} = 2k_1^2 + k_2^2$
Expected signal is QPSK	$C_{40,x}^{\alpha=f_{ci}} = k_1^2 = k^2$	$C_{40,x}^{\alpha=f_{ci}} = k_1^2 + 2k_2^2$	$C_{40,x}^{\alpha=f_{ci}} = k_1^2 + k_2^2$
Expected signal is 16 QAM	$C_{40,x}^{\alpha=f_{ci}} = k_1^2 = 0.68k^2$	$C_{40,x}^{\alpha=f_{ci}} = 0.68k_1^2 + 2k_2^2$	$C_{40,x}^{\alpha=f_{ci}} = 0.68k_1^2 + k_2^2$

We can see that if the received power is fixed as k, $C_{40,x}^{\alpha=f_{ci}}$ varies with the interference power k_2 and the modulation modes. If the expected signal is BPSK modulation, and the FOCC without interference is larger than the FOCC with interference, then an appropriate threshold can be set to distinguish whether there is interference. If $C_{40,x}^{\alpha=f_{ci}}$ is greater than the threshold, there is no interference, and vice versa. Considering the noise effect, the threshold σ_1 can be designed according to the noise power, and the interference detection is given by

$$\text{Expected signal is } BPSK \Rightarrow \begin{cases} H_0, \left(C_{40,x}^{\alpha=f_{ci}} > 2k^2 - \sigma_1\right) \\ H_1, (otherwise) \end{cases} \tag{9}$$

In Equation (9), H_1 represents the presence of interference, and H_0 represents no interference. Considering the noise influence and the decision distance in the first row of Table 1, the threshold σ_1 in Equation (9) is set to 0.5 in the following algorithm simulation.

If the expected signal is QPSK modulation, the FOCC without interference $C_{40,x}^{\alpha=f_{ci}} = k^2$ is not always the largest. Similarly, a threshold σ_2 can be designed, and the interference detection is given by

$$\text{Expected signal is } QPSK \Rightarrow \begin{cases} H_0, \left(k^2 - \sigma_2 < C_{40,x}^{\alpha=f_{ci}} < k^2 + \sigma_2\right) \\ H_1, (otherwise) \end{cases} \tag{10}$$

The meanings of H_0 and H_1 are the same as those described after Equation (9). The threshold σ_2 in Equation (10) is set to 0.1 in the following algorithm simulation.

The procedure for deriving the FOCCs for BPSK/QPSK/16 QAM modulated signals is given above. For received signals with multipath Doppler effects, the signal model is shown in Equation (2), and the FOCC of the received signal with multipath Doppler effects is given by

$$\begin{aligned} C_{40,x}^{\alpha}(0,0,0) &= \left\langle \left(\sum_{i=1}^{2} \sum_{l=1}^{2} h_i s_i(t) e^{j2\pi f_{di} t} \right)^4 e^{-j8\pi\alpha t} \right\rangle_t - 3 \left\langle \left(\sum_{i=1}^{2} \sum_{l=1}^{2} h_i s_i(t) e^{j2\pi f_{di} t} \right)^2 e^{-j4\pi\alpha t} \right\rangle_t^2 \\ &= \left\langle \sum_{i=1}^{2} \sum_{l=1}^{2} {h_i}^4 A^4 a^4(t) e^{j8\pi(f_{ci}+f_{di}-\alpha)t} \right\rangle_t - 3 \left\langle \sum_{i=1}^{2} \sum_{l=1}^{2} {h_i}^2 A^2 a^2(t) e^{j4\pi(f_{ci}+f_{di}-\alpha)t} \right\rangle_t^2 \end{aligned} \tag{11}$$

From Equation (11), it can be seen that when the cyclic frequency is equal to the sum of the carrier frequency and the Doppler frequency, there is a non-zero value of the FOCC of the received signal. Since the Doppler frequency is much smaller than the carrier frequency, i.e., $f_c + f_d \approx f_c$, we see that the FOCC value is not zero when the cyclic frequency is equal to the carrier frequency. However, due to the multipath, the amplitude of FOCC changes, and simultaneously, the signal power decreases, but the overall nature of the FOCC does not change. In summary, the presence of a multipath Doppler effect in the received signal does not significantly alter the nature of the FOCC, making it feasible to use the FOCC for interference detection, and subsequent simulations have confirmed this conclusion.

3.4. Computational Complexity

In this subsection, the computational complexity of the proposed method is analyzed and compared with the traditional interference detection algorithms based on second-order statistical properties. The time complexity, which describes the number of elementary operations needed to perform a method, is used to measure the computational complexity. In general, we focus on the asymptotic behavior of the computational workload as the input size increases. The time complexity $T(N)$ is usually expressed as $T(N) = O(N)$, where N is the input size. To be impartial, the time complexity of the three methods is analyzed below.

The computational complexity of all algorithms includes statistics calculation complexity and interference decision complexity. The computational complexity comparison of the three methods is shown in Table 2. For the proposed FOCC-based algorithm, the statistics calculation process is shown in Equation (5), and the input sizes are the number of symbols in one data frame and the length of the cyclic frequency. If N is the number of symbols, and M is the length of the cyclic frequency, then the statistics calculation complexity is $O(MN^5)$. To detect an interference, a suitable threshold is preset, and the interference decision complexity is $O(M)$. For traditional second-order statistical methods, the energy detector in Ref. [10] and the frequency-domain CME algorithm in Ref. [12] are considered. In the energy detector, the energy of the received signal is calculated; then, the statistics calculation complexity is $O(N^2)$. To detect an interference, the noise variance is estimated to set the threshold, and the interference decision complexity is $O(N^2)$. In the frequency-domain CME algorithm, the fast Fourier transform (FFT) of the received signal is performed, and the statistics calculation complexity is $O(N\log N)$. To detect an interference, the continuous mean elimination with P iterations is performed, and the interference decision complexity is $O(PN)$.

Table 2. Computational complexity comparison of three methods.

Methods	Statistics Calculation	Interference Decision
The proposed algorithm	$O(MN^5)$	$O(M)$
Energy detector in Ref. [10]	$O(N^2)$	$O(N^2)$
CME algorithm in Ref. [12]	$O(N\log N)$	$O(PN)$

In practice, the variables mentioned above usually satisfy $500 \leq N \leq 1000$, $100 \leq M \leq 200$, and $100 \leq P \leq 500$. From Table 2, we summarize that the proposed FOCC-based algorithm has higher statistics calculation complexity and lower interference decision complexity than the traditional methods. Although the total computational complexity is increased, it is acceptable in the application of rail transit wireless communication interference detection.

4. Simulation Experiments and Result Analysis

In this section, the proposed interference detection algorithms are evaluated using three simulation experiments, where the performance metric is the correct rate of the detection algorithms. In each experiment, 1000 independent trials are performed, and the correct rate is defined as the ratio of the number of correct trials to the total number of trials.

In Experiment A, the correct rate of the proposed adjacent-frequency interference detection algorithm is simulated for different numbers of interference signals. The traditional energy detector method in Ref. [10] and the frequency-domain CME algorithm in Ref. [12] are also tested for comparison. In Experiment B, the correct rate of the proposed same-frequency interference detection algorithm is tested. In the rail transit scenario, the wireless signals will experience multipath propagation and the Doppler effect, due to reflections in station buildings and high-speed train movement. Therefore, the algorithm performances in a wireless channel with multipath and Doppler are evaluated in Experiment C. All the simulation experiments are implemented on a PC using MATLAB R2016a.

Experiment A. Correct rate of the adjacent-frequency interference detection algorithm

In this experiment, the algorithm described in Section 3.2 is simulated, and three source signals, with carrier frequencies 5 MHz/5.01 MHz/5.02 MHz and modulation modes BPSK/QPSK/16 QAM, are considered. To evaluate the performance of different numbers of interference signals, the received signal is mixed with any two or three source signals, where one is the expected source and the others are interference sources. In our simulation, there are 800 bits in one data frame, and the symbol rate is set to 0.5 Mbps for all the source signals. Assuming that the signal power is $P_s = 1$, and the SNR range is [−2:0.5:1, 2:1:8], the simulation results are obtained by averaging over 500 independent data frames. To illustrate the effect of different combinations of signals, three sets of signals with the same modulation and two sets of signals with different modulations were selected for testing.

The results in Figure 2 show that the FOCC values at the carrier frequencies are determined by the modulation modes of the signals. On the other hand, the FOCC values at the non-carrier frequencies depend on the noise and bit sequences. In order to estimate the number of carriers more accurately, the FOCC at the non-carrier frequencies is reduced by averaging over multiple calculations, assuming that the noise and bit sequences for each source have a random equal probability distribution.

The case of two sources, one being the expected signal and the other being the interference signal, is considered. The correct rate of the proposed adjacent-frequency interference detection algorithm is shown in Figure 3a. It is seen that the correct rate of the proposed method is greater than 90% when the SNR is higher than 2 dB, and the expected signal is BPSK or QPSK. The correct rate decreases at the same SNR when both the expected signal and the interference signal are 16 QAM because the value of the FOCC is affected by the modulation order. It is easier to see that the lower the modulation order, the higher the value of the FOCC. When the case of three sources, one being the expected signal and the other two being the interference signals, is considered, the correct rate of the proposed adjacent-frequency interference detection algorithm is shown in Figure 3b. It can be seen that as the number of overlapping signals increases, the correct rate of the proposed method decreases due to the greater influence of the interference term. The correct rate is 100% for BPSK mixed signals, and the results for QPSK and 16 QAM are 90% and 75%, respectively.

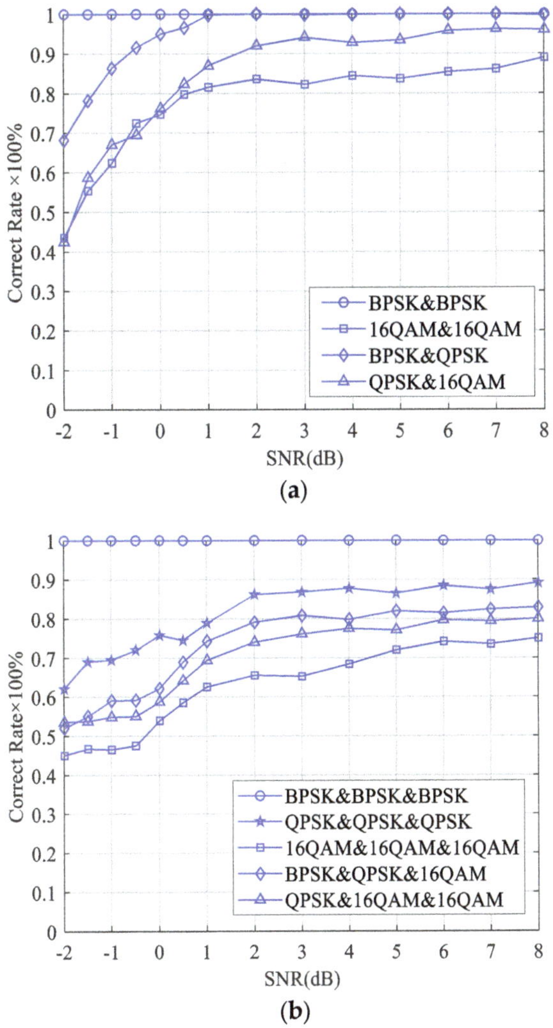

Figure 3. Correct rate of the adjacent-frequency interference detection algorithm. (**a**) One expected signal and one interference signal. (**b**) One expected signal and two interference signals.

It is worth noting that the curves in Figure 3 are obtained by averaging over 500 independent trials. In each trial, the expected signal and interference signal are randomly selected from the source signals. Therefore, it may not be easy to distinguish the expected source and interference source in the legend of Figure 3.

Because the energy detector is only suitable for interference detection during the quiet period when the expected signal is not present, in order to fairly compare the proposed algorithm with existing algorithms, we consider the situation that the interference signal occurs with a probability of 0.5, and the expected signal does not exist. The simulation results of the proposed adjacent-frequency interference detection algorithm, the time-domain energy detector in Ref. [10], and the frequency-domain CME algorithm in Ref. [12] are shown in Figure 4 for comparison.

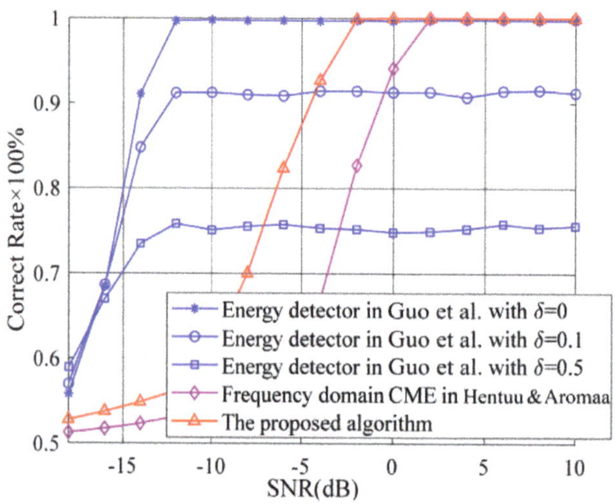

Figure 4. Correct rate comparison with only the interference signal. Guo et al., 2015 [10]; Hentuu & Aromaa, 2002 [12].

The decision threshold of the energy detector is set based on the noise variance estimation, and the estimation error will deteriorate the detection performance. Assuming that the error of the noise variance is denoted by δ, the correct rate curves of the energy detector with $\delta = 0$, $\delta = 0.1$ and $\delta = 0.5$ are given in Figure 4 by the blue solid lines. We can see that the energy detector has a very good detection performance when $\delta = 0$. As the noise estimation error increases, the correct rate decreases significantly due to the deviation of the decision threshold. For both the frequency-domain CME algorithm and the proposed algorithm, the correct rates are greater than 50% even in the low-SNR region because they can always correct when there is no interference. The correct rate of the proposed algorithm reaches 100% when SNR = −2 dB, which is 4 dB lower than the result of the frequency-domain CME algorithm.

As we can see from Figure 4, the proposed method has a poor performance compared to a traditional energy detector in the low-signal-to-noise ratio region. However, the energy detector is only suitable for detecting the interference signal during the quiet period, and the algorithm performance is deteriorated by the noise variance estimation error. The proposed method has a wider applicability and does not require estimation of noise variance.

Experiment B. Correct rate of the same-frequency interference detection algorithm

In this experiment, the algorithm described in Section 3.3 is simulated. Three source signals, with the same carrier frequency of 5 MHz are considered. The received signal contains one expected signal and no more than two interference signals. The power of each signal is random, and the sum power of the received signal is normalized to 1. The other experimental conditions are identical to Experiment A, and the results are shown in Figure 5.

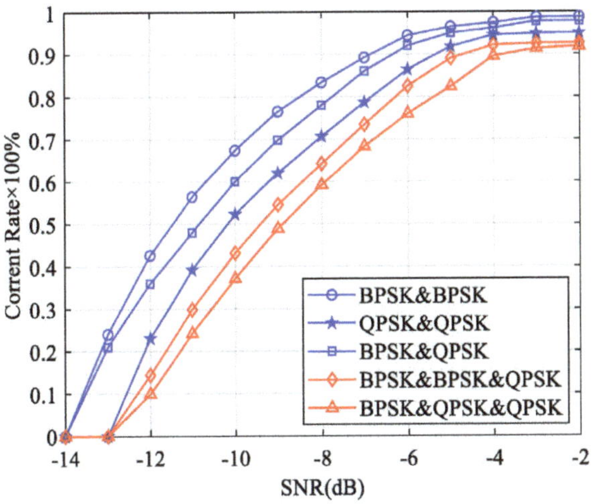

Figure 5. Correct rate of the same-frequency interference detection algorithm.

We can see from Figure 5 that the correct rate of the same-frequency interference detection is greater than 90% when the SNR is higher than −4 dB, which is better than the results of the adjacent-frequency interference detection in Figure 3. This is because the adjacent-frequency interference signals have a different frequency component compared with the expected signal, resulting in an interference term between signals. The interference term will cause the FOCC to have a non-zero value at the non-cyclic frequency, which will affect the spectral line detection at the cyclic frequencies. In the case of same-frequency interference, the FOCC of the received signal at the cyclic frequency is not affected by other frequency components, and the relationship between the received signal power and the FOCC is used to limit the influence of the interference term.

Experiment C. Interference detection in multipath and Doppler environments

In the rail transit scenario, such as a subway or high-speed train, the wireless channel is mostly single-path or two-path. Meanwhile, the received signal suffers from the Doppler effect when the train is running at high speed [25]. Therefore, it is useful to analyze the performance of interference detection algorithms in multipath and Doppler environments. In this experiment, a two-path propagation channel with $h_1 = 1$, $h_2 = 0.6$, $\tau_1 = 0$, $\tau_2 = 1$ and the speed of the train of 30 m/s is considered, and one expected source and one interference source exist with the same parameters as in the previous experiments.

The FOCC of the adjacent-frequency interference case with BPSK and QPSK modulation is shown in Figure 6a, where the power of both signals is normalized. Compared with the results of the single-path case in Figure 2a, the amplitude decreases at carrier frequencies and increases at other frequencies due to the multipath effect. The carrier frequencies have changed due to the Doppler effect, but the changes are only 10^{-7} of the carrier frequencies and are not obvious in the figure. The FOCC of the same-frequency interference case with BPSK and QPSK modulation is shown in Figure 6b. Compared with the result in Figure 3a, it can be seen that the FOCC is affected by the multipath propagation and almost not affected by the Doppler.

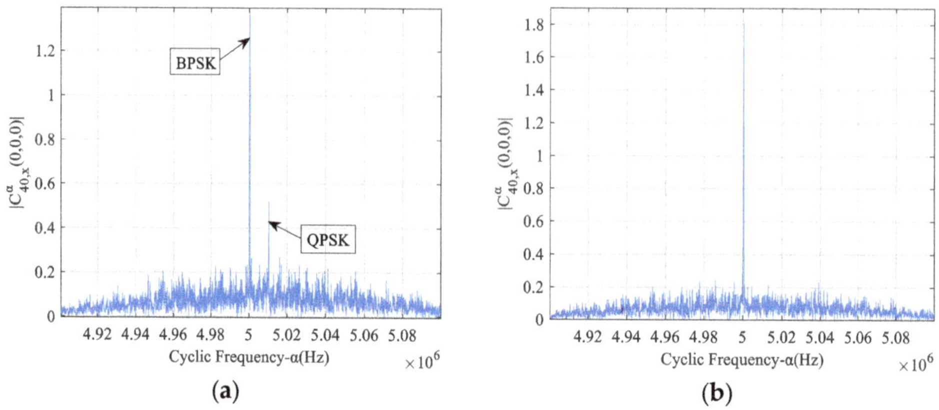

Figure 6. FOCC of received signal with multipath and Doppler effect. (**a**) The case of adjacent-frequency interference. (**b**) The case of same-frequency interference.

The correct rate of the adjacent-frequency interference detection algorithm in multipath and Doppler environments is shown in Figure 7, with single-path results shown for comparison. It can be seen that the correct rate in a multipath channel is worse than the results in a single path. The reason is that the amplitude of the FOCC at the carrier frequencies decreases due to the power reduction of each signal component when there is a multipath. And the amplitude increases in non-carrier frequencies due to multipath interference, which results in a decrease in the accuracy of the detection of the number of carriers. For low-order modulation, such as BPSK&QPSK, the correct rate decreases from 100% to 91% at SNR = 3 dB. For high-order modulation, such as QPSK&16 QAM, the correct rate decreases from 95% to 89% at the same SNR. These results indicate that the adjacent-interference detection algorithm based on the FOCC is also applicable under multipath and Doppler channels.

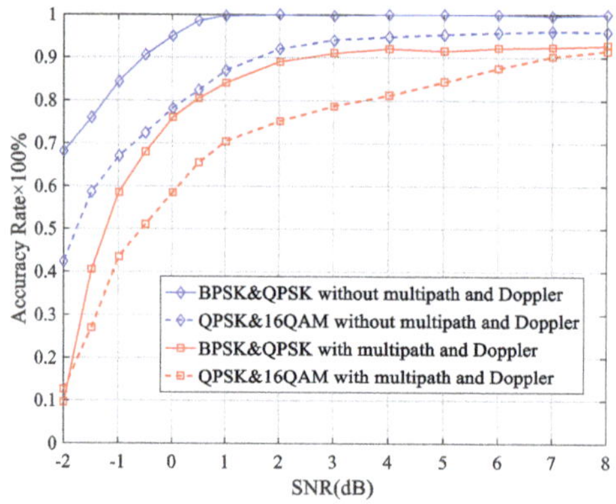

Figure 7. Correct rate of the adjacent-frequency interference detection algorithm in multipath and Doppler environments.

The correct rate of the same-frequency interference detection algorithm in multipath and Doppler environments is shown in Figure 8. It shows that the multipath has a large impact on the same-frequency interference detection algorithm when compared to the results in Figure 5. This is because the signal components propagated by the multipath can be considered as new sources of interference, and the relationship between the FOCC and the received power is changed. Even so, the correct rate is more than 80% when the SNR is greater than 3 dB and can reach 90% when the SNR is greater than 7 dB. This ensures that the algorithm is available in practice.

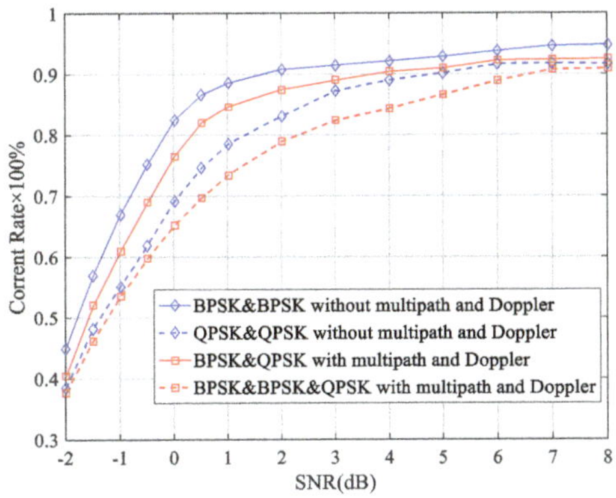

Figure 8. Correct rate of the same-frequency interference detection algorithm in multipath and Doppler environments.

5. Conclusions

For rail transit wireless communication systems, the interference signal has a great impact on the normal operation of trains. Minor interference will lead to a decrease in throughput, and severe interference will lead to communication interruption and train shutdown. In this paper, two detection algorithms are proposed for both adjacent-frequency interference and same-frequency interference, and simulation results validate their effectiveness. To reduce the influence of noise, the FOCC of the received signal is analyzed, and the theoretical derivation for different modulation modes, including BPSK, QPSK and 16 QAM, is performed. The results show that adjacent-frequency interference can be detected by identifying the number of spectral lines that is higher than a preset threshold, and the correct rate is more than 90% when one interference is present and the SNR is greater than 2 dB. Same-frequency interference can be detected based on the relationship between the FOCC and received power, and the correct rate is more than 90% when one interference is present and the SNR is greater than −4 dB. Compared with the traditional second-order statistical methods, the proposed method can detect both adjacent-frequency interference and same-frequency interference. In addition, the performance of both algorithms is affected by multipath and Doppler propagation. However, the correct rate can reach 90% when the SNR is greater than 7 dB. Thus, interference detection based on the FOCC is proved to be effective for the wireless communication of rail transit systems.

Author Contributions: Conceptualization, P.W. and J.Y.; Data curation, P.W., J.Y. and L.W.; Formal analysis, J.Y.; Funding acquisition, J.Y.; Investigation, P.W., J.Y. and Y.P.; Methodology, P.W., J.Y. and Y.P.; Project administration, L.W.; Resources, P.W.; Software, Y.P. and S.Z.; Supervision, P.W. and J.Y.; Validation, P.W., J.Y., Y.P. and L.W.; Visualization, P.W., J.Y. and Y.P.; Writing—original draft, J.Y., Y.P. and S.Z.; Writing—review and editing, P.W., J.Y. and Y.P. All authors have read and agreed to the published version of the manuscript.

Funding: This work was supported in part by the National Natural Science Foundation of China under Grant 62073258 and in part by the Natural Science Basic Research Program of Shaanxi Province under Grant 2020JM-456 and Grant 2021JQ-487.

Institutional Review Board Statement: Not applicable.

Informed Consent Statement: Not applicable.

Data Availability Statement: Not applicable.

Conflicts of Interest: The authors declare no conflict of interest.

References

1. Cheng, P.F.; Wang, H.Y.; Yang, G.J. Integration of digital economy and rail transportation industry: Realistic logic, path and countermeasures. *Bus. Stud.* **2022**, *29*, 84–97.
2. He, J.H. Application of 800MHz wireless digital trunking communication system in Qinhuangdao port. *Port Technol.* **2015**, *1*, 33–36.
3. Guan, M.D.; Liu, J.H.; Cai, B. 2.4 G and 5.8 G wireless communication systems used in subways. *Railr. Commun. Signal Eng. Technol.* **2008**, *1*, 37–39.
4. Dai, K.P. Anti-interference research of LTE-M vehicle-ground wireless communication system. *Railw. Commun. Signal* **2017**, *53*, 47–50.
5. Pochet, J.; Baro, S.; Sandou, G. Automatic train supervision for a CBTC suburban railway line using multi-objective optimization. In Proceedings of the 2017 IEEE 20th International Conference on Intelligent Transportation Systems (ITSC), Yokohama, Japan, 16–19 October 2017; pp. 1–6.
6. Du, H.; Wen, C.; Li, W. A new method for detecting and early-warning in-band interference of the GSM-R network. In Proceedings of the 2017 IEEE 17th International Conference on Communication Technology (ICCT), Chengdu, China, 27–30 October 2017; pp. 800–804.
7. Zhang, Z.; He, X.; Yuan, H. A robust parking detection algorithm against electric railway magnetic field interference. *IEEE Trans. Intell. Transp. Syst.* **2020**, *21*, 882–893. [CrossRef]
8. Mohine, S.; Bansod, B.S.; Kumar, P. Time domain energy detector approach for detection of moving ground vehicle. In Proceedings of the 2019 IEEE 5th International Conference for Convergence in Technology (I2CT), Pune, India, 29–31 March 2019; pp. 1–4.
9. Gismalla, E.H.; Alsusa, E. Performance analysis of the periodogram-based energy detector in fading channels. *IEEE Trans. Signal Process.* **2011**, *59*, 3712–3721. [CrossRef]
10. Guo, H.; Jiang, W.; Luo, W. A modified energy detector for random signals in Gaussian noise. *IEEE Commun. Lett.* **2015**, *19*, 1358–1361. [CrossRef]
11. Jin, E.S.; Park, H. A novel frequency domain preamble detection for OFDM based WLAN systems. In Proceedings of the 2020 International Conference on Information and Communication Technology Convergence (ICTC), Jeju Island, Republic of Korea, 21–23 October 2020; pp. 1494–1498.
12. Henttu, P.; Aromaa, S. Consecutive mean excision algorithm. In Proceedings of the IEEE Seventh International Symposium on Spread Spectrum Techniques and Applications, Prague, Czech Republic, 2–5 September 2002; pp. 450–454.
13. Vartiainen, J. Always one/zero malicious user detection in cooperative sensing using the FCME method. In Proceedings of the 2012 7th International ICST Conference on Cognitive Radio Oriented Wireless Networks and Communications (CROWNCOM), Stockholm, Sweden, 18–20 June 2012.
14. Li, G.; Liu, N.; Zhang, C. Research and design of a transmission system for time-frequency-domain electromagnetic method. *IEEE Access* **2019**, *7*, 153999–154007. [CrossRef]
15. Xing, L.; Wen, Y.; Thomas, D.W.P. A Joint time-frequency analytical method for electromagnetic interference in railway GNSS system. In Proceedings of the 2020 International Symposium on Electromagnetic Compatibility, EMC EUROPE, Rome, Italy, 23–25 September 2020; pp. 1–4.
16. Yadav, S.K.; George, N.V. Coarray MUSIC-Group delay: High-resolution source localization using non-uniform arrays. *IEEE Trans. Veh. Technol.* **2021**, *70*, 9597–9601. [CrossRef]
17. Zhang, Z.C. New Wigner distribution and ambiguity function based on the generalized translation in the linear canonical transform domain. *Signal Process.* **2016**, *118*, 51–61. [CrossRef]
18. Mhiri; Korso, M.E.; Breloy, A. Multifrequency array calibration in presence of radio frequency interferences. *Signal Process. Off. Publ. Eur. Assoc. Signal Process. (EURASIP)* **2022**, *199*, 108613–108622.
19. Huang, L.; Wu, S.; Li, X. Reduced-rank MDL method for source enumeration in high-resolution array processing. *IEEE Trans. Signal Process.* **2007**, *55*, 5658–5667. [CrossRef]
20. Fernández-Menduiña, S.; Krahmer, K.; Leus, G. Computational array signal processing via modulo non-linearities. *IEEE Trans. Signal Process.* **2022**, *70*, 2168–2179. [CrossRef]
21. Liu, M.Q. Research on Key Technologies for Signal Detection and Modulation Identification in Non-Cooperative Communication. Ph.D. Thesis, Xidian University, Xi'an, China, 2013.
22. Zhang, L.H.; Qiu, T.S.; Li, J.C. A blind source separation method based on fourth-order cyclic cumulant. *Signal Process.* **2011**, *27*, 1412–1415.
23. Zhan, Y.; Zeng, G.; Duan, C. A high order statistics based multipath interference detection method. *IET Commun.* **2021**, *15*, 1586–1596. [CrossRef]
24. Sun, C.Q.; Liu, K.; Wang, P.Q. Blind frequency offset estimation algorithm for the same-frequency 16QAM signal based on higher-order cumulants. *Electron. Meas. Technol.* **2019**, *42*, 85–90.
25. Bao, C.F. Research on Doppler Frequency Shift Estimation Algorithm for High-Speed Railroad Wireless Communication System. Master's Thesis, Lanzhou Jiaotong University, Lanzhou, China, 2014.

sensors

MDPI

Article

An Overview of Cognitive Radio Technology and Its Applications in Civil Aviation

Ruikang Zheng [1], Xuan Li [2,*] and Yudong Chen [1]

[1] China Airport Planning & Design Institute Co., Ltd. Northwest Branch, Xi'an 710075, China; zhengrk@cacc.com.cn (R.Z.); xbchenyd@cacc.com.cn (Y.C.)
[2] School of Information Engineering, East China Jiaotong University, Nanchang 333000, China
* Correspondence: lixuan@ecjtu.edu.cn

Abstract: This paper provides an overview of cognitive radio technology and its applications in the field of civil aviation. Cognitive radio technology is a relatively new and emerging field that allows for dynamic spectrum access and efficient use of spectrum resources. In the context of civil aviation, cognitive radio technology has the potential to enable more efficient use of the limited radio spectrum available for communication and navigation purposes. This paper examines the current state of cognitive radio technology, including ongoing research and development efforts, regulatory issues, and potential challenges to widespread adoption. The potential applications of cognitive radio technology in civil aviation are also explored, including improved spectrum utilization, increased safety and security, and enhanced situational awareness. Finally, the paper concludes with a discussion of future research directions and the potential impact of cognitive radio technology on the future of civil aviation. It is hoped that this paper will serve as a useful resource for researchers, engineers, and policy makers interested in the emerging field of cognitive radio technology and its potential applications in the field of civil aviation.

Keywords: cognitive radio; applications and trends; challenges; civil aviation

1. Introduction

Since the concept of cognitive radio (CR) was proposed by Joseph Mitola III in 1998, it has attracted lots of attention in scientific research organizations and institutions. In recent years, due to the increasing shortage of radio spectrum resources, CR technology came into being. Through the perception and analysis of radio spectrum, cognitive radio can access the idle spectrum resource opportunity for data transmission, improve spectrum utilization, and reduce spectrum waste. Therefore, cognitive radio technology is widely used in military, commercial, and civil aviation, and other fields.

However, with the continuous development of cognitive radio technology, there still exist some problems and challenges. For example, further research is needed on how to improve the accuracy of spectrum sensing and how to achieve dynamic spectrum access. In addition, the development and research of the integration of cognitive radio and existing wireless communication systems are also an important research direction.

In order to solve the above challenges, lots of efforts have been made in the CR research area, including basic theory, radio frequency (RF) front end, spectrum sensing, adaptive modulation wave technology, smart antenna, etc. Furthermore, studies have also explored multiple input multiple output (MIMO) and orthogonal frequency division multiplexing (OFDM) technologies to improve the efficiency and reliability.

As a radio communication system based on artificial intelligence technology, CR can automatically analyze different frequencies through the cognition of radio signals and realize intelligent switching and automatic modulation of radio signals. In this light, it has been widely used in many fields, including aviation, automobiles, smart homes, Internet of

Things, and its potential application scenarios are very wide, which are mainly discussed in this paper.

The basic principles of cognitive radio technology can be summarized as follows: first, the system needs to use certain sensors and computer hardware to collect and process radio signals. Then, it also has to process and convert the signal data collected by sensors and computer hardware in order to achieve automatic recognition and analysis of the signal. Finally, automatic selections of the system involve the corresponding modulation method, demodulation method, frequency, and so on based on the recognized signal type.

In the application of cognitive radio technology, different frequencies and signals need to be identified and analyzed, which requires special processing and conversion of radio signals. This processing and conversion require the use of special sensors and computer hardware, so the cost of cognitive radio technology is high. In addition, cognitive radio technology needs to comply with specific laws and regulations in the application process, which also requires system manufacturers and users to comply with relevant regulations and standards to ensure the stability and safety of the system.

The applications of CR technology in civil aviation, which are discussed in this paper, improve spectral efficiency, communication reliability, and flexibility. CR can perceive and monitor spectrum usage and identify idle frequency bands or shareable spectrum resources. Firstly, the available spectrum can be managed and planed effectively, ensuring that spectrum resources between different communication systems have no interference with each other. Secondly, the available spectrum can be dynamically allocated to different communication systems and users with real-time requirements. Finally, spectrum sharing among multiple communication systems can also be achieved, thereby improving spectrum utilization efficiency and system capacity.

In the field of civil aviation, the application of cognitive radio technology will be widely promoted. For example, airports, airlines, and navigation equipment manufacturers need to monitor and manage radio signals to ensure safe communication and flight operations. This kind of monitoring and management needs to identify and analyze radio signals, and cognitive radio technology is able to meet this demand.

It can be seen that CR technology can play a crucial role in the civil aviation field by enabling intelligent monitoring and management of radio signals. It will significantly enhance the safety and efficiency of flight operations, providing a reliable guarantee for smooth and secure flights. Moreover, cognitive radio technology introduces innovative ideas and technical solutions for intelligent monitoring and management of wireless electrical signals, paving the way for advanced advancements in this domain.

The organization of this article is as follows: Section 2 provides an overview of the fundamentals of cognitive radio technology, including the architecture of a cognitive radio network. In Section 3, we present an overview of the existing enabling techniques for CR communications. Section 4 comprehensively analyzes the current research in cognitive radio. The applications and trends of cognitive radio networks are discussed in Section 5. Section 6 highlights the specific challenges that cognitive radio faces in today's context. The application of cognitive radio in the field of civil aviation is discussed in Section 7. Finally, Section 8 concludes this paper, summarizing the key findings and contributions.

2. Fundamentals

2.1. Main Characteristics of Cognitive Radio

Wireless networks face a significant challenge of limited energy and bandwidth, which restricts the quality of service and channel capacity. To address this issue, researchers are exploring new communication and networking paradigms that can intelligently and efficiently utilize these scarce resources. Cognitive radio is a crucial technology that can enable future communications and networking to utilize network resources in a flexible and efficient manner. Different from traditional communication approaches, CR can adapt the operating parameters such as transmission power, frequency, and modulation type according to the changes in the surrounding radio environment [1]. Cognitive capability

is the unique characteristic that allows CR devices to acquire necessary information from the radio environment, such as the transmitted waveform, RF spectrum, communication network type, geographic information, locally available resources and services, user needs, security policy, and so on. Once CR devices gather the needed information, it can dynamically adjust the transmission parameters based on the sensed environment changes to achieve optimal performance [2]. The authors of [1,2] considered CR as an opportunistic or dynamic use of existing bands that addresses the analysis of radio environment and estimation of channel state.

2.2. Functions of Cognitive Radio

Cognitive radio operates on a typical duty cycle that involves detecting spectrum idle space, selecting optimal frequency bands, coordinating spectrum access with other users, and vacating the frequency when a primary user (PU) appears. To support this cognitive cycle, the spectrum sensing, management, and sharing are utilized.

Spectrum sensing allows CR to detect spectrum idle space, which is an unused frequency band that can be utilized by secondary users (SUs). When PUs start using the licensed spectrum again, CR can detect their activity through sensing and vacate the frequency to avoid harmful interference.

After detecting the spectrum idle space, the spectrum management function enables SUs to choose the best frequency band and hop among multiple bands to meet various Quality of Service (QoS) requirements [3]. For example, when a PU reclaims their frequency band, the SU can direct their transmission to other available frequencies, based on factors such as noise and interference levels, path loss, channel error rate, and holding time. A brief process of CR is shown in Figure 1.

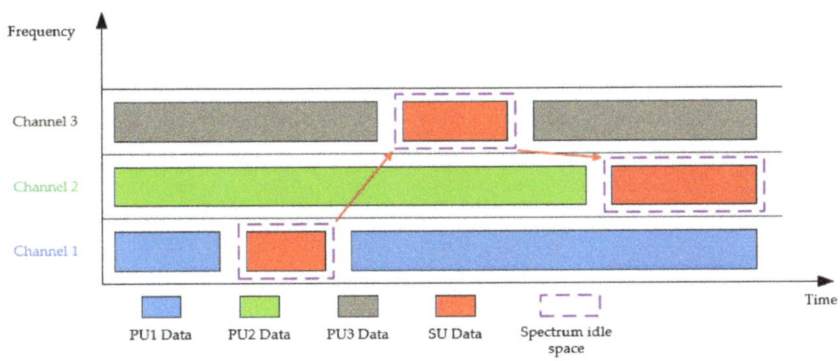

Figure 1. Functional diagram of CR.

Effective spectrum allocation and sharing mechanisms are crucial in dynamic spectrum access to achieve high spectrum efficiency. When SUs coexist with PUs in a licensed band, the interference level due to secondary spectrum usage should be limited by a certain threshold since PUs own the spectrum rights. When multiple SUs share a frequency band, their access should be coordinated to avoid collisions and interference.

2.3. Infrastructure of Cognitive Radio Network

Cognitive radio technologies enable SUs to utilize temporally unused licensed spectrum bands owned by PUs. In a CR network architecture, there are both primary and secondary networks. The secondary network comprises a set of SUs with or without a secondary base station. These users can access the licensed spectrum only when it is not occupied by a PU. The opportunistic spectrum access of SUs is coordinated by a secondary base station, which serves as a hub of the secondary network and is equipped with CR functions. If several secondary networks share a common spectrum band, a central network entity, known as a spectrum broker [4], can coordinate their spectrum usage for efficient and fair spectrum sharing.

The primary network consists of a set of PUs and one or more primary base stations. PUs are authorized to use certain licensed spectrum bands under the coordination of primary base stations. Their transmission should not be interfered with by secondary networks. PUs and primary base stations are generally not equipped with CR functions. Therefore, if a secondary network shares a licensed spectrum band with a primary network, the secondary network needs to immediately detect the presence of a PU and direct its transmission to another available band to avoid interfering with primary transmission.

Cognitive communications, through the sensing, detecting, and monitoring of the surrounding RF environment, increase spectrum efficiency and support higher bandwidth service. Moreover, the real-time autonomous decisions for efficient spectrum sharing reduce the burdens of centralized spectrum management. Therefore, CRs can be employed in many applications.

3. Techniques

3.1. Spectrum Sensing Techniques

Spectrum sensing is a crucial mechanism for obtaining information about the occupancy of the primary spectrum. It aims to identify spectral gaps in different domains such as time, frequency, space, polarization, and angular domains. To achieve this, the secondary transmitter must have an RF chain equipped with sensing capability, which enables it to determine the presence or absence of PUs using various signal processing techniques. Some of the commonly used spectrum sensing techniques include energy detection (ED), feature detection, matched filter-based detection, autocorrelation-based detection, covariance-based detection, and eigenvalue-based detection [5–7].

3.2. Spectrum Management

Dynamic spectrum management is a technique that involves the efficient allocation and reallocation of spectrum resources to optimize their utilization while preventing interference. Cognitive radio devices are capable of dynamically assigning and reassigning spectrum resources based on their availability and the requirements of users, while taking into consideration the existence of PUs. This approach can enhance spectrum efficiency and ensure effective spectrum sharing between primary and SUs [8].

3.3. Spectrum Sharing

This technique involves sharing spectrum resources among multiple users, including both primary and SUs. Cognitive radio devices can share spectrum resources by using techniques such as time division multiple access (TDMA), frequency division multiple access (FDMA), and code division multiple access (CDMA) [9,10].

3.4. Resource Allocation

The technique of resource allocation involves the efficient management of the limited radio resources, including power, bandwidth, and time, among different users to enhance the overall system performance. Resource allocation can be performed centrally or in a distributed manner. With the use of intelligent techniques, cognitive radio can optimize spectrum usage and achieve efficient resource allocation, resulting in improved system performance [11,12].

3.5. Adaptive Modulation and Coding

Adaptive modulation and coding are a technique used to optimize performance and minimize interference by adapting modulation and coding schemes. Cognitive radio devices are capable of adjusting their transmission parameters, such as modulation and coding, based on the quality of the channel and the needs of the users. By employing adaptive modulation and coding techniques, cognitive radio devices can optimize the use of the available spectrum and improve overall system performance.

CR technology is applied to manage radio signals and provide relevant information on communication protocols and security in cognitive radio networks [13,14], which can be adopted to ensure the safety of communication and flight operations in civil aviation. Ref. [15] provided insights into the practical applications and deployments of cognitive networks, including the potential usage in civil aviation for improved communication and operations. The authors of [16,17] discussed the future research directions and challenges in CR technology, which may provide potential applications in the field of civil aviation. However, the security of CR also deserves attention. Ref. [18] focused on security and privacy protocols specifically designed for cognitive wireless sensor networks.

4. Research Status of Cognitive Radio

In [19], the researchers presented a compact, low-cost dual-port filtered antenna system for interleaved cognitive radio applications. The filtered antenna system consists of a frequency bandwidth reconfigurable narrowband filtering antenna (RNF) and a wideband filtering antenna (WF), both of which have a sharp suppression in the stopband. RNF switches between two narrowbands, including WiMAX, 5G, and WLAN frequency bands. The working frequency of WF covers a certain range, which is sufficient to cover the two frequency bands of RNF. The proposed filter produces a flat response in the passband, with two sharp transmission zeros at the stopband.

A two-unit MIMO antenna for cognitive radio MIMO applications was proposed in [20] to avoid the complexity involved in reconfigurable antennas and improve spectral efficiency. The proposed MIMO antenna possesses salient features, such as polarization diversity and performing a maximum of four communication tasks when all the white spaces are detected.

With the discussion of parametric analysis and design and results of the frequency and pattern reconfigurable antenna (FPRA), a new compact reconfigurable antenna was presented in [21]. FPRA has a circular patch with a circular stub on top and a slotted defective ground structure (DGS) on the bottom. In frequency reconstruction mode, FPRA can switch between ultra-wide bandwidth (UWB, 3.05~11.45 GHz), multiband (4 GHz, 6.5 GHz, and 9.2 GHz), and narrowband (7.1 GHz) modes. In UWB and narrowband operations, the radiation pattern is almost omnidirectional, while in broadband situations, the pattern can be reconfigured without affecting the operating frequency band. The reflection coefficient of FPRA is less than −15 dB, and the gain is greater than 2 dBi. The proposed FPRA was processed and validated. The almost constant group delay within the UWB frequency range confirms the excellent time-domain characteristics of FPRA.

A new method to identify unfilled PU spectrum was proposed in [22], which analyzed the collaboration between users using machine learning methods. Learning methods are applied to construct classifiers, appropriate fusion algorithms are selected for the environment under consideration, and efficient out of band sensing is performed. They studied the impact of fading on perceptual performance and found that with the presence of fading, perceptual performance decreases, which is consistent with early research results. From the simulation results, it can be inferred that Weibull fading is superior to all other fading models. Achieving a 1% miss detection probability in the Rayleigh channel yields a false alarm probability of almost 0.8, while achieving the same miss detection probability in the Weibull channel yields a false alarm probability of less than 0.1, which is very beneficial for both indoor and outdoor scenarios. Here, using the time series prediction learning method, a hidden Markov model is used for numerical analysis of the PU channel condition. It is evident that the prediction performance has reached 100% as the result of using the Weibull fading model for a period of 200 ms when compared to the Rayleigh model which achieves only 84.5% accuracy in prediction.

In [23], the author proposed a power allocation method for underlay energy acquisition cognitive radio networks based on supermodel games. An operable market-driven cognitive radio scheme was presented to improve spectrum utilization in radio and television frequency [24].

The early radio navigation equipment can only obtain the single function of the ship's position at sea by receiving radio signals, but with the development of information technology, the radio navigation system equipment in shipping has been effectively upgraded and updated, better meeting the requirements of modern navigation [25].

Ref. [26] introduced the key technologies of radio applications: wireless spectrum, adaptive transmission, cross layer design. The specific applications of GNU radio integration in mobile communication is also proposed, which are described one by one from high-speed rail and military aspects, so as to highlight the functions of cognitive radio technology and explore its development prospects.

As a conclusion, the research directions and contributions of the latest works mentioned above are shown in Table 1.

Table 1. The latest research directions and contributions.

Reference	Research Field	Method	Contribution
Ref. [19]	Filtered antenna system	Dual-port filtered antenna	Flat response in the passband with two sharp transmission zeros at the stopband
Ref. [20]	MIMO antenna system	Two-element MIMO antenna	Possesses salient features, e.g., polarization diversity
Ref. [21]	FPRA for UWB and CR application	A new compact reconfigurable antenna	Suitable for UWB and CR application
Ref. [22]	Idle spectrum prediction	Time series forecasting learning method	Promotion in the prediction performance
Ref. [23]	Power allocation in CR	Supermodel game power allocation method	Energy acquisition allocation scheme
Ref. [24]	CR application	A novel market-driven CR scheme	Improves spectrum utilization in radio and television frequency
Ref. [25]	CR communication and networking	CR schemes in maritime navigation	Proves the possibility of CR in maritime navigation
Ref. [26]	GNU radio and CR application	GNU radio integration in mobile communication	Highlights CR technology and explores its development prospects

5. Applications and Trends in Cognitive Radio Networks

With the continuous development of cognitive radio networks (CRNs), people have gradually deepened their understanding of its development direction. At present, the development directions of cognitive radio technology mainly focus on the following aspects (see Figure 2).

5.1. 5G Communication System

With the emergence of the 5G era, conventional wireless communication technologies are no longer adequate to meet future demands. Cognitive network technology is increasingly gaining traction as a dominant force in the future development trend. As an emerging network communication technology, cognitive networks will have a significant impact on the future of communication. The development direction of cognitive radio in the context of 5G communication systems is focused on achieving higher data transmission rates, better coverage, and more flexible and diverse communication modes [27].

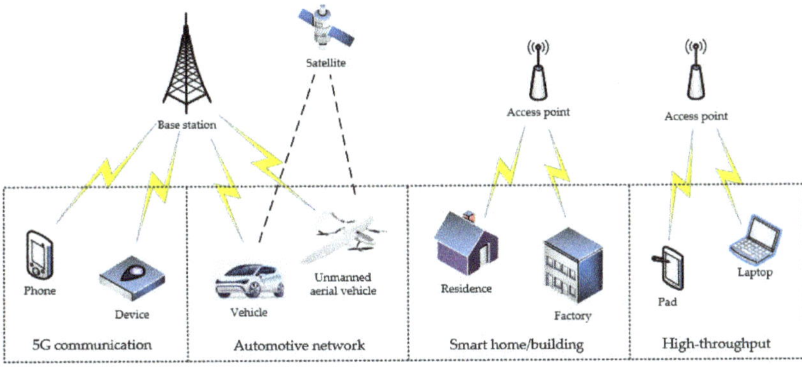

Figure 2. Development directions of CR networks.

5.2. Mobility and Automotive Networks

The use of multichannel operations in wireless access for vehicular environments (WAVE) has been proposed in the IEEE 1609.4 standard. The WAVE system employs the 75 MHz spectrum in the 5.9 GHz band, with one control channel and six service channels. However, there is still a scarcity of spectrum which poses a challenge for the WAVE system. The application of cognitive radio technology in WAVE has been studied to address this problem [28–30]. Some initial work on cognitive radio-enabled vehicular communications has already been conducted. Vehicular wireless sensor networks are becoming a popular network paradigm for collecting monitoring information in urban environments. In this field, CRNs are likely to play a significant role, and protocols for highway safety using CRNs have already been proposed [31,32], although more research is needed in this area.

5.3. High-Throughput Applications

Delivering multimedia applications, such as on-demand or live video streaming, audio, and still images, over resource-constrained CRNs is a significant challenge due to their high bandwidth requirements. Refs. [33–35] further discuss the advantages of CRNs in these high-throughput application scenarios. Similarly, other CRN applications, including those used in hospitals, vehicular networks, tracking, and surveillance, have varying data densities and require significant bandwidth, low delay, and high bursts of data. However, because SUs in CRNs can access multiple channels whenever necessary and available, CRNs are well suited for bandwidth-intensive applications.

5.4. Smart Home or Building Applications

A wide range of indoor applications, such as intelligent buildings, home monitoring systems, factory automation, and personal entertainment, require a dense CRN environment to ensure an adequate Quality of Service (QoS). However, conventional CRNs face significant challenges in achieving reliable communication in indoor areas due to the extremely crowded industrial, scientific, and medical (ISM) bands [36]. To overcome these challenges, CRNs offer a promising solution. By using cognitive radio technology, CRNs can mitigate the limitations of traditional indoor CRN applications, thereby improving their efficiency and effectiveness.

5.5. Immediate Monitoring Applications

Immediate monitoring applications, such as traffic monitoring, biodiversity mapping, habitat monitoring, environmental monitoring, crop and livestock monitoring, irrigation, underwater CRNs, vehicle and inventory tracking, disaster relief operations, as well as bridge and tunnel monitoring require reliable and low-latency communication with minimum channel access. For some real-time surveillance applications, even a small delay can be critical, and they therefore require high reliability. However, in multihop CRNs, a delay can occur due to link failure if the channel condition is poor. In contrast, CR nodes can switch to another channel with a better condition if they find one. To increase the channel bandwidth, CRNs can use channel aggregation and multiple channels concurrently, which can be beneficial for real-time surveillance applications [37,38].

5.6. Military and Security-Related Applications

Conventional cognitive radio networks find application in numerous military and public security scenarios. Some of these applications include chemical, biological, radiological, and nuclear (CBRN) attack detection and investigation, command and control operations, gathering information for battle damage evaluation, battlefield surveillance, intelligence assistance, targeting, and more. In hostile regions, adversaries may employ jamming signals to interfere with radio communication channels. In such situations, CRNs can switch to different frequency bands to avoid the band with the jamming signal. Furthermore, certain military applications require large bandwidths, minimal channel access, and low communication delays, making CRNs an optimal choice for these purposes.

Authors have achieved some research results in cognitive wireless networks and cognitive vehicular networks [39,40]. There are also some novel research directions in applying cognitive radio networks in civil aviation such as air–ground communication, cognitive radar, etc.

In general, the future development direction of cognitive radio technology is mainly to achieve more efficient communication and integration with other communication technologies.

The future development direction of cognitive radio mainly includes the following aspects:

Firstly, incorporating artificial intelligence (AI) technology into cognitive radio enables signal adaptation, autonomous sensing, and intelligent modulation, leading to enhanced system performance. Furthermore, integrating predictive, analytical, and reasoning capabilities into the cognitive radio system can augment its intelligence and efficiency.

Secondly, high-speed transmission is crucial. With the introduction and continuous upgrading of various new wireless technologies, the demand for higher transmission rates has surged. Consequently, cognitive radio technology must enable faster data transmission rates while ensuring data accuracy and reliability [41].

Thirdly, security is a crucial consideration in cognitive radio systems as the number of wireless devices increases. To ensure the stability and reliability of the system, more attention and research are required to strengthen the security protection of cognitive radio systems in the future.

6. Challenges Faced by Cognitive Radio

Cognitive radio refers to the use of radio technology to obtain, process, and interpret the information transmitted and processed by electronic equipment and communication systems.

The challenges faced by cognitive radio are as follows:

6.1. Limitations on Frequency Range

The limitation of the cognitive radio frequency range is that human cognition and perception are limited by the radio frequency range. In modern communication technology, the use of radio frequencies has become an unavoidable problem, and these frequency ranges have a profound impact on human perception and cognition.

The frequency spectrum of radio waves has a wide range from 3 kHz to 300 GHz, which includes many frequencies required by modern communication technologies, such as Wi-Fi, Bluetooth, mobile data, satellite communication, etc. However, these frequencies impose some limitations on human perception and cognition [42,43].

Firstly, the frequency of radio waves may result in different frequency ranges received by devices exposed to radio waves, which may affect their perception and cognition. For example, the frequency range of Wi-Fi radio waves is between 2.4 GHz and 2.45 GHz, but different Wi-Fi devices may have different frequency ranges, which may lead to different wireless access points (APs) connecting to the network, thereby affecting the perception and cognition of the devices.

Secondly, the frequency of radio waves may also affect the perception and cognition of the eyes and brain. The shorter the wavelength of a radio wave, the better its penetration ability. This means that high-frequency radio waves can penetrate fog and obstacles, while shortwave radio waves are more easily received by radar antennas. However, the wavelength of high-frequency radio waves is relatively short, so the human eye's perception and cognition of shortwave radio waves may be affected, which may lead to high-frequency radio wave devices being considered ineffective.

Finally, the frequency of radio waves may also affect human cognition and culture. In some cultures, radio waves are considered an important cultural element, such as the radio and television frequency bands in China and some radio frequency bands in other countries, such as the 1.1 GHz and 1.8 GHz frequency bands in the UK, which are considered to have cultural significance. The use of these frequency bands may affect people's cognition and culture.

Therefore, as human cognition and perception ability are limited by the radio frequency range and different radio frequencies may affect human perception and cognition, so different frequency ranges are required.

6.2. Increase in Complexity

Cognitive radio technology, also known as cognitive communication, is a cross field research area involving radio technology and artificial intelligence technology. The development and widespread application of this technology have led to a deeper understanding of communication and computing methods, while also increasing the complexity of radio technology. The development of cognitive radio technology is inseparable from the support of artificial intelligence technology. The application of artificial intelligence technology makes the training of algorithms, data structures, and models more efficient, and also makes the process of machine learning applied to cognitive radio more complex [44,45].

There are papers and patents involving the applications of CR technology in different fields. For example, Ref. [46] proposed a dual-user relay protocol based on energy harvesting and cognitive radio technology, which achieved energy efficiency via relay communication. Ref. [47] compared the application potential and performance differences in different CR technologies in the FM broadcast frequency band. Ref. [48] proposed a novel channel allocation method to improve spectrum utilization efficiency with an OFDM-based CR technology. Ref. [49] introduced a CR technology utilizing a multiphase downconverter to efficiently receive multiple signals. A research report from Ain Shams University had reported a CR technology utilizing major L-band ranging devices for aviation communication [50]. These related works provided valuable knowledge and technical solutions for the academic and industrial communities.

The development of cognitive radio has also increased the complexity of radio technology. The development and improvement of radio technology changes with the change in cognitive radio technology. For example, antenna array, modulation mode, demodulation mode, signal power transmission, etc. are closely related to cognitive radio technology [51].

Finally, cognitive radio technology also needs to identify and understand the signal, which requires compression, quantization, noise removal, reverberation removal, shock removal, and other operations on the signal. These operations require in-depth research and analysis of the signal in order to extract useful information [52].

To sum up, the development of cognitive radio technology enables people to have a deeper understanding of communication and computing and also increases the complexity of radio technology. The development of cognitive radio technology makes the cross field research between computer and radio technology more complex and efficient [53–55].

The research of cognitive radio technology needs the cross knowledge of computers, artificial intelligence, radio technology, and other fields. The cross research in these fields not only increases the complexity of technology but also promotes the progress and application of technology [56,57].

6.3. Efficient Data Storage and Processing

In cognitive radio, data storage and processing are very important. Traditional data storage requires a lot of disk space and professional equipment support, which is unrealistic for cognitive radio applications with high real-time requirements. Cognitive radio data storage and processing use efficient algorithms and hardware support, making data storage and processing more efficient [58].

The high efficiency of cognitive radio data storage is because it uses distributed storage technology to store data on multiple nodes, so as to achieve efficient data storage and processing. Compared with traditional data storage methods, cognitive radio data storage can access and process data faster, while reducing the number and complexity of storage devices [59].

Cognitive radio data storage also uses adaptive algorithms, which can automatically select suitable storage nodes and storage methods according to the real-time needs, size, and mode of data and other factors. This algorithm can reduce the workload of storage devices and improve their response speed and efficiency [60].

In conclusion, the efficiency of cognitive radio data storage is the key reason for its wide application in the field of radio communication. By using distributed storage technology, efficient algorithms and hardware support and other technical means, cognitive radio data storage can efficiently store and process data, providing strong technical support for the development and application of cognitive radio communication technology [61–63].

6.4. Security Issues

Cognitive radio is a new radio communication technology, which can use artificial intelligence and machine learning algorithms to design and generate radio signals, so as to achieve extensive radio coverage and low-power wireless sensor networks. In cognitive radio technology, information and data are encoded as radio signals and transmitted through radio waves, which can be received and processed by different devices [64].

However, the security of cognitive radio also deserves attention. With more and more devices using cognitive radio technology, how to handle and protect the security of these devices becomes particularly important. Among the security issues of cognitive radio, the following are the main issues:

- Interference and deception.

The transmission and reception of cognitive radio signals require low power consumption and short-distance communication, so they are vulnerable to interference and deception. These interferences and deceptions can be caused by the visibility, interference, or deception of other devices, networks, or technologies, as well as specific frequency noise and other interferences. Therefore, when using cognitive radio, it is necessary to ensure that the equipment and network are secure to avoid interference and deception [65–67].

- Data integrity.

Cognitive radio devices can receive and transmit data, which may be damaged or tampered with. In order to protect data integrity, data encryption and integrity verification must be implemented. This will ensure that only the correct devices can access and transmit data and prevent data from being tampered with [68,69].

- Unknown risks.

The use of cognitive radio technology involves unknown risks and vulnerabilities, which may pose a threat to the security of equipment. Therefore, it is necessary to test the security of devices and networks and identify and fix vulnerabilities and potential security risks, in order to ensure the security of devices [70].

- Lack of standardization.

At present, cognitive radio technology lacks standardization. This means that devices from different manufacturers may have different safety and functionality, and there is a lack of unified testing and evaluation standards. This may pose a threat to the security and stability of the devices, as well as prevent them from communicating with the network or working properly [71].

Cognitive radio technology has great potential and broad application prospects, but security issues must be addressed and solved to ensure the safety and correct use of equipment. This includes implementing standardization, testing and verifying the integrity and safety of equipment, as well as developing unified safety standards and testing processes. This will ensure the security of devices and networks and enable a wider range of cognitive radio applications [72,73].

6.5. Standardization Issues

CR standardization is the core issue of the current technical field. Standardization organizations and relevant research units are working closely together to develop unified standards and testing methods to secure the rapidly developing fields of CR technology. In particular, Ref. [74] pointed out that cognitive radio technology needs to be standardized in multiple aspects, including spectrum management, channel selection, interference detection and management, security, etc. Meanwhile, due to the complexity and diversity of CR technology, the development of standards also should fully consider the differences and special demands of application scenarios.

Cognitive radio is a method of designing, constructing, and testing radio systems using artificial intelligence and machine learning methods. In cognitive radio, the acquisition, processing, and interpretation of signals rely on computer vision and natural language processing technologies. These technologies can enable artificial intelligence systems to recognize, analyze, and interpret human language and signals, so as to better complete the task of cognitive radio [75,76].

Due to the complexity and innovation of cognitive radio systems, standardization is particularly important to ensure that all systems can work in the same environment and meet the needs of users.

At present, the standardization of cognitive radio involves many aspects, such as the layout of radio channels, communication protocols, signal processing algorithms, equipment design, and so on. Although various standardization organizations have already carried out some work, there are still many challenges, such as: different standardization organizations may have different standard definitions, standard testing methods, standard implementation methods, etc., which may lead to inconsistencies between different systems. In addition, the research and development of cognitive radio need a lot of resources, and the workload of standardization organizations will increase accordingly, so more cooperation and support are needed [77].

Standardization organizations and relevant research units need to work closely together to develop unified standards and test methods to ensure that all systems can work in the same environment, meet user needs, and promote the development of cognitive radio technology. It is also necessary to strengthen relevant research results and experimental work to better guide the practice of standardization work. Only in this way can we truly realize the popularization and development of cognitive radio technology and create more communication and cognitive capabilities for humankind [78,79].

6.6. Environmental Adaptability Issues

Cognitive radio is a new technology, which involves the conversion of radio signals into cognitive signals so that computers and humans can recognize and understand these signals. This technology has applications in multiple fields, such as medical, military, transportation, and communication. The problem of environmental adaptability of cognitive radio refers to how humans recognize and interpret radio signals in the environment [80].

Cognitive radio technology requires powerful computers and complex algorithms to analyze and understand radio signals. These computers require sufficient memory to process a large amount of data and high-speed memory to store the parsing results. Algorithms also require strong computing power to process large amounts of data in order to convert complex radio signals into cognitive signals [81].

The problem of environmental adaptability of cognitive radio is related to human cognitive processes, and it needs to consider how the human brain processes information. When analyzing radio signals, it is necessary to consider factors such as frequency, strength, and phase of the radio signal in order to convert this information into cognitive information. It is necessary to consider how different human brains process information differences in order to achieve consistency in analytical results among different individuals [82].

The environmental adaptability of cognitive radio needs to consider complex radio signals, which requires strong computing power of analytical algorithms and powerful processing power of computers to process large amounts of data. Humans need to adapt to different radio signals in the cognitive radio environment in order to convert this information into cognitive information and different signal sources are needed in the environment so that different people have different results of analytic signals [83].

The environmental adaptability of cognitive radio needs to consider how the human brain processes information, and different signal sources and resolution algorithms are needed to achieve consistency of resolution results between different human individuals and switch signal sources in different environments [84].

In a word, cognitive radio's environmental adaptability needs to consider complex radio signals, requires humans to adapt to different radio signal sources and resolution algorithms, and requires switching signal sources in different environments. The application of this technology will bring many changes to humanity, including applications in medical, military, transportation, and communication fields, while also bringing more efficient communication and more accurate diagnosis and treatment to humanity [85].

7. Applications of Cognitive Radio in Civil Aviation

7.1. Applications in Civil Aviation

In civil aviation, with proper definition and working principles, CR technology can be applied to monitor and manage radio signals, ensuring the safety of communication and flight operations [86].

The increase in air traffic leads to the increasing use of radio spectrum for aeronautical communications in civil aviation, which is expected to be more and more congested. Considering the advantages of cognitive radio in resource-constrained situations, researchers have begun to discuss the applications of cognitive radio in civil aviation such as air–ground communication, air–air communication, unmanned aerial systems, etc. [87]. Furthermore, cognitive radio technology, along with other radio technologies, can be widely used in the following aspects in Figure 3.

7.1.1. Communication System

The application of CR in civil aviation communication systems refers to the use of new radio technologies and equipment to make it easier and safer for passengers and crew to use communication equipment, including mobile phones, laptops, e-books, etc. In the civil aviation communication system, the application of CR can make flight communication more efficient, safe, and reliable [88].

With the continuous development of modern communication technology, people's understanding of radio is also constantly improving. Radio technology plays an increasingly important role in modern communication, with applications including radio beacons, radio measurements, radio reception and transmission, satellite communication, and more. CR technology is based on the improvement of people's radio awareness and, through the use of new radio technology and equipment, passengers and crew can use communication equipment more easily and safely, including mobile phones, laptops, e-books, etc. [89].

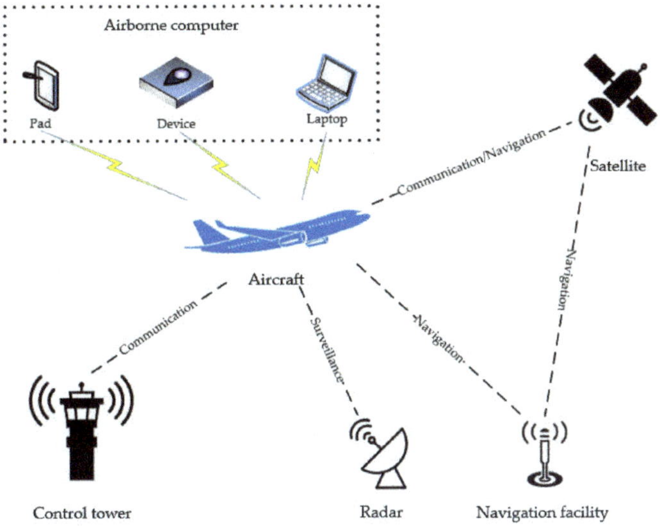

Figure 3. Application scenarios of CR in civil aviation.

In civil aviation communication systems, the application of CR can improve communication efficiency and security. The flight communication system needs to handle different types and quantities of communication devices, such as mobile phones, laptops, e-books, etc. CR can make these devices easier to manage and recognize, while providing better security [90].

Airport wireless beacons are an important component of airport communication systems. They are used for transmitting and receiving wireless network signals, making it easier for passengers and crew members to connect to the airport wireless network and access various applications and resources during flights. CR can replace traditional radio beacons because it uses higher-frequency radio waves, making it easier to connect to wireless networks while providing better signal-to-noise ratio and stability. A radio direction finding system is a type of radio equipment that uses radio waves of a specific frequency to detect and locate radio equipment. In civil aviation communication systems, CR can replace traditional radio direction finding systems by using high-frequency radio waves to detect and locate various communication devices. CR can better detect and locate handheld radio devices, including mobile phones and other communication devices. The airport wireless communication system is an important component of the flight communication system. CR can be used to connect airport wireless communication systems, making it easier for passengers and crew to access flight communication systems, such as flight information, entertainment systems, and flight communication beacons. CR can also provide safer and more efficient communication, enabling passengers and crew members to more effectively access information and maintain flight communication systems [91,92].

The application of CR in civil aviation communication system is an important part of airport communication system and flight communication systems. CR can make it easier and safer for passengers and crew to use communication equipment, improving the efficiency and safety of airport communication systems and flight communication systems [93].

7.1.2. Satellite Communication

Cognitive radio is a new type of radio technology. Different from traditional analog signal radio, cognitive radio uses digital signal technology for communication, which can automatically identify and adapt to different communication needs, including voice communication, data communication, and video communication, and can communicate in different wireless communication systems, such as satellite communication, mobile communication, and terrestrial radio communication.

In the field of civil aviation, cognitive radio technology is widely used to ensure the safety and smoothness of flights. The radio systems of civil aviation airports and airlines are usually maintained and upgraded by professional radio engineers who need to be trained and certified in cognitive radio technology in order to master the latest technologies and tools and ensure the reliability and safety of the radio system [92].

Cognitive radio technology is applied to flight communication, including voice communication, data communication, and video communication. For example, on a flight, communication between the captain and passengers usually uses satellite phones for real-time communication. Since the aircraft will generate noise in flight, the captain's voice will be interfered with by the noise. Therefore, the captain needs to use cognitive radio technology to record and confirm the call to ensure the quality and safety of the call.

The radio system at the airport is used to monitor the status, flight speed, and dangerous situations encountered by the aircraft, such as the fuel condition, attitude, and altitude of the aircraft. The airport radio system also needs to use cognitive radio technology for monitoring and communication, so that flight and airport management personnel can understand the flight situation in real time.

The application of cognitive radio technology in the radio systems of civil aviation and airports can improve the reliability and security of communication, ensure the safety and smoothness of flights, and ensure the safety and interests of passengers and airport managers.

7.1.3. Navigation

Cognitive radio is a radio communication technology based on machine learning and cognitive computing. Its goal is to optimize and improve the radio communication system by making use of human cognitive and language abilities to enable radio devices to automatically recognize and receive signals. The application of cognitive radio technology in civil aviation navigation can bring the following advantages:

- Improvement of navigation accuracy.

Traditional GPS navigation uses satellite signals, which are influenced by factors such as the Earth's curvature and satellite signal attenuation, resulting in navigation errors. Cognitive radio technology uses cognitive computing and machine learning algorithms to identify and correct satellite signals in real time and improve navigation accuracy.

- Reduction of wireless channel interference.

Wireless channels are an important part of communication systems, but they are often affected by interference from other communication signals, which poses a threat to the integrity and accuracy of the signal. Cognitive radio technology can automatically identify and avoid interference and improve the stability and accuracy of radio channels, thus ensuring the accuracy of navigation.

- Improvement of aircraft safety.

During aircraft flight, navigation equipment must ensure flight safety. Traditional navigation methods require manual signal confirmation and operation, which are susceptible to human errors and fatigue. The cognitive radio technology can be used by the crew to confirm the radio signal through simple operation, so as to reduce the occurrence of human errors and improve flight safety [91].

- Improvement of aircraft communication.

A traditional aircraft communication system is mainly composed of a mechanical navigation system and radio navigation system. The mechanical navigation system is vulnerable to bad weather and mechanical failure. The communication of the radio navigation system is vulnerable to signal interference and interruption, resulting in reduced reliability of navigation information. Cognitive radio technology can improve the communication between the aircraft mechanical navigation system and radio navigation system while maintaining the stability of radio signals.

The application of cognitive radio technology in civil aviation navigation can improve navigation accuracy, reduce radio channel interference, improve aircraft safety and improve aircraft communication, and lay a solid foundation for accurate flight, safe flight, and reliable operation of flights.

7.1.4. Airborne Computers

Airborne computers are the core computer systems used by airlines on aircraft. Cognitive radio technology is used to receive signals needed by airborne computers and convert them into formats that can be processed by computers. For example, airlines can use cognitive radio technology to receive data required by airborne computers, such as flight plans, weather data, etc., so that airborne computers can process these data and execute tasks.

Cognitive radio technology plays a vital role in the field of civil aviation. With the development of computer and digital communication technology, cognitive radio technology is constantly updated to meet the needs of flight communication and improve the quality and efficiency of flight communication.

The application of cognitive radio technology in the field of civil aviation involves communication systems, satellite communications, navigation and airborne computers. With the continuous development of technology, cognitive radio technology is becoming more and more important and influential in the field of civil aviation [94].

7.2. Development Trend of Cognitive Radio in Civil Aviation

Cognitive radio technology in the field of civil aviation has received extensive attention and research. In the future, this technology will be more widely applied in the field of civil aviation, with the following main development trends:

7.2.1. Higher Security and Reliability

Cognitive radio technology can improve flight safety by identifying and tracking the surrounding signal environment. In the future, cognitive radio technology will adopt more security mechanisms and adaptive algorithms to improve its reliability in response to unexpected situations such as communication interruption.

7.2.2. Multimodal Communication

In the field of civil aviation, an aircraft must communicate with ground control consoles, other flights, weather stations, and other aircraft simultaneously, thus achieving multimodal communication is necessary. Cognitive radio technology will respond to this demand and support multiple communication modes to improve the utilization of the radio spectrum.

7.2.3. Higher Spectrum Utilization

One of the development trends in the future is to improve the spectral efficiency. In the case of scarce spatial spectrum resources, improving the spectral efficiency of cognitive radio technology has become a major goal. Combining OFDM and MIMO technology with cognitive radio technology can achieve more efficient spectrum utilization.

7.2.4. Advanced Algorithms

With the improvement of computing power and processing speed, cognitive radio technology can support more complex and advanced algorithms and models to solve some complex problems that have not been solved in the existing technology. For example, machine learning, artificial intelligence, and other technologies will be widely used in cognitive radio technology to improve its performance and accuracy [95].

7.2.5. Integration with Other Technologies

Cognitive radio technology will be combined with other technologies to support more applications. For example, combined with aviation control center technology, cognitive radio technology can improve the communication efficiency between different flights and provide more secure and reliable technical support for flight scheduling.

In conclusion, cognitive radio technology will play an important role in the field of civil aviation. With the development of technology and the increase in application scenarios, cognitive radio technology will show a broader and flexible development prospect.

7.3. Future Prospects of Cognitive Radio in Civil Aviation

Based on the development trend of cognitive radio technology, we can predict that it will play an important role in the field of civil aviation. First of all, using cognitive radio technology, aircraft communication, navigation, monitoring, control and other systems can achieve autonomous, cooperative, reliable, and efficient operation. Secondly, cognitive radio technology can process the sensing information obtained by the aircraft with high accuracy, efficiency, and speed, so as to realize prediction, control, and intervention operations and improve the performance and safety of the aircraft. At the same time, cognitive radio technology can also effectively reduce congestion, delay, and other problems in the civil aviation field and improve the overall operation efficiency and service quality through intelligent data analysis and optimization.

In the future, the application of cognitive radio technology in the field of civil aviation will be more popular and widespread. First, with the gradual maturity and improvement of related technologies, the performance of cognitive radio systems will be greatly improved, which can better meet the needs of practical applications. Secondly, the application scope of cognitive radio technology will continue to expand, covering more fields and scenarios, including control and command, airport operation, aircraft maintenance, etc. In the future, more application innovations may emerge, continuously improving the overall operational efficiency and safety of the civil aviation system.

However, the application of cognitive radio technology in the field of civil aviation still faces some challenges and problems. First, with the continuous development of civil aviation and the continuous expansion of flight control scale, the channel resource utilization and spectrum coexistence of cognitive radio technology will become more and more complex. Secondly, the application of cognitive radio technology needs to optimize and coordinate the resource allocation and scheduling of different systems and needs to solve many technical problems such as information security, equipment standards, mutual interference, etc.

Therefore, the future direction of development should be to pay more attention to the research and innovation of new technologies and develop more efficient, accurate, and intelligent cognitive radio systems to meet the needs and challenges of flight network operation. While achieving functional upgrades, we also need to develop unified standards and specifications, promote technical coordination and information sharing in the field of civil aviation, integrate and optimize civil aviation resources, and achieve more efficient, safe, and reliable operation of the civil aviation network.

8. Conclusions

In the future development of cognitive radio, there are many trends and prospects. First of all, research in cognitive radio will pay more attention to the development of intelligence and strive to continuously improve its intelligence level, so that it can more accurately perceive the radio environment and more accurately adjust the use of radio resources. Secondly, research in cognitive radio will seek better technology combinations to improve the utilization and efficiency of its radio spectrum.

In addition, the future development direction of cognitive radio will also involve its application in the field of civil aviation. The civil aviation industry, as a crucial field for air traffic safety, has high requirements for the safety and reliability of radio usage. The future development direction of cognitive radio will be closely combined with the civil aviation field to improve the reliability and safety level of its communication and navigation.

Finally, cognitive radio will continue to face various challenges and problems, such as insufficient spectrum resources, interference, etc. Therefore, in the future development of cognitive radio, it is necessary to constantly strengthen technical research and improve the technical level to meet these challenges and problems. The future development of cognitive radio will pay more attention to technological innovation and intelligent development and use radio in a more efficient, safer, and more reliable way by combining various advanced technologies. At the same time, the future development of cognitive radio also needs to strengthen the solution of existing problems and constantly solve new problems to promote its continuous development and growth.

Author Contributions: Conceptualization, R.Z.; methodology, R.Z.; investigation, X.L.; resources, Y.C.; supervision, Y.C.; funding acquisition, X.L. All authors have read and agreed to the published version of the manuscript.

Funding: This work was supported by the National Natural Science Foundation of China (61861018).

Institutional Review Board Statement: Not applicable.

Informed Consent Statement: Not applicable.

Data Availability Statement: The data used to support the findings of this study are included within the article.

Conflicts of Interest: The authors declare no conflict of interest.

References

1. FCC. *Spectrum Policy Task Force Report*; FCC Doc. ET Docket No. 02-135; FCC: Washington, DC, USA, 2002.
2. Haykin, S. Cognitive radio: Brain-empowered wireless communications. *IEEE J. Sel. Areas Commun.* **2005**, *23*, 201–220. [CrossRef]
3. Akyildiz, I.F.; Lee, W.-Y.; Vuran, M.C.; Mohanty, S. NeXt generation/dynamic spectrum access/cognitive radio wireless networks: A survey. *Comput. Netw.* **2006**, *50*, 2127–2159. [CrossRef]
4. Raman, C.; Yates, R.D.; Mandayam, N.B. Scheduling variable rate links via a spectrum server. In Proceedings of the First IEEE International Symposium on New Frontiers in Dynamic Spectrum Access Networks (DySPAN), Baltimore, MD, USA, 8–11 November 2005; pp. 110–118.
5. Axell, E.; Leus, G.; Larsson, E.G.; Poor, H.V. Spectrum Sensing for Cognitive Radio: State-of-the-Art and Recent Advances. *IEEE Signal Process. Mag.* **2012**, *29*, 101–116. [CrossRef]
6. Yucek, T.; Arslan, H. A survey of spectrum sensing algorithms for cognitive radio applications. *IEEE Commun. Surv. Tuts.* **2009**, *11*, 116–130. [CrossRef]
7. Akyildiz, I.F.; Lo, B.F.; Balakrishnan, R. Cooperative spectrum sensing in cognitive radio networks: A survey. *Phys. Commun.* **2011**, *4*, 40–62. [CrossRef]
8. Akyildiz, I.F.; Lee, W.-Y.; Vuran, M.C.; Mohanty, S. A survey on spectrum management in cognitive radio networks. *IEEE Commun. Mag.* **2008**, *46*, 40–48. [CrossRef]
9. Ghasemi, A.; Sousa, E.S. Fundamental limits of spectrum-sharing in fading environments. *IEEE Trans. Wirel. Commun.* **2007**, *6*, 649–658. [CrossRef]
10. Ban, T.W.; Choi, W.; Jung, B.C.; Sung, D.K. Multi-user diversity in a spectrum sharing system. *IEEE Trans. Wirel. Commun.* **2009**, *8*, 102–106. [CrossRef]
11. Naeem, M.; Anpalagan, A.; Jaseemuddin, M.; Lee, D.C. Resource Allocation Techniques in Cooperative Cognitive Radio Networks. *IEEE Commun. Surv. Tutor.* **2013**, *16*, 729–744. [CrossRef]
12. Ahmad, A.; Ahmad, S.; Rehmani, M.H.; Hassan, N.U. A Survey on Radio Resource Allocation in Cognitive Radio Sensor Networks. *IEEE Commun. Surv. Tutor.* **2015**, *17*, 888–917. [CrossRef]
13. Meghanathan, N. A survey on the communication protocols and security in cognitive radio networks. *Int. J. Commun. Netw. Inf. Secur.* **2013**, *5*, 19. [CrossRef]
14. Hossain, E.; Niyato, D.; Kim, D.I. Evolution and future trends of research in cognitive radio: A contemporary survey. *Wirel. Commun. Mob. Comput.* **2015**, *15*, 1530–1564. [CrossRef]
15. Mauri, J.L.; Ghafoor, K.Z.; Rawat, D.B.; Perez, J.M.A. (Eds.) *Cognitive Networks: Applications and Deployments*; CRC Press: Boca Raton, FL, USA, 2014.
16. Khan, A.A.; Rehmani, M.H.; Saleem, Y. Neighbor discovery in traditional wireless networks and cognitive radio networks: Basics, taxonomy, challenges and future research directions. *J. Netw. Comput. Appl.* **2015**, *52*, 173–190. [CrossRef]

17. Marinho, J.; Monteiro, E. Cognitive radio: Technology survey and future research directions. In Proceedings of the ANACOM Conference Proceeding, Lisbon, Portugal, 23–24 September 2010.
18. Sen, J. A survey on security and privacy protocols for cognitive wireless sensor networks. *arXiv* **2013**, arXiv:1308.0682.
19. Hasan, Y.M.; Abdullah, A.S.; Alnahwi, F.M. Dual-Port Filtenna System for Interweave Cognitive Radio Applications. *Iran. J. Sci. Technol. Trans. Electr. Eng.* **2022**, *46*, 943–958. [CrossRef]
20. Srikar, D.; Anuradha, S. A new two-element MIMO antenna system for cognitive radio applications. *Circuit World* **2021**, *48*, 111–125. [CrossRef]
21. Niture Deeplaxmi, V.; Mahajan, S.P. A Compact Reconfigurable Antenna for UWB and Cognitive Radio Applications. *Wirel. Pers. Commun.* **2022**, *125*, 3661–3679. [CrossRef]
22. Mathai, V.; Indumathi, P. Spectral Vacancy Prediction Using Time Series Forecasting for Cognitive Radio Applications. *Intell. Autom. Soft Comput.* **2022**, *33*, 1729–1746. [CrossRef]
23. Liu, S. Power allocation method of supermodel game in cognitive radio networks. *Fujian Comput.* **2023**, *39*, 26–30.
24. Xu, W. Reflection and exploration on the application of cognitive radio in radio and television frequency. *Digit. Media Res.* **2022**, *39*, 34–40.
25. Zhou, W. On the Application of Radio Communication Navigation in Maritime Navigation—Review of cognitive radio Communication and Networking: Principles and Applications. *Mod. Radar* **2022**, *44*, 116.
26. Li, N. Research on the application of GNU radio based cognitive radio in mobile communication. *Inf. Commun.* **2020**, *5*, 224–225.
27. Wang, W. Research on Spectrum Sensing Technology Based on Energy Detection in Cognitive Radio. Master's Thesis, Nanjing Information Engineering University, Nanjing, China, 2022.
28. Ghandour, A.J.; Fawaz, K.; Artail, H. Data Delivery Guarantees in Congested Vehicular Ad Hoc Networks Using Cognitive Networks. In Proceedings of the IEEE IWCMC 2011, Istanbul, Turkey, 4–8 July 2011; pp. 871–876.
29. Di Felice, M.; Doost-Mohammady, R.; Chowdhury, K.R.; Bononi, L. Smart radios for smart vehicles: Cognitive vehicular networks. *IEEE Veh. Technol. Mag.* **2012**, *7*, 26–33. [CrossRef]
30. Rawat, D.B.; Zhao, Y.; Yan, G.; Song, M. CRAVE: Cognitive Radio Enabled Vehicular Communications in Heterogeneous Networks. In Proceedings of the IEEE Radio and Wireless Symposium (RWS 2013), Austin, TX, USA, 20–23 January 2013; pp. 190–192.
31. El Mougy, A.; Ibnkahla, M. A Cognitive WSN Framework for Highway Safety based on Weighted Cognitive Maps and Q-learning. In Proceedings of the Second ACM International Symposium on Design and Analysis of Intelligent Vehicular Networks and Applications, (DIVANet'12), Paphos, Cyprus, 21–25 October 2012; pp. 55–62.
32. Aalamifar, F.; Vijay, G.; Abedi Khoozani, P.; Ibnkahla, M. Cognitive Wireless Sensor Networks for Highway Safety. In Proceedings of the First ACM International Symposium on Design and Analysis of Intelligent Vehicular Networks and Applications (DIVANet '11), Miami Beach, FL, USA, 31 October–4 November 2011; pp. 55–60.
33. Misra, S.; Reisslein, M.; Xue, G. A survey of multimedia streaming in wireless sensor networks. *IEEE Commun. Surv. Tutor.* **2008**, *10*, 18–39. [CrossRef]
34. Akyildiz, I.F.; Melodia, T.; Chowdhury, K.R. A survey on wireless multimedia sensor networks. *Comput. Netw.* **2007**, *51*, 921–960. [CrossRef]
35. Akyildiz, I.F.; Melodia, T.; Chowdhury, K.R. Wireless multimedia sensor networks: Applications and testbeds. *Proc. IEEE* **2008**, *96*, 1588–1605. [CrossRef]
36. Zhou, G.; Stankovic, J.A.; Son, S. Crowded Spectrum in Wireless Sensor Networks. In Proceedings of the Third Workshop on Embedded Networked Sensors (EmNets 2006), Cambridge, MA, USA, 30–31 May 2006.
37. Gao, B.; Yang, Y.; Park, J.M. Channel Aggregation in Cognitive Radio Networks with Practical Considerations. In Proceedings of the IEEE International Conference on Communications (ICC), Kyoto, Japan, 5–9 June 2011; pp. 1–5.
38. Joshi, G.P.; Kim, S.W. An Enhanced Synchronized MAC Protocol for Cognitive Radio Networks. In Proceedings of the 7th International Conference on Wireless Communications, Networking and Mobile Computing (WiCOM), Wuhan, China, 23–25 September 2011; pp. 1–4.
39. Li, X.; Zhou, R.; Zhou, T.; Liu, L.; Yu, K. Connectivity Probability Analysis for Green Cooperative Cognitive Vehicular Networks. *IEEE Trans. Green Commun. Netw.* **2022**, *2022*, 1553–1563. [CrossRef]
40. Sheng, M.; Li, X.; Wang, X.; Xu, C. Topology Control with Successive Interference Cancellation in Cognitive Radio Networks. *IEEE Trans. Commun.* **2017**, *65*, 37–48. [CrossRef]
41. Premkumar, M.; Sathiyapriya, S.; Arun, M.; Sachan, V. Medical Signal Processing via Digital Filter and Transmission Reception Using Cognitive Radio Technology. *Trait. Du Signal* **2022**, *39*, 1357–1362. [CrossRef]
42. Babu, C.R.; Balakrishnan, A.; Ramana, K.; Singh, S.; Ra, I.-H. Elite-CAM: An Elite Channel Allocation and Mapping for Policy Engine over Cognitive Radio Technology in 5G. *Sensors* **2022**, *22*, 5011. [CrossRef]
43. Antonova, V.M.; Balakin, K.A.; Grechishkina, N.A.; Klygin, D.S.; Kuznetsov, N.A. Application of the Hypergraphs Theory to Solve the Problem of Optimizing Network Performance in Civil Aviation. *J. Commun. Technol. Electron.* **2022**, *67*, 717–721. [CrossRef]
44. Lu, J. Data Protection Software for Civil Aviation Control Flight Information System Based on FPE Algorithm. *Secur. Commun. Netw.* **2022**, *2022*, 4150660. [CrossRef]
45. Shao, Q.; Yuan, J. Study on the disposal strategy of civil aviation passenger collective events based on evolutionary game theory. *Phys. A Stat. Mech. Its Appl.* **2022**, *598*, 127341. [CrossRef]

46. Lin, Z.; Li, G.; Li, J. Two-User Relay Protocol Based on Energy Harvesting and Cognitive Radio Techniques. *Wirel. Pers. Commun.* **2021**, *121*, 3061–3083. [CrossRef]
47. Hamed, I.; Elmenyawi, M.A.; Hassan, A.Y. Comparative study between cognitive radio techniques in FM broadcasting band. *IOP Conf. Ser. Mater. Sci. Eng.* **2021**, *1172*, 012009. [CrossRef]
48. Yilmazel, R.; Inanç, N. A Novel Approach for Channel Allocation in OFDM Based Cognitive Radio Technology. *Wirel. Pers. Commun.* **2021**, *120*, 307–321. [CrossRef]
49. Naval Information Warfare Center Pacific. Cognitive Radio Technique for Efficiently Receiving Multiple Signals Using Polyphase Downconverter Channelizers. U.S. Patent USPTO 10,763,906, 1 September 2020.
50. Investigators from Ain Shams University. Report New Data on Information and Data Loss and Recovery "Cognitive Radio Techniques for Utilizing the Primary L-band Distance Measuring Equipment for Aeronautical Communications". *Inf. Technol. Newsweekly* **2020**, *52*, 104–110.
51. Cicioğlu, M.; Çalhan, A. Channel aware wireless body area network with cognitive radio technology in disaster cases. *Int. J. Commun. Syst.* **2020**, *33*, e4565. [CrossRef]
52. Samala, S.; Mishra, S.; Singh, S.S. Spectrum Sensing Techniques in Cognitive Radio Technology: A Review Paper. *J. Commun.* **2020**, *15*, 577–582. [CrossRef]
53. Xinsheng, Z.; Zhou, X.; Zhao, L.; Yu, P.; Dong, Q.; Yang, J.; Liu, J.; Zhao, Y. Research on Dynamic Allocation Strategy of KU Satellite Spectrum Based on Cognitive Radio Technology. *J. Phys. Conf. Ser.* **2020**, *1575*, 012031.
54. Ajitha, C.; Jaya, T. Cognitive radio technology with reduced PAPR and complexity for IoT-based OFDM networks. *Peer-to-Peer Netw. Appl.* **2020**, *13*, 2065–2077. [CrossRef]
55. Rathee, G.; Ahmad, F.; Kurugollu, F.; Azad, M.A.; Iqbal, R.; Imran, M. CRT-BIoV: A Cognitive Radio Technique for Blockchain-Enabled Internet of Vehicles. *IEEE Trans. Intell. Transp. Syst.* **2020**, *22*, 4005–4015. [CrossRef]
56. Elaty, E.A.; Zekry, A.; El Agooz, S.; Helaly, A.M. Cognitive Radio Techniques for Utilizing the Primary L-Band Distance Measuring Equipment for Aeronautical Communications. *IEEE Access* **2020**, *8*, 124812–124823. [CrossRef]
57. Veeramakali, T. Cognitive Radio Technology Based Spectrum Allocation for D2d-U Communication. *Int. J. Eng. Adv. Technol.* **2019**, *9*, 520–522. [CrossRef]
58. Panda, S. Cognitive Radio Techniques over Conventional Radio Systems. *Int. J. Eng. Adv. Technol.* **2019**, *9*, 3330–3333. [CrossRef]
59. Neeta, S.; Jyoti, S. Queueing analysis of mobile adhoc networks equipped with cognitive radio technology. *ICTACT J. Commun. Technol.* **2018**, *9*, 1875–1879. [CrossRef]
60. Ajmery, S.; Zhao, L.; Xavier, F. Efficient Resource Allocation in Device-to-Device Communication Using Cognitive Radio Technology. *IEEE Trans. Veh. Technol.* **2017**, *66*, 10024–10034.
61. Polak, L.; Kalller, O.; Klozar, L.; Sebesta, J.; Kratochvil, T. Influence of the LTE System using Cognitive Radio Technology on the DVB-T2 System using Diversity Technique. *Autom. Časopis Autom. Mjer. Elektron. Računarstvo Komun.* **2017**, *57*, 496–505. [CrossRef]
62. Advancements in Cognitive Radio Technologies—Microelectronics TOE. M2 Presswire. 2017. Available online: https://www.businesswire.com/news/home/20170112005782/en/Advancements-in-Cognitive-Radio-Technologies{-}{-}-Microelectronics-TOE{-}{-}-Research-and-Markets (accessed on 15 May 2023).
63. Le Trong, N.; Chen, W.L.; Hwa, C.H. Standardization and Security for Smart Grid Communications Based on Cognitive Radio Technologies—A Comprehensive Survey. *IEEE Commun. Surv. Tutor.* **2017**, *19*, 423–445.
64. Parnika, D.; Singh, S. Journey of mobile generation and cognitive radio technology in 5G. *Int. J. Mob. Netw. Commun. Telemat. (IJMNCT)* **2016**, *6*, 19.
65. Khandaker, M.R.A.; Wong, K.-K. Guest Editorial for the Special Issue on Self-Optimizing Cognitive Radio Technologies. *Phys. Commun.* **2016**, *19*, 81–83. [CrossRef]
66. Daidj, N.; Salahaldin, L. The impact of cognitive radio technology on mobile network operators' interorganizational networks and business models. *J. High Technol. Manag. Res.* **2016**, *27*, 101–109. [CrossRef]
67. Sharma, S.K.; Chatzinotas, S.; Ottersten, B. In-line interference mitigation techniques for spectral coexistence of GEO and NGEO satellites. *Int. J. Satell. Commun. Netw.* **2016**, *34*, 11–39. [CrossRef]
68. Zhu, J.; Song, Y.; Jiang, D.; Song, H. Multi-Armed Bandit Channel Access Scheme with Cognitive Radio Technology in Wireless Sensor Networks for the Internet of Things. *IEEE Access* **2016**, *4*, 4609–4617. [CrossRef]
69. Kloc, M.; Riess, S.; Brendel, J.; Linz, S.; Gardill, M.; Weigel, R.; Fischer, G. Let's Make Them Cognitive Cognitive Radio Technology Applied to Professional Wireless Microphone Systems. *IEEE Microw. Mag.* **2016**, *17*, 70–78. [CrossRef]
70. Lee, W.; Jung, B.-C.; Ban, T. Cooperative Spectrum Sensing for Cognitive Radio Technology Considering Heterogeneous Primary User. *J. Korean Inst. Inf. Commun. Eng.* **2015**, *19*, 1546–1553. [CrossRef]
71. Saleem, Y.; Rehmani, M.H.; Zeadally, S. Integration of Cognitive Radio Technology with unmanned aerial vehicles: Issues, opportunities, and future research challenges. *J. Netw. Comput. Appl.* **2015**, *50*, 15–31. [CrossRef]
72. Communication Equipment Companies. Terminal Registration Method, Device, and System. U.S. Patent US9622205B2, 11 April 2015.
73. Sharma, S.K.; Bogale, T.E.; Chatzinotas, S.; Ottersten, B.; Le, L.B.; Wang, X. Cognitive Radio Techniques Under Practical Imperfections: A Survey. *IEEE Commun. Surv. Tutorials* **2015**, *17*, 1858–1884. [CrossRef]

74. Khattab, A.; Bayoumi, M.A. An overview of IEEE standardization efforts for cognitive radio networks. In Proceedings of the 2015 IEEE International Symposium on Circuits and Systems (ISCAS), Lisbon, Portugal, 24–27 May 2015; pp. 982–985. [CrossRef]
75. Penagos, H.P.; Bohórquez, M.C.; Rodríguez, D.E.; Maldonado, D.O.; Sánchez, C.L.; Galarza, C.M. Cognitive radio technology in the UHF band. *Tecnura* **2014**, *18*, 138–151.
76. Wei, Z.; Yang, D.; Sang, L. Dynamic system level frequency spectrum allocation scheme based on cognitive radio technology. *China Commun.* **2014**, *11*, 84–91. [CrossRef]
77. Song, M.; Chang, S. A Cognitive Radio Technology for Retransmission in Cellular-Based Emergency Alert Broadcast Systems. *J. Broadcast Eng.* **2014**, *19*, 415–418. [CrossRef]
78. Full Spectrum Inc. System and Method for the Delivery of High Speed Data Services over Dedicated and Non-Dedicated Private Land Mobile Radio (PLMR) Channels Using Cognitive Radio Technology. US Patent 8619804, 31 December 2013.
79. Anwar, M.M. Cognitive Radio Technology for Improving Capacity of Mobile Communication Systems in Palestine. *Palest. Tech. Univ. Res. J.* **2014**, *6*, 19–34.
80. Kumar, M.R.; Kumar, S.M. A Survey on Rural Broadband Wireless Access Using Cognitive Radio Technology. *IEEE 802.22 Wirel. Reg. Area Networks. Univers. J. Commun. Netw.* **2013**, *1*, 27–31.
81. Reyes, H.; Kaabouch, N. Improving the Reliability of Unmanned Aircraft System Wireless Communications through Cognitive Radio Technology. *Commun. Netw.* **2013**, *5*, 225–230. [CrossRef]
82. Baldini, G.; Holland, O.; Stavroulaki, V.; Tsagkaris, K.; Demestichas, P.; Polydoros, A.; Karanasios, S.; Allen, D. The evolution of cognitive radio technology in Europe: Regulatory and standardization aspects. *Telecommun. Policy* **2013**, *37*, 96–107. [CrossRef]
83. Abadie, A.; Wijesekera, D. An approach for risk assessment and mitigation in cognitive radio technologies. *Int. J. Inf. Priv. Secur. Integr.* **2013**, *1*, 344–359.
84. Giorgetti, A.; Sithamparanathan, K. *Cognitive Radio Techniques: Spectrum Sensing, Interference Mitigation, and Localization*; Artech: Boston, MA, USA; London, UK, 2012.
85. Deock-Ho, H. Improvements of UWB System Performance Based on Cognitive Radio Technology in Indoor LOS Wireless Channel Model. *Korea Inst. Inf. Technol. Mag.* **2013**, *11*, 49–55.
86. What Is Cognitive Radio (CR) and How Does It Work? Available online: https://www.techtarget.com/searchnetworking/definition/cognitive-radio (accessed on 15 May 2023).
87. Jacob, P.; Sirigina, R.P.; Madhukumar, A.S.; Prasad, V.A. Cognitive Radio for Aeronautical Communications: A Survey. *IEEE Access* **2016**, *4*, 3417–3443. [CrossRef]
88. Zhang, Z.; Long, K.; Wang, J. Self-organization paradigms and optimization approaches for cognitive radio technologies: A survey. *IEEE Wirel. Commun.* **2013**, *20*, 36–42. [CrossRef]
89. Yao, J.; Huang, L. Analysis on the Application of Cognitive Radio Technology in Electric Power Communication. In Proceedings of the 2012 International Conference on Power Electronics Engineering and Computer Technology (PEECT 2012), Austin, TX, USA, 5–10 May 2012; pp. 218–222.
90. Rohde, J.; Toftegaard, T.S. Adapting Cognitive Radio Technology for Low-Power Wireless Personal Area Network Devices. *Wirel. Pers. Commun.* **2011**, *58*, 111–123. [CrossRef]
91. Marcus, M.J. Can cognitive radio technology help solve some difficult spectrum management issues by creating "virtual guardbands"? *IEEE Wirel. Commun.* **2011**, *18*, 5. [CrossRef]
92. Fitch, M.; Nekovee, M.; Kawade, S.; Briggs, K.; MacKenzie, R. Wireless service provision in TV white space with cognitive radio technology: A telecom operator's perspective and experience. *IEEE Commun. Mag.* **2011**, *49*, 64–73. [CrossRef]
93. The Wireless Innovation Forum. *Wireless Innovation Forum Approves Report on Quantifying the Benefits of Cognitive Radio Technologies*; The Software Defined Radio Forum, Inc./Wireless Innovation Forum: White Rock, BC, Canada, 2011.
94. Atlabachew, M.; Tuko, M.A. Investigation of Adaptive Beamforming Algorithms for Cognitive Radio Technology. *Zede J.* **2011**, *28*, 31–47.
95. Laddomada, M.; Chen, H.H.; Daneshgaran, F.; Mondin, M.; Sadjadpour, H. Spectrum Sharing and Sensing for Future Broadband Networks: The Cognitive Radio Technology. *Int. J. Digit. Multimedia Broadcast.* **2010**, *2010*, 898470. [CrossRef]

sensors MDPI

Article

Free-Rider Games for Cooperative Spectrum Sensing and Access in CIoT Networks

Kejian Jiang [1], Chi Ma [2], Ruiquan Lin [2], Jun Wang [2], Weibing Jiang [2] and Haifeng Hou [1,*]

[1] Guangdong Planning and Designing Institute of Telecommunications Co., Ltd., Guangzhou 510630, China; jkj@gpdi.com
[2] College of Electrical Engineering and Automation, Fuzhou University, Fuzhou 350108, China; 200120077@fzu.edu.cn (C.M.); rqlin@fzu.edu.cn (R.L.); wangjunfzu@fzu.edu.cn (J.W.); 210110008@fzu.edu.cn (W.J.)
* Correspondence: houhaifeng@gpdi.com

Abstract: With the rapid development of technologies such as wireless communications and the Internet of Things (IoT), the proliferation of IoT devices will intensify the competition for spectrum resources. The introduction of cognitive radio technology in IoT can minimize the shortage of spectrum resources. However, the open environment of cognitive IoT may involve free-riding problems. Due to the selfishness of the participants, there are usually a large number of free-riders in the system who opportunistically gain more rewards by stealing the spectrum sensing results from other participants and accessing the spectrum without spectrum sensing. However, this behavior seriously affects the fault tolerance of the system and the motivation of the participants, resulting in degrading the system's performance. Based on the energy-harvesting cognitive IoT model, this paper considers the free-riding problem of Secondary Users (SUs). Since free-riders can harvest more energy in spectrum sensing time slots, the application of energy harvesting technology will exacerbate the free-riding behavior of selfish SUs in Cooperative Spectrum Sensing (CSS). In order to prevent the low detection performance of the system due to the free-riding behavior of too many SUs, a penalty mechanism is established to stimulate SUs to sense the spectrum normally during the sensing process. In the system model with multiple primary users (PUs) and multiple SUs, each SU considers whether to free-ride and which PU's spectrum to sense and access in order to maximize its own interests. To address this issue, a two-layer game-based cooperative spectrum sensing and access method is proposed to improve spectrum utilization. Simulation results show that compared with traditional methods, the average throughput of the proposed TL-CSAG algorithm increased by 26.3% and the proposed method makes the SUs allocation more fair.

Keywords: cognitive IoT; cooperative spectrum sensing; dynamic spectrum access; game theory; energy harvesting

1. Introduction

In recent years, there has been widespread interest in the automation revolution in the industrial, infrastructure, and healthcare fields through the implementation of the Internet of Things (IoT) [1]. However, available wireless spectrum resources are extremely limited, and the large-scale deployment of IoT devices will greatly exacerbate the shortage of spectrum resources and the problem of interference between devices. A large amount of data shows that the utilization of spectrum in the communication field is not high, and many authorized spectrums are not fully utilized, resulting in a huge waste of spectrum resources [2]. To address these challenges, Cognitive Radio (CR) provides an effective solution. In a CR system, Secondary Users (SUs) can access the authorized spectrum of Primary Users (PUs) with little impact on the PUs. Applying CR technology to IoT networks can effectively improve their spectrum utilization efficiency. Therefore, the design and optimization of Cognitive IoT (CIoT) networks based on CR have received considerable attention [3].

Spectrum sensing is a key technology in CR to address the problem of insufficient spectrum resources by reusing underutilized spectrum. However, the detection performance of spectrum sensing may deteriorate significantly due to uncertainties such as noise and channel conditions [4]. Cooperative Spectrum Sensing (CSS) can improve the sensing performance of fading and shadowed channels by combining the results of multiple SUs' global sensing [5]. Specifically, due to factors such as obstruction, SUs may be located in the shadow of PU transmission signals and unable to detect their signals, which results in interference with PU's communication. CSS can effectively address the hidden terminal problem by allowing multiple SUs to cooperate and greatly reducing the impact of shadowed areas on sensing. Furthermore, dynamic spectrum access can greatly improve the spectrum utilization efficiency between SUs and PUs [6]. Dynamic spectrum access technology allows SUs to effectively utilize spectrum resources by "leasing" or "sharing" idle spectrum resources, improving spectrum utilization, and also having significant advantages in fairness and efficiency. This technology can be used either in the overlay mode, where the SUs opportunistically utilize the unused spectrum of the PUs, or in the underlay mode, where both PUs and SUs transmit simultaneously under SINR constraints in the same frequency band [7].

Due to the limited battery capacity of IoT devices, it is difficult for them to run for a long time. Energy Harvesting (EH) allows devices to harvest energy from the environment to refill batteries [8]. For any type of energy source, the amount of energy transferred from the energy source to the energy harvester is not deterministic but random. Collecting energy through Radio Frequency (RF) signals can avoid the influence of harsh environments and light, and compared with solar energy and wind energy, the energy harvesting process is more stable. Therefore, RF EH technology has become an important direction for researchers to study the energy efficiency of CIoT [9]. However, since selfish SUs can harvest more energy for system consumption by free-riding during spectrum sensing slots, the application of EH will exacerbate the free-riding behavior of selfish SUs, leading to degraded system detection performance. Therefore, the application of EH technology needs to consider how to deal with the free-riding problem to ensure more stable and reliable system performance.

In recent years, people have been more committed to developing distributed intelligent collaboration systems due to reasons of scalability and high performance [10]. These systems consist of distributed processing nodes with autonomous learning capabilities and usually include a large number of free-riding nodes [11]. However, these free-riding nodes will result in fewer nodes providing resources, and the entire system will face the problem of insufficient utilization of resources, and the purpose of sharing will gradually disappear. In CSS, SUs report their sensing results to the FC by control channel and the final channel state is determined by the FC. However, SUs may be selfish, and they may adopt a free-riding strategy when multiple SUs occupy different sub-bands of a PU's spectrum and can occasionally overhear the sensing results of other SUs. Free-riders tend to wait for other SUs to sense them first and take advantage of other SUs' results so that they can reserve more time for their own data transmission. Therefore, due to the existence of free-riding behavior, there are two types of SUs with different sensing attitudes in the system: Contributing Users (CUs) who participate in cooperative sensing and Free-riding Users (FUs) who are dishonest. CUs increase the amount of information, but regardless of how much they contribute, the added information is shared equally among participants. However, if the cost of contribution exceeds the marginal benefit, SUs will be tempted not to contribute. Free-riders severely reduce the fault tolerance and content availability of the peer-to-peer system. In cooperative systems, the key to solving this problem is to establish incentive mechanisms that aim to improve utility by incentivizing nodes to become more cooperative. Ref. [12] proposed a game-theoretic framework for participating in cooperation and free-riding in federated learning to support spectrum sensing for NextG communications.

2. Relate Work

Spectrum sensing is the basis and key to achieving spectrum sharing, improving the efficiency of spectrum utilization, and ensuring the security of spectrum resources in CR technology. So far, many papers have applied spectrum sensing technology in IoT field. For the issue of energy efficiency in spectrum sensing, Ref. [13] investigated how to use spectrum sensing technology to improve the energy efficiency of CIoT networks, and proposed a new two-way information exchange algorithm and optimal transmit power allocation technique to achieve higher dynamic spectrum sensing capability and data throughput. In addition, cooperative spectrum sensing has been studied extensively. Ref. [14] proposes an integrated CSS and access control model to maximize the throughput of the cognitive industrial IoT by jointly optimizing the sensing time, the number of sensing nodes, and the transmit power of each node. Dynamic spectrum access is an important research direction in CIoT networks. It provides an effective way for existing radio networks to optimize system performance, enhance resource utilization, and improve system reliability. In ref. [15], the benefits of cooperative communication and cognitive radio networks (CRN) are merged to meet IoT networks' needs. In this work, the hybrid overlay-underlay CRN is employed to guarantee both SU stability and an acceptable total throughput. In ref. [16], spectrum aware Ad hoc on-demand distance vector routing protocol is proposed without fixed base stations, aiming to address the routing needs in future IoT environments.

There are now numerous papers focusing on free-riding behavior. In ref. [17], Adar et al. analyzed the user traffic in the Gnutella system and found that there were a large number of free-riding users in the system, with only about 28% of the users sharing files. To address the problem of the behavior of free-riders in the system, Ref. [18] proposes a peer-to-peer protocol with adaptive and self-organizing topology to punish malicious peers and free-riders by considering the trustworthiness. In ref. [19], machine learning based malicious signal detection is employed for cognitive radio networks. In [20], To enhance cooperation among agents, one of the main goals of the multi-agent system is to solve the possible free-riding problem. To address this problem, the authors propose a novel strategy that allows agents to manipulate the utility of each adversary simultaneously, further promoting mutual cooperation among all agents. Although the above literature has emphasized the users' free-riding problem, it does not consider the scheduling problem of dynamic spectrum access techniques for SUs to further improve the system's utility.

Game theory [21] is an effective tool for analyzing optimal behavior among rational decision makers. Game theory has been widely used in distributed CR networks to effectively improve system performance and to help SUs make better use of spectrum resources. With the increasingly complex interaction behavior of IoT devices [22], single-layer networks can no longer meet the needs of the system, and the study of multilayer networks has become a key to the development of many fields. To explore the cooperative behavior among rational individuals, researchers have combined complex networks and game theory [23], which is more helpful to study the interaction behavior problem among users in complex networks. Ref. [24] explored an important tradeoff of CSS among SUs in cognitive radio networks and proposed a distributed coalition formation game based on genetic algorithms to select the optimal coalition leader to motivate SUs to actively participate. Ref. [25] proposed a potential game-based spectrum access and power control method to solve the problem of how to achieve optimal throughput in a multibeam cognitive satellite communication network. Furthermore, game theory is also widely applied to solving the free-rider problem. In ref. [26], the free-riding problem was first modeled as a public goods game. In ref. [27], the authors considered an evolutionary game model that allows SUs to choose between free-riding and normal sensing strategies.

Although [12,26,27] considered two sensing attitudes of SUs, they did not address the scheduling problem of SUs when sensing multiple PU channels, which did not work well in multi-channel cooperative spectrum sensing. In the case where SUs cooperatively sense multiple PU channels, most SUs may tend to sense the same PU channel due to geographical location and other factors, resulting in a serious shortage of SUs sensing other PU channels, which greatly affects the detection performance. This may lead to a waste of spectrum resources and even affect the normal communication of PUs. Therefore, it is necessary to consider using dynamic spectrum access to solve the scheduling problem of SUs in this case.

Different from past research, this paper focuses on the problem of SUs' free-riding behavior in Energy Harvesting CIoT (EH-CIoT) networks. A punishment mechanism is established to stimulate SUs to sense the spectrum normally during the sensing process. For the system model with multiple PUs and SUs, a two-layered game-based cooperative spectrum sensing and access method is proposed to dynamically coordinate the strategies of SUs. Simulation results show that the proposed method significantly improves the overall throughput of the system, and the allocation of SUs in the coalition is more fair compared with traditional methods.

The rest of this paper is organized as follows: Section 3 describes the system model and time slot structure for cooperative spectrum sensing and access and gives the optimization problem modeling of this paper. Section 4 proposes a two-layer game to solve the proposed problems in Section 3. In Section 5, simulation results are presented. Finally, the conclusion is made at last in Section 6.

3. System Model and Problem Description

3.1. System Models

Consider an EH-CIoT network covered by M PU bands, which share channels with N SUs in overlay mode. When the PU bands are not occupied, each SU can only sense and access one band in the same time slot due to the hardware limitation. As shown in Figure 1, when SUs upload their sensing results, there is a possibility that the sensing results will be listened to by other users, and the listeners will decide to become CUs or FUs according to their utility. In each time slot, PU bands are sensed simultaneously. The SUs are divided into M coalitions for sensing different PU bands, and each SU can freely choose the sensing attitudes (to become CU or FU) and the band of which PU to access. In each coalition, the SU with the highest detection probability is considered the Coalition Head (CH). CH plays the role of FC in the coalition.

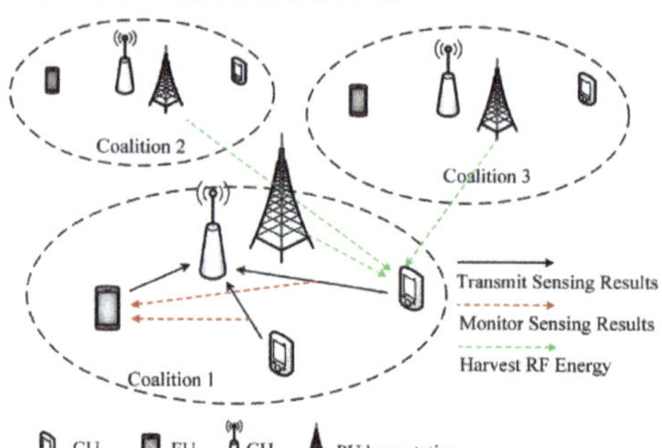

Figure 1. System model.

Due to its simplicity and low cost, the energy detector is chosen in this paper as the spectrum sensing technique to sense the PU. The noise is assumed to be an independent, identically distributed random process with a zero mean and a variance σ^2. In the Rayleigh fading environment, the detection probability and false alarm probability of the i-th secondary user SU_i detecting the state of the j-th primary user PU_j channel are:

$$P_{d,i,j} = e^{-\frac{\theta_j}{2}} \sum_{n=0}^{m-2} \frac{1}{n!} \left(\frac{\theta_j}{2}\right)^n + \left(\frac{1+\bar{\gamma}_{i,j}}{\bar{\gamma}_{i,j}}\right)^{m-1} \times \left[e^{-\frac{\theta_j}{2(1+\gamma_{i,j})}} - e^{-\frac{\theta_j}{2}} \sum_{n=0}^{m-2} \frac{1}{n!} \left(\frac{\theta_j \bar{\gamma}_{i,j}}{2(1+\bar{\gamma}_{i,j})} \right)^n \right] \quad (1)$$

and

$$P_{f,i,j} = \frac{\Gamma\left(m, \frac{\theta_j}{2}\right)}{\Gamma(m)} \quad (2)$$

where θ_j is the detection threshold of PU_j, m is the time-bandwidth product, and $\bar{\gamma}_{i,j}$ denotes the average signal-to-noise ratio of the received signal from PU_j to SU_i, $\Gamma(.,.)$ is the incomplete gamma function, and $\Gamma(.)$ is the gamma function. When detecting the PU channel, the probability of missing detection is:

$$P_{m,i,j} = 1 - P_{d,i,j} \quad (3)$$

In the coalition Ω, since the control channel transmits sensing results of 0 and 1, the corresponding error reporting probability [28] can be expressed as follows:

$$P_{e,i,k} = \frac{1}{2}\left(1 - \sqrt{\frac{\bar{\gamma}_{i,k}}{1+\bar{\gamma}_{i,k}}}\right) \quad (4)$$

where $\bar{\gamma}_{i,k}$ is the average SNR for bit reporting between SU_i and CH_k. The decision fusion criterion at CHs is the OR verdict criterion. By cooperative sensing, the detection probability and the false alarm probability of the coalition Ω are respectively given as follows:

$$P_{d,e}^{or} = 1 - \prod_{i \in \Omega} [P_{m,i}(1 - P_{e,i,k}) + (1 - P_{m,i})P_{e,i,k}] \quad (5)$$

and

$$P_{f,e}^{or} = 1 - \prod_{i \in \Omega} \left[\left(1 - P_f\right)(1 - P_{e,i,k}) + P_f P_{e,i,k} \right] \quad (6)$$

where P_f, $P_{m,i}$ and $P_{e,i,k}$ are given by Equations (2)–(4).

3.2. Time Slot Structure

In this paper, the harvested energy is derived from the RF energy transmitted by the PU. As shown in Figure 2, the sensing result of the sensing phase directly affects the action of the SU in the next phase. If the PU is present, the SU harvests energy in the data transmission phase, and if the PU is not present, the SU transmits data in the data transmission phase.

CUs follow the time frame structure shown in Figure 2a, which can be divided into three phases: energy harvesting phase, the spectrum sensing phase, and the data transmission phase. FUs follow the time frame structure shown in Figure 2b and can be divided into two phases, energy harvesting phase and the data transmission phase. Compared with CUs, FUs are not involved in spectrum sensing, so they have more time to harvest energy and transmit data.

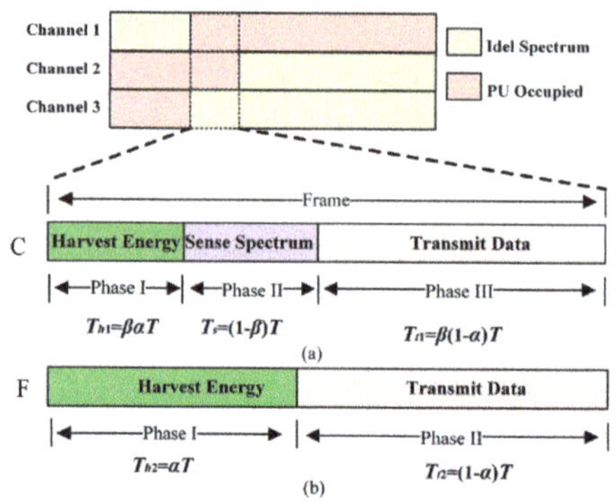

Figure 2. Time slot structure of the proposed method.

If the PU is present, according to the time slot structure, the energy harverted by CUs and FUs can be calculated separately as follows:

$$\begin{aligned} E_{hc}^{i}(\mathrm{C}) &= \sum_{j=1}^{M}\left(1-P_{H_1}^{j}\right)Q_i^j(T_{h1}+T_{t1}) \\ &= \sum_{j=1}^{M}\left(1-P_{H_1}^{j}\right)Q_i^j(\beta T) \end{aligned} \tag{7}$$

and

$$\begin{aligned} E_{hc}^{i}(\mathrm{F}) &= \sum_{j=1}^{M}\left(1-P_{H_1}^{j}\right)Q_i^j(T_{h2}+T_{t2}) \\ &= \sum_{j=1}^{M}\left(1-P_{H_1}^{j}\right)Q_i^j T \end{aligned} \tag{8}$$

If the PU is absent, according to the time slot structure, the energy harvested by CUs and FUs can be calculated separately as follows:

$$\begin{aligned} E_{hc}^{i}(\mathrm{C}) &= \sum_{j=1}^{M} P_{H_1}^{j} Q_i^j T_{h1} \\ &= \sum_{j=1}^{M} P_{H_1}^{j} Q_i^j(\beta\alpha T) \end{aligned} \tag{9}$$

and

$$\begin{aligned} E_{hc}^{i}(\mathrm{C}) &= \sum_{j=1}^{M} P_{H_1}^{j} Q_i^j T_{h1} \\ &= \sum_{j=1}^{M} P_{H_1}^{j} Q_i^j(\beta\alpha T) \end{aligned} \tag{10}$$

where $P_{H_1}^j$ is the probility of the presence of PU$_j$, α and β are the weights, and $0 < \alpha, \beta < 1$, T is the duration of a time slot, Q_i^j is the energy arrival rate at SU$_i$, given by:

$$Q_i^j = P_{p_j} h_{j,i} \tag{11}$$

where P_{p_j} is the transmission power of PU$_j$ and $h_{j,i}$ is the path loss of PU$_j$ and SU$_i$.

When PU_j is absent, and not detected by SU_i , the average throughput of SU_i is

$$R_{H_0} = T_t(1 - P_f)\log_2\left(1 + \frac{P_{s_i}h_{j,i}}{\sigma^2}\right) \quad (12)$$

When PU_j is present, and not detected by SU_i, the average throughput of the SU_i is

$$R_{H_1} = T_t(1 - P_d)\log_2\left(1 + \frac{P_{s_i}h_{j,i}}{\sum_j^M P_{p_j}g_0 + \sigma^2}\right) \quad (13)$$

Let P_{H_0} denotes the probability that the PU is absent, then the total throughput of SU_i is

$$\begin{aligned} R_i =& P_{H_0}R^j_{H_0} + \left(1 - P^j_{H_0}\right)R_{H_1} \\ =& T_t\left[P^j_{H_0}\left(1 - P_f\right)\log_2\left(1 + \frac{P_{s_i}h_{j,i}}{\sigma_i^2}\right)\right. \\ &\left. + \left(1 - P^j_{H_0}\right)(1 - P_d)\log_2\left(1 + \frac{P_{s_i}h_{j,i}}{\sum_j^M P_{p_j}g_0 + \sigma_i^2}\right)\right] \end{aligned} \quad (14)$$

In dynamic spectrum access, it is required that the SUs'operation should not conflict or interfere with the PUs, and P_d should be very close to one. Moreover, we usually have $\sum_j^M P_{p_j}g_0 + \sigma_i^2 > \sigma_i^2$ due to the interference from the PUs to the SUs, the second term on the RHS of (14) is much smaller than the first term on the RHS of (14). To simplify the formulation, the total throughput of SU_i access PU_j channel can be expressed as follows:

$$\begin{aligned} R_i &= P^j_{H_0}R_{H_0} \\ &= T_tP^j_{H_0}\left(1 - P_f\right)\log_2\left(1 + \frac{P_{s_i}h_{j,i}}{\sigma_i^2}\right) \end{aligned} \quad (15)$$

where $P^j_{H_0}$ denotes the probability that the PU_j is not present and R_{H_0} denotes the average throughput of SU_i when PU_j is not present. T_t is the duration of the data transmission phase in a time slot. According to Equation (15), the throughput R_i^{C} of CU_i can be expressed as follows:

$$R_i^{\mathrm{C}} = T_{t1}P^j_{H_0}\left(1 - P_f\right)\log_2\left(1 + \frac{P_{s_i}h_{j,i}}{\sigma_i^2}\right) \quad (16)$$

where $T_{t1} = \beta(1-\alpha)T$, P_{s_i} is the transmission power of SU_i. Similarly, the throughput of FU_i can be expressed as follows:

$$R_i^{\mathrm{F}} = T_{t2}P^j_{H_0}\left(1 - P_f\right)\log_2\left(1 + \frac{P_{s_i}h_{j,i}}{\sigma_i^2}\right) \quad (17)$$

where $T_{t2} = (1-\alpha)T$.

3.3. Optimization Problem Modeling

The SUs aim to accomplish a joint task, which is to protect the PU from interference, achieve a required detection threshold for fusion detection probability through CSS, and obtain high throughput by accessing the PU spectrum. CUs can eavesdrop on sensing results and have more time for their own data transmission. However, if a sufficient number of SUs do not sense the PU channel, all of them may get very low throughput. Therefore, the SUs need to try different strategies in each time period and learn the optimal strategy from their strategic interactions.

The utility of SUs consists of four components: the benefit obtained based on throughput, the benefit obtained from harvesting energy, the cost of consuming energy during data transmission, and the penalty for free-riders based on the coalition detection probability. Considering that FUs do not contribute to the coalition detection probability, a penalty needs to be imposed on them. The penalty function can be expressed as follows:

$$\chi(P_d^j) = \lambda \min\left\{1, -\log S\left(P_d^j\right)\right\} \tag{18}$$

where λ is a predetermined parameter defining the harshness of the penalty, and P_d^j is the detection probability of detecting PU_j. The sigmoid function for the satisfaction degree of the detection performance is calculated as follows:

$$S\left(P_d^j\right) = \frac{1}{1 + e^{-\rho\left(P_d^j - \widetilde{P}_d\right)}} \tag{19}$$

where $\widetilde{P}_d$ is the predefined requirement for the uncertainty, and ρ decides on the steepness of the satisfactory curve. The utility functions of CUs and FUs can be defined as follows:

$$U_i(\mathrm{C}) = f_c\left(R_i^{\mathrm{C}}\right) - E_{t1} + E_{hc}^i(\mathrm{C}) \tag{20}$$

and

$$U_i(\mathrm{F}) = f_c\left(R_i^{\mathrm{F}}\right) - E_{t2} - \chi\left(P_d^j\right) + E_{hc}^i(\mathrm{F}) \tag{21}$$

where the first term $f_c(x)$ on the RHS of Equations (20) and (21) is the satisfaction function of each SU with respect to the achievable throughput, and for simplicity we choose to set $f_c(x) = \mu x$, where μ is the uniform unit parameter for the SUs to convert the throughput into the corresponding benefit, E_{t1} and E_{t2} are the energy consumed by CUs and FUs during data transmission. Then Equations (20) and (21) can be expressed as follows:

$$\begin{aligned} U_i(\mathrm{C}) = {} & \mu\beta(1-\alpha)TP_{H_0}^j\left(1-P_f\right)\log_2\left(1+\frac{P_{s_i}h_{j,i}}{\sigma_i^2}\right) \\ & - P_{H_0}^j P_{s_i}\beta(1-\alpha)T \\ & + \sum_{j=1}^{M}\left[P_{H_0}^j Q_i^j(\beta T) + \left(1-P_{H_0}^j\right)Q_i^j(\beta\alpha T)\right] \end{aligned} \tag{22}$$

and

$$\begin{aligned} U_i(\mathrm{F}) = {} & \mu(1-\alpha)TP_{H_0}^j\left(1-P_f\right)\log_2\left(1+\frac{P_{s_i}h_{j,i}}{\sigma_i^2}\right) \\ & - P_{H_0}^j P_{s_i}(1-\alpha)T - \lambda\min\{1, -\log S(P_{d,i})\} \\ & + \sum_{j=1}^{M} Q_i^j\left[P_{H_0}^j T + \left(1-P_{H_0}^j\right)(\alpha T)\right] \end{aligned} \tag{23}$$

To improve the utility of SUs, it is crucial to increase the system throughput and enhance the cooperative detection probability to reduce penalties, as indicated Equations (20) and (21), and it can be inferred that system throughput is primarily influenced by the false alarm probability. Increasing the false alarm probability can improve the system's throughput. In order to improve SUs' throughput, the problem of whether SUs participate in cooperative sensing and which coalitions SUs choose to access is formulated as an optimization problem. Equations (22) and (23) need to be optimized simultaneously, and the problem formulation is given as follows:

$$\max_{\{P_d^j\},\{P_f^j\}} \quad \boldsymbol{U}_i(\mathrm{C}), \boldsymbol{U}_i(\mathrm{F}) \quad \forall i = 1,2,\ldots,\mathrm{N} \tag{24a}$$

$$\text{s.t. } 0 \leq \beta \leq 1 \tag{24b}$$

$$0 \leq \alpha \leq 1 \tag{24c}$$

4. Two-Layer Cooperative Sensing and Access Game (TL-CSAG)

As shown in Figure 3, based on evolutionary game and hedonic coalition game models, an iterative algorithm for solving the game problem mentioned above is proposed in this section. In this paper, the cooperative spectrum sensing and access mechanism is modeled as a two-layer game, where the problem of whether SUs participate in cooperative sensing is modeled as an evolutionary game and the problem of which coalitions SUs choose to access is modeled as a hedonic coalition game. In this game, the SUs are considered participants, and U_i is considered a utility function of SU_i. The SUs' sensing attitude preference in the coalitions can be described as follows:

$$\boldsymbol{o}(t) = \begin{bmatrix} \boldsymbol{o}_1(t) \\ \boldsymbol{o}_2(t) \\ \vdots \\ \boldsymbol{o}_N(t) \end{bmatrix} = \begin{bmatrix} o_{1,1}(t) & o_{1,2}(t) & \ldots & o_{1,M}(t) \\ o_{2,1}(t) & o_{2,2}(t) & \ldots & o_{2,M}(t) \\ \vdots & \vdots & \ddots & \vdots \\ o_{N,1}(t) & o_{N,2}(t) & \ldots & o_{N,M}(t) \end{bmatrix} \tag{25}$$

where $o_{i,j}(t)$ denotes the sensing attitude preference of SU_i in coalition Ω_j in the t-th time slot, $o_{i,j}(t) \in \{0,1\}$. $o_{i,j}(t) = 1$ indicates that SU_i will choose to sense PU_j channel in coalition Ω_j, and $o_{i,j}(t) = 0$ indicates that SU_i will choose to be a free-rider in coalition Ω_j. The SU_i's strategy is denoted by $s_i = (\boldsymbol{o}_i, \omega_{i,j})$ and $\omega_{i,j}$ denotes the SU_i choose the coalition Ω_j. To provide a clearer depiction of how SUs can modify their utility through specific actions, its utility function can be rewritten as follows:

$$U_i(s_i) = U_i(\boldsymbol{o}_i, \omega_{i,j}) = \begin{cases} U_i(\mathrm{C}), & \text{if } o_{i,j} = 1 \\ U_i(\mathrm{F}), & \text{if } o_{i,j} = 0 \end{cases} \tag{26}$$

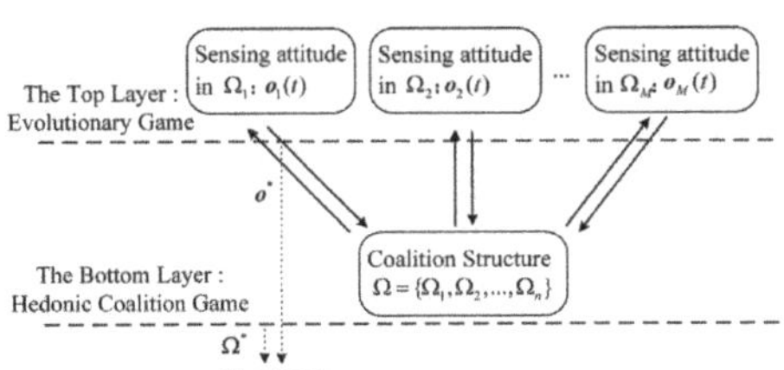

Figure 3. The structure of TL-CSAG model.

Each SU maximize its utility by optimizing its sensing attitude and then enter a coalition. Thus, the optimization problem is formulated as follows:

$$\begin{aligned} &\max_{\{\boldsymbol{o}_i\},\{\omega_{i,j}\}} && U_i(\boldsymbol{o}_i, \omega_{i,j}) \quad \forall i = 1,\ldots,N; \forall j = 1,\ldots,M \\ &\text{s.t.} && \begin{cases} 0 \leqslant \beta \leqslant 1 \\ 0 \leqslant \alpha \leqslant 1 \end{cases} \end{aligned} \tag{27}$$

From Equations (24)–(27), it can be seen that the strategies of both problems are interdependent. Therefore, the interaction of SUs can be formulated as a noncooperative game. Under the condition of Nash Equilibrium (NE), each SU's strategy cannot get better utility without changing the strategy of other SUs. To comprehensively consider the feasibility and convergence of the two-level game algorithm, the definition of NE is given as follows:

Definition 1. *Let* $\boldsymbol{o}^* = (\boldsymbol{o}_1^*, \ldots, \boldsymbol{o}_N^*)$ *and* $\boldsymbol{\omega}^* = (\omega_1^*, \ldots, \omega_N^*)$ *be the NEs of the sensing attitude strategy and the coalition formation strategy derived by the proposed TL-CSAG, respectively.* $\boldsymbol{s}^* = ((\boldsymbol{o}_1^*, \omega_1^*), \ldots, (\boldsymbol{o}_N^*, \omega_N^*))$, *is an NE of TL-CSAG, and it can be given by:*

$$U_i(s_i^*, s_{-i}^*) \geq U_i(s_i, s_{-i}^*), \forall s_i^* \in s^*, i \in \mathcal{N} \tag{28}$$

where $s_i^* = (\boldsymbol{o}_i^*, \omega_i^*)$, s_{-i}^* *is the stable strategy for all SUs except* SU_i.

4.1. The Top Layer Game: Sensing Attitude Strategy in the Coalition

This subsection analyzes the evolution of SUs' sensing attitudes using the idea of evolutionary game theory. In each time slot, SUs calculate their own utility. If SU_i's utility $\bar{U}_{ij}(\text{C})$ for participating in sensing in Ω_j is higher than the average utility $\bar{U}_{ij}$, then in the next time slot, the probability of SU_i participating in sensing will increase. To describe the evolution of the sensing attitude of SU_i in Ω_j, the following differential equations [29] are constructed as follows:

$$\dot{\mathcal{P}}_i^j(\text{C}) = \frac{\mathcal{P}_i^j(\text{C}, t+1) - \mathcal{P}_i^j(\text{C}, t)}{\mathcal{P}_i^j(\text{C}, t)} = \eta_i[\bar{U}_i^j(\text{C}) - \bar{U}_i^j] \tag{29}$$

where η_i is the adjustment step size determined by SU_i. $\mathcal{P}_i^j(\text{C}, t)$ is the probability that in the t-th time slot SU_i participates in sensing in Ω_j.

In the next time slot, the probability of SU_i participation in sensing in Ω_j can be calculated as follows:

$$\mathcal{P}_i^j(\text{C}, t+1) = \mathcal{P}_i^j(\text{C}, t) + \eta_i\left[\bar{U}_i^j(\text{C}) - \bar{U}_i^j\right]\mathcal{P}_i^j(\text{C}, t) \tag{30}$$

Equation (30) describes the dynamic process of SU_i choosing its sensing attitude, where the sum of the probabilities of participating in sensing and free-riding is equal to 1.

This paper first study the game of two SUs in Ω_j, i.e., $\Omega_j = \{g_1, g_2\}$. The payoff table of the two SUs is shown in Table 1 according to Equations (19) and (20), where $A = 1 - P_f^{\text{C}}$, $B_i = 1 - P_{fi}$, $D_i = \mu(1-\alpha)TP_{H_0}^j \log_2\left(1 + \frac{P_{s_i} h_{j,i}}{\sigma_i^2}\right)$, $E_i = \sum_{j=1}^{\text{M}}[P_{H_0}^j Q_i^j T + (1 - P_{H_0}^j Q_i^j \alpha T)]$, $F = P_{H_0}^j P_{p_j}(1 - \alpha T)$, $W_i = P_{H_0}^j P_{s_i}(1 - \alpha T) + \lambda \min\{1, -\log S(P_{d,i})\}$.

Table 1. Payoff table of a two-user.

	C	F
C	$\beta(AD_1 + E_1 - F), \beta(AD_2 + E_2 - F)$	$\beta(B_1D_1 + E_1 - F), B_1D_2 - W_2 + E_2$
F	$B_2D_1 - W_1 + E_1, \beta(B_2D_2 + E_2 - F)$	$0, 0$

Let x_1 and x_2 denote the probability that g_1 and g_2 take attitude "C", respectively, and $V_i = E_i - F$, then the expected payoff $\bar{U}_{g_1}(\text{C})$ while g_1 chooses to contribute is expressed as follows:

$$\bar{U}_{g_1}(\text{C}) = \beta(AD_1 + V_1)x_2 + \beta(B_1D_1 + V_1)(1 - x_2) \tag{31}$$

The mean utility function of g_1 is given by:

$$\begin{aligned}\bar{U}_{g_1} &= \beta(AD_1 + V_1)x_1x_2 + \beta(B_1D_1 + V_1)x_1(1 - x_2) \\ &\quad + (B_2D_1 + E_1 - W_1)(1 - x_1)x_2\end{aligned} \tag{32}$$

Therefore, the replicator dynamics of the two SUs using Equation (26) will be:

$$\mathcal{G}(x_1) = x_1(1 - x_1)(G_1x_2 + H_1) \tag{33}$$

and

$$\mathcal{F}(x_2) = x_2(1 - x_2)(G_2x_1 + H_2) \tag{34}$$

where $H_i = \beta(B_iD_i + V_i)$, $G_1 = \beta(AD_1 - B_1D_1) - (B_2D_1 + E_1 - W_1)$ and $G_2 = \beta(AD_2 - B_2D_2) - (B_1D_2 + E_2 - W_2)$. According to the conditions of equilibrium, we have $\dot{x_1} = 0$ and $\dot{x_2} = 0$, then, we get four equilibrium points: $(0,0)$, $(0,1)$, $(1,0)$, $(1,1)$, and the mixed strategy equilibrium $(-\frac{E_2}{G_2}, -\frac{E_1}{G_1})$. The Jacobian matrix is formed by taking the partial derivatives of Equations (38) and (39), then we can obtain:

$$J_m = \begin{bmatrix} \frac{\partial \mathcal{G}(x_1)}{\partial x_1} & \frac{\partial \mathcal{G}(x_1)}{\partial x_2} \\ \frac{\partial \mathcal{F}(x_2)}{\partial x_1} & \frac{\partial \mathcal{F}(x_2)}{\partial x_2} \end{bmatrix} \tag{35}$$

where

$$\begin{cases} \frac{\partial \mathcal{G}(x_1)}{\partial x_1} = (1 - 2x_1)(G_1x_2 + H_1) \\ \frac{\partial \mathcal{G}(x_1)}{\partial x_2} = G_1x_1(1 - x_1) \\ \frac{\partial \mathcal{F}(x_2)}{\partial x_1} = G_2x_2(1 - x_2) \\ \frac{\partial \mathcal{F}(x_2)}{\partial x_2} = (1 - 2x_2)(G_2x_1 + H_2) \end{cases} \tag{36}$$

From the stability conditions of the Jacobian, the system converges when $\det(J_m) > 0$ and $\mathrm{tr}(J_m) < 0$. The stability conditions for the four equilibrium points are:

(1) When $\beta < \frac{D_2B_1+V_2-W_2}{D_2A+V_2-W_2}$, the strategies of g_1 and g_2 converge to (C, F).

(2) When $\beta < \frac{D_1B_2+V_1-W_1}{D_1A+V_1-W_1}$, the strategies of g_1 and g_2 converge to (F, C).

(3) When $\beta > \frac{D_2B_1+V_2-W_2}{D_2A+V_2-W_2}$ and $\beta > \frac{D_1B_2+V_1-W_1}{D_1A+V_1-W_1}$, the strategies of g_1 and g_2 converge to (C, C).

(4) When $\beta < \frac{D_2B_1+V_2-W_2}{D_2A+V_2-W_2}$ and $\beta < \frac{D_1B_2+V_1-W_1}{D_1A+V_1-W_1}$, the strategies of g_1 and g_2 converge to (C, F) or (F, C) based on the initial adopted strategies.

The above mentioned has demonstrated the $\boldsymbol{o}^*$ of the two SUs. However, due to the different utility functions of different SUs, it is difficult to obtain $\boldsymbol{o}^*$ in the case of multiple SUs. Through repeated games, SUs can autonomously adapt to the changing environment, and each SU has a clear sensing strategy.

4.2. The Bottom Layer Game: Coalition Formation Strategy

Regardless of what sensing attitude is adopted or which coalition is accessed, it will directly affect the utility of SUs. In the bottom layer game, the method used by SUs to solve the problem of which coalition to access is modeled as a hedonic coalition game. Next, the relevant definitions and related theorems are stated and proved.

Definition 2. *(Switch Rule): Given a partition $\Pi = \{\Omega_1, \ldots, \Omega_m, \ldots \Omega_M\}$ of SUs' set $\mathcal{N}$, $SU_i \in \Omega_m$ decides to leave it current coalition Ω_m and join another coalition $\Omega_{m'} \in \Pi$, where $m \neq m'$, if and only if $\Omega_{m'} \cup \{i\} \succ_i \Omega_m$, where $\succ_i$ is the preference relation of SU_i. As a result, $\{\Omega_m, \Omega_{m'}\} \to \{\Omega_m \backslash \{i\}, \Omega_{m'} \cup \{i\}\}$.*

In order to evaluate the preferences of SU_i over its own sets of possible coalitions, the concept of preference relation is introduced [30] as follows:

$$\Omega_{m'} \succeq_i \Omega_m \Leftrightarrow u_i^{\Omega_{m'}} \geq u_i^{\Omega_m} \tag{37}$$

where the relationship of $\Omega_{m'} \succeq_i \Omega_m$ means that SU_i prefers to join coalition $\Omega_{m'}$ over coalition Ω_m. The $u_i^{\Omega_m}$ which is the preference function of SU_i in Ω_m can be expressed as follows:

$$u_i^{\Omega_m} = \begin{cases} U_i^{\Omega_m}(\boldsymbol{o}_{i,m}, \Omega_m), & \text{if } \Omega_m \notin h(i) \\ -\infty, & \text{otherwise} \end{cases} \tag{38}$$

where $U_i^{\Omega_m}$ is the utility of SU_i in coalition Ω_m and $h(i)$ is the history set of SU_i. The history set contains the coalitions that SU_i has joined prior to formation of the current partition Π. In general, considering the history set can accelerate the convergence of the system.

According to the preference relationship of SU in Equation (34), it can be considered that when the history set of SU_i is not considered:

$$\Omega_{m'} \succ_i \Omega_m \Leftrightarrow U_i(\boldsymbol{o}_{i,m'}, \Omega_{m'}) > U_i(\boldsymbol{o}_{i,m}, \Omega_m) \tag{39}$$

Theorem 1. *In TL-CSAG, the convergence rate of the sensing attitude strategy of SUs always precedes that of the coalition formation strategy. That is, the formation of $\boldsymbol{o}^*$ precedes the formation of $\boldsymbol{\omega}^*$.*

Proof. When the sensing attitude strategy does not reach $\boldsymbol{o}^*$, according to Equation (26), the utility function of SUs has two states, and the utility is unstable. According to the description of Equation (39), when SU_i exchanges its coalition, it may have $U_i(\boldsymbol{o}_{i,m'}, \Omega_{m'}) > U_i(\boldsymbol{o}_{i,m}, \Omega_m)$ and will join coalition $\Omega_{m'}$. However, in the subsequent time slots, due to the change of $\boldsymbol{o}^*$, $U_i(\boldsymbol{o}_{i,m'}, \Omega_{m'}) > U_i(\boldsymbol{o}_{i,m}, \Omega_m)$ may occur, and SU_i will switch coalition again. □

After $\boldsymbol{o}$ reaches $\boldsymbol{o}^*$, SUs determines their possible strategy of obtaining high returns $\boldsymbol{o}^*$, and will focus on the distribution of SUs in the coalition. When a division Π^* is a NE, it means that there is no coalition that makes the SU strictly like to join, while the other coalitions are not hurt by the formation of this new coalition. The NE formed by the hedonistic coalition is specifically defined as follows:

Definition 3. *A parition $\Omega^* = \{\Omega_1^*, \Omega_2^*, \ldots, \Omega_M^*\}$ is NE if $\forall SU_i \in \Omega_j^*$ with $\forall j \in \mathcal{M}, \Omega_j^* \succeq_i \Omega_l \cup \{i\}, \forall l \in \mathcal{M}$.*

Theorem 2. *Once the sensing attitudes of SUs reache the stability strategy $\boldsymbol{o}^*$, the final coalition will also achieve the stability strategy Π^*.*

Proof. After each switching operation, SUs will obtain higher utility in the new coalition. Given the number of channels M in the CIoT network and the number of SUs N, the total number of different partitions is M^N, which is a finite number. Thus, from any given initial partition Π_0, the switching operation always terminates at some point after a finite number of iterations, where the coalition structure converges to the final partition Π^*. □

4.3. Algorithm Steps

Based on the analysis of the above two parts, the whole process of the proposed cooperative spectrum sensing and access algorithm based on the two-layer game can be obtained. During the initialization phase, each coalition selects the SU with the highest detection probability among all SUs as its CH. Furthermore, once a CH is selected by a coalition, it cannot be chosen by any other coalition, and it must always have a cooperative attitude. Furthermore, $\boldsymbol{o}^*$ of each coalition's sensing attitude strategy is obtained by the evolution among the SUs. On the basis of $\boldsymbol{o}^*$, the ω^* formed by SUs can be obtained through the switch rules, and the specific steps are shown in Algorithm 1.

Algorithm 1 The proposed two-layer cooperative sensing and accsee game

Require: Location of the SU and the PU
Ensure: Stable sensing attitudes $\boldsymbol{o}^*$; Stable sensing coalitions Π^*;
1: Initialization: Set iterations $t = 1$, initial sensing strategy probability $\mathcal{P}_0 = 50\%$, initialize parameter μ and η; Randomly initialize the attitudes of SUs $\boldsymbol{o}_0$; Randomly and evenly distribute the SUs to each coalition Π_0;
2: **while** $s \neq s^*$ and $t \in 1 : \text{MAX}$ **do**
3: $t = t + 1$;
4: Randomly select two coalitions Ω_n and Ω_m from the coalitions set Π , and then switch $\text{SU}_i \in \Omega_n$ to the coalition Ω_m, etc., $\{\Omega_n, \Omega_m\} \rightarrow \{\Omega_n', \Omega_m'\} = \{\Omega_n \backslash \{i\}, \Omega_m \bigcup \{i\}\}$;
5: SU_i chooses a strategy $o_{i,m}$ with probability $\mathcal{P}_i^m(C, t)$;
6: Calculate the utility $U_i(o_{i,n}, \Omega_n)$ of SU_i, and calculate the utility $U_i(o_{i,n}, \Omega_m')$ of the coalition Ω_m' at which SU_i is exchanged;
7: SU_i decides whether to join a new coalition based on Equation (37);
8: Update $\mathcal{P}_i^n(C, t)$ and $\mathcal{P}_i^m(C, t)$ based on Equations (29) and (30);
9: **while** $\boldsymbol{o} = \boldsymbol{o}^*$ **do**
10: Set $t = t + 1$;
11: Repeat steps 4, 6, 7;
12: Set $\boldsymbol{o} \leftarrow \boldsymbol{o}^*$;
13: **end while**
14: **end while**

5. Simulation Analysis and Evaluation

In this section, MATLAB is used to simulate the performance of the proposed method. The spatial environment is established in a 1 km $\times$ 1 km planar coordinate system with 3 PUs and 15 SUs randomly distributed in the coordinate system; the sampling frequency is $f_s = 1$ MHz, the bandwidth is 2 MHz and the time of each frame is T = 10 ms. The rest of the simulation parameters are shown in Table 2. The method proposed in [31] is used here as a comparison, which is denoted by "Con". The random algorithm randomizes the sensing attitudes and the coalitons of all SUs, which are denoted by "Random". In the following section, the final state distribution of SUs, the relationship between detection probability and cooperation probability, which represents the proportion of contributors out of 15 SUs, the factors affecting cooperation probability, the fairness of the SUs distribution in the coalition, and finally the performance of the three algorithms in terms of throughput are investigated.

Table 2. Simulation parameters.

Parameter	Meaning	Value
m	time bandwidth product	5
λ	the parameter to determine the value of penalty	0.02
μ	the parameter to determine the value of revenue	10
η	adjustment step size	3
σ^2	Gaussian noise variance	$1 \times 10^{(-9)}$ W
P_{p_j}	PU transmission power	0.1 W
P_{s_i}	SU transmission power	0.1 W
P_{H_0}	probability that the PU is absent	0.9
β	weighting factor	0.9
α	weighting factor	0.5
ρ	penalty gradient coefficient	20

The initial coalition structure and the final coalition structure obtained from TL-CSGA are shown in Figure 4a,b, respectively. In these two figures, each colored box represents an SU, and the boxes in the same column belong to the same coalition. As can be seen from Figure 4b, SUs are finally distributed in three coalitions, and FUs are all distributed in coalition Ω_2. When the CIoT network is stable, P_f and P_d for each coalition will be fixed. As FUs do not contribute to the two parameters, switching their coalitions after the system is stable will not affect the interests of other SUs. Eventually, all FUs will be concentrated in the coalition with the highest benefit. Since the number of CUs will affect P_f, it is not advisable for each coalition to have too many CUs; otherwise, it will reduce the throughput of the system. Generally, CUs will be roughly evenly distributed in each coalition.

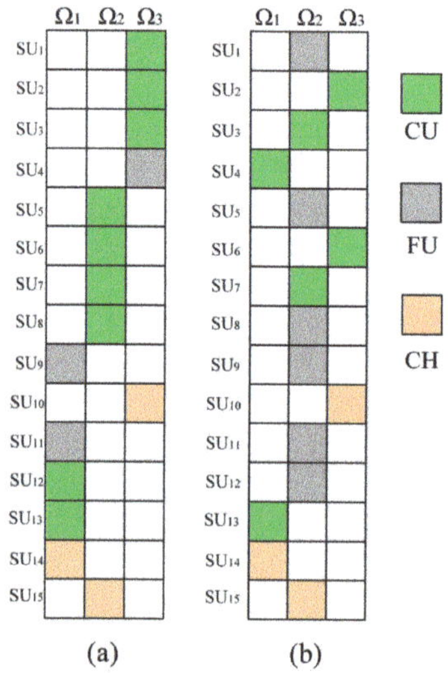

Figure 4. (**a**) Initial and (**b**) Final coalition structures.

Figure 5 shows the relationship curves between the cooperation probability and the detection probability. As can be seen from Figure 5, with a certain number of SUs, the average detection probability of coalitions increases as the cooperation probability increases. In CSS, to avoid interference from SUs to PUs, the coalition detection probability P_d needs to reach a certain threshold of θ_{th}, i.e., $P_d > \theta_{th}$. As a certain number of SUs participate, when the cooperation probability reaches a certain value, the coalition detection probability P_d can meet the requirement of $P_d > \theta_{th}$. This paper assumes that the coalition detection probability needs to meet $P_d > 0.95$.

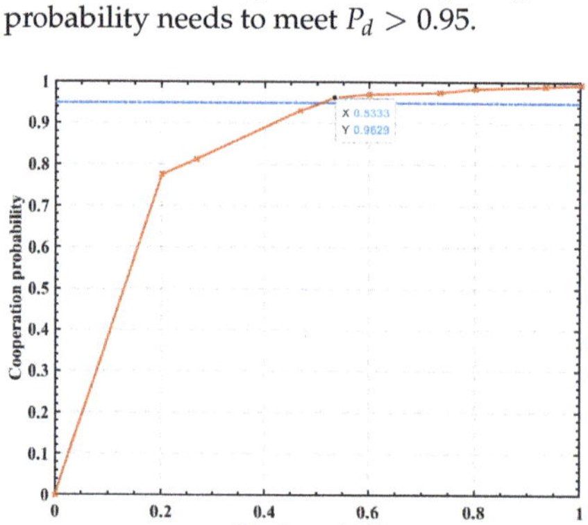

Figure 5. The relationship between cooperation probability and detection probability.

Figure 6 shows the effect of λ and $\widetilde{P}_d$ on the probability of cooperation with the 15 SUs. It can be seen that the increased value of λ is more likely to prompt SUs to adopt a cooperative sensing attitude. Moreover, it may also increase the probability of SUs cooperation with increasing $\widetilde{P}_d$ values.

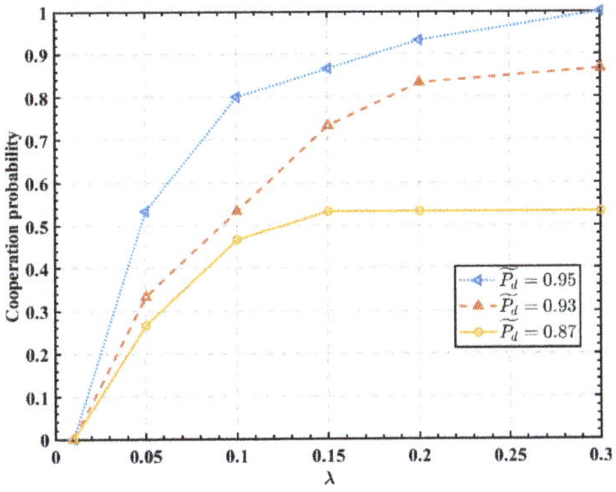

Figure 6. Effect of the parameters λ and $\widetilde{P}_d$ on the cooperation probability.

Figure 7 shows the effect of ρ, $\widetilde{P}_d$, and detection probability P_d on the penalty function. As shown in Figure 7, the higher the detection probability of the coalition, the smaller the penalty to FUs in the coalition. For the same detection probability, decreasing $\widetilde{P}_d$ will reduce the penalty to FUs. In Figure 6, the effect of $\widetilde{P}_d$ on the cooperation probability is essentially the effect of $\widetilde{P}_d$ on the penalty function, which in turn affects the cooperation probability of SUs.

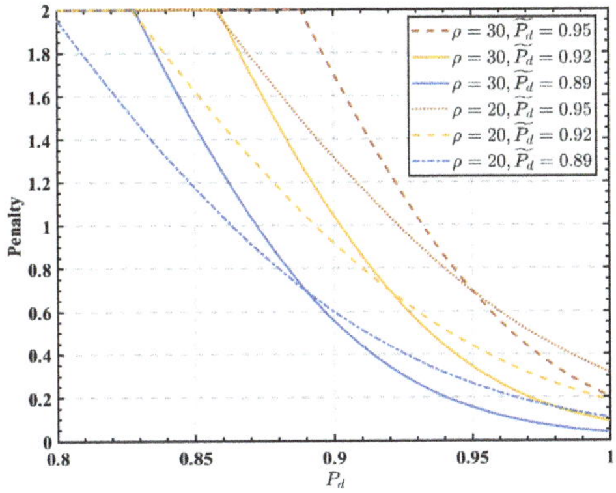

Figure 7. Effect of the coalition detection probability on the user penalty.

Figure 8 shows the effect of the energy harvested by FUs and the weight factor β on the cooperation probability of the system. In Figure 8, it can be observed that the more energy harvested by FUs, the lower the cooperation probability. The reason is that free-riders can obtain high profits by harvesting more RF energy. Similarly, a lower weight factor β leads to a lower probability of cooperation. Under a certain amount of energy harvesting by FUs, increasing β can improve energy harvesting benefits, reduce consumption during spectrum sensing, and improve data transmission efficiency. The simulation results show that the EH technology can provide more benefits to selfish SUs, and its application may exacerbate their free-riding behavior.

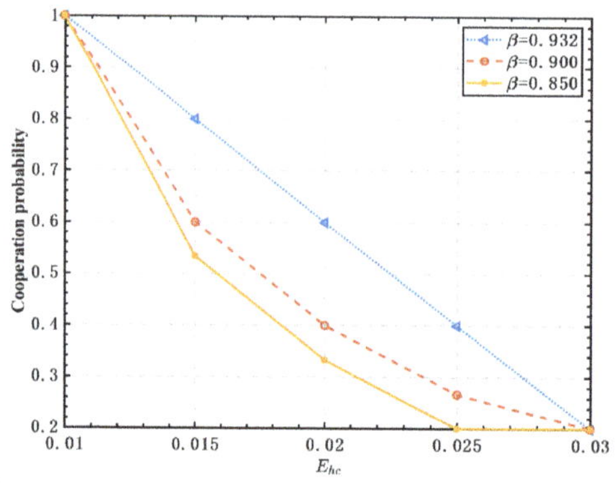

Figure 8. Effect of the energy harvested by FUs on the cooperation probability.

In order to evaluate the fairness of SUs' allocation of each coalition, Jain's fairness is introduced [32], which is defined as follows:

$$\mathcal{F} = \frac{(\sum_{i=1}^{n} x_i)^2}{n \sum_{i=1}^{n} x_i^2} \tag{40}$$

Jain's fairness evaluates the fairness of a set of values, where there are n coalitions and x_i is the value allocated to SU$_i$, and it is located in $[1/N, 1]$. $1/N$ corresponds to the minimum fair allocation where only one SU obtains a non-zero value, and 1 corresponds to the maximum fair allocation where all SUs receive the same value.

By adjusting the cooperation probability to ensure $P_d > 0.95$, the fairness of different algorithms in SUs' allocation was analyzed. The fairness indices of the three coalitions in Figure 4b are shown in Table 3, which lists the detection probabilities, false alarm probabilities, and corresponding fairness $\mathcal{F}$ of the three coalitions using TL-CASG and "Con" algorithms. It can be seen that the fairness index of the TL-CASG algorithm is superior to that of the "Con" algorithm in terms of both detection probability and false alarm probability. This is because the "Con" algorithm only considers the distance between SUs and PUs when forming coalitions without considering the influence of SUs in other coalitions globally.

Table 3. Coalition fairness analysis.

Coalition	P_d-TL-CSAG	P_d-Con	P_f-TL-CSAG	P_f-Con
Ω_1	0.9729	0.9653	0.2849	0.3751
Ω_2	0.9846	0.9934	0.1611	0.1718
Ω_3	0.9613	0.9583	0.2139	0.4139
Fairness $\mathcal{F}$	0.9999	0.9997	0.9495	0.9002

The relationship between throughput and the number of iterations of three algorithms is compared in Figure 9. By modeling the problem of which PU channel is sensed by SUs, the proposed algorithm can fully utilize the sensing capability of SUs in the multi-PU model and improve the overall system throughput. After reaching an equilibrium state in the "Con" algorithm and TL-CSAG, the average throughput of the proposed TL-CSAG algorithm increased by 78% and 26.3%, respectively, compared with the "Random" algorithm and "Con" algorithm.

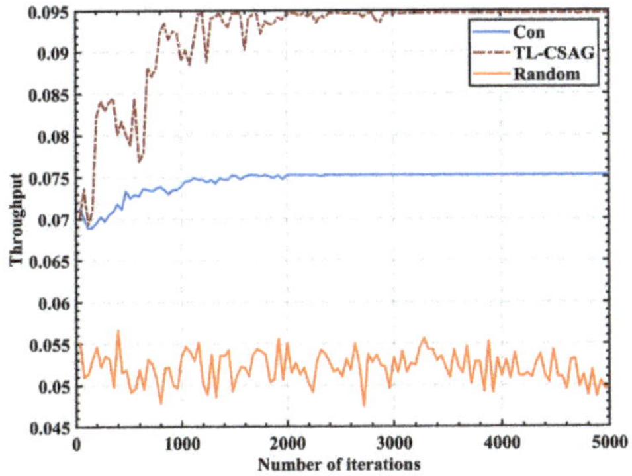

Figure 9. Curve of throughput with number of iterations.

6. Conclusions

On the basis of the EH-CIoT model, this paper considers the free-riding problem of SUs. In order to prevent low system detection performance due to excessive SUs' free-riding behavior, this paper establishes a penalty mechanism to stimulate SUs to sense the PU spectrum normally. To address the issue of whether SUs should free-ride and which PU spectrum to sense and access to gain higher benefits, a two-layer game-based cooperative spectrum sensing and access method is proposed to improve spectrum utilization and the fairness of the proposed method. Simulation results show that after the system reaches equilibrium, CUs are uniformly distributed among the three coalitions, but FUs are in the coalition with the highest profit. In the case of a fixed number of SUs, as the probability of cooperation increases, the average detection probability of coalitions also increases. At the same time, this paper also studied the impact of different parameters on the probability of cooperation, and simulation results confirmed that the more energy harvested, the lower the probability of SUs adopting a cooperative sensing attitude. Furthermore, Jain's fairness concept is introduced to evaluate the higher fairness of the coalition in SUs' allocation in our algorithm. Finally, this paper studied the impact of using different algorithms on throughput, and the results show that the proposed algorithm improves throughput by 78% and 26.3% compared with traditional algorithms. In the future study, we will further explore the secrecy and energy-efficient resource allocation for our proposed network. In the future study, we will further explore the application of free-riders in mobile networks and optimize algorithm models to achieve faster iteration speeds.

Author Contributions: Conceptualization, K.J. and J.W.; methodology, K.J. and C.M.; software, C.M., J.W. and R.L.; validation, J.W., R.L. and H.H.; formal analysis, K.J., H.H., C.M. and R.L.; investigation, K.J. and C.M.; resources, K.J., H.H. and C.M.; data curation, K.J., H.H. and W.J.; writing—original draft preparation, C.M.; writing—review and editing, C.M., W.J. and H.H.; visualization, C.M., R.L. and J.W.; supervision, K.J. and H.H.; project administration, K.J., C.M. and H.H.; funding acquisition, H.H. All authors have read and agreed to the published version of the manuscript.

Funding: This work was supported in part by the Natural Science Foundation of China under Grants No. 61871133 and in part by the Industry-Academia Collaboration Program of Fujian Universities under Grants No. 2020H6006.

Institutional Review Board Statement: Not applicable.

Informed Consent Statement: Not applicable.

Data Availability Statement: Not applicable.

Conflicts of Interest: The authors declare no conflict of interest.

References

1. Al-Fuqaha, A.; Guizani, M.; Mohammadi, M.; Aledhari, M.; Ayyash, M. Internet of things: A Survey on Enabling Technologies, Protocols, and Applications. *IEEE Commun. Surv. Tutor.* **2015**, *17*, 2347–2376. [CrossRef]
2. Voicu, A.M.; Simić, L.; Petrova, M. Survey of Spectrum Sharing for Inter-Technology Coexistence. *IEEE Commun. Surv. Tutor.* **2018**, *21*, 1112–1144. [CrossRef]
3. Giuliano, A.; Hilal, W.; Alsadi, N.; Gadsden, S.A.; Yawney, J. A Review of Cognitive Dynamic Systems and Cognitive IoT. In Proceedings of the 2022 IEEE International IOT, Electronics and Mechatronics Conference (IEMTRONICS), Toronto, ON, Canada, 1–4 June 2022; pp. 1–7.
4. Awin, F.; Abdel-Raheem, E.; Tepe, K. Blind Spectrum Sensing Approaches for Interweaved Cognitive Radio System: A Tutorial and Short Course. *IEEE Commun. Surv. Tutor.* **2019**, *21*, 238–259. [CrossRef]
5. Liu, X.; Zheng, K.; Chi, K. Cooperative Spectrum Sensing Optimization in Energy-harvesting Cognitive Radio Networks. *IEEE Trans. Wirel. Commun.* **2020**, *19*, 7663–7676. [CrossRef]
6. Hu, F.; Chen B.; Zhu, K. Full Spectrum Sharing in Cognitive Radio Networks toward 5G: A Survey. *IEEE Access* **2018**, *6*, 15754–15776. [CrossRef]
7. Liu, X.; Hu, S.; Li, M.; Lai, B. Energy-Efficient Resource Allocation for Cognitive Industrial Internet of Things with Wireless Energy Harvesting. *IEEE Trans. Ind. Inform.* **2021**, *17*, 5668–5677. [CrossRef]
8. Kaswan, A.; Jana P.K.; Das, S.K. A Survey on Mobile Charging Techniques in Wireless Rechargeable Sensor Networks. *IEEE Commun. Surv. Tutor.* **2022**, *24*, 1750–1779. [CrossRef]
9. Vu, H.S.; Nguyen, N.; Ha-Van, N.; Seo, C.; Le, M.T. Multiband Ambient RF Energy Harvesting for Autonomous IoT Devices. *IEEE Microw. Wirel. Compon. Lett.* **2020**, *30*, 1189–1192. [CrossRef]
10. Gharib, A.; Ejaz W.; Ibnkahla, M. Distributed Spectrum Sensing for IoT Networks: Architecture, Challenges, and Learning. *IEEE Internet Things J.* **2021**, *4*, 66–73. [CrossRef]
11. Pal, D.; Khethavath, P.; Thomas, J.P.; Mangla, U. KeyPIn-mitigating The Free Rider Problem in the Distributed Cloud Based on Key, Participation, and Incentive. *J. Cloud Comput.* **2022**, *11*, 8. [CrossRef]
12. Sagduyu Y.E. Free-Rider Games for Federated Learning with Selfish Clients in NextG Wireless Networks. In Proceedings of the 2022 IEEE Conference on Communications and Network Security (CNS), Austin, TX, USA, 3–5 October 2022; pp. 365–370.
13. Ansere, J.A.; Han, G.; Wang, H. A Reliable Energy Efficient Dynamic Spectrum Sensing for Cognitive Radio IoT Networks. *IEEE Internet Things J.* **2019**, *6*, 6748–6759. [CrossRef]
14. Liu, X.; Jia, M.; Zhou, M. Integrated Cooperative Spectrum Sensing and Access Control for Cognitive Industrial Internet of Things. *IEEE Internet Things J.* **2023**, *10*, 1887–1896. [CrossRef]
15. Moayedian, N.S.; Salehi, S.; Khabbazian, M. Fair resource allocation in cooperative cognitive radio iot networks. *IEEE Access* **2020**, *8*, 191067–191079. [CrossRef]
16. Cicioğlu, M.; Çalhan, A.; Miah, M.S. An Effective Routing Algorithm for Spectrum Allocations in Cognitive Radio Based Internet of Things. *Concurr. Comput.* **2022**, *34*, e7368. [CrossRef]
17. Adar, E.; Huberman, B.A. Free Riding on Gnutella. *First Monday* **2000**, *5*, 32–35. [CrossRef]
18. Condie, T.; Kamvar, S.D.; Garcia-Molina, H. Adaptive Peer-to-Peer Topologies.In Proceedings of the Fourth International Conference on Peer-to-Peer Computing, Zurich, Switzerland, 27 August 2004; pp. 53–62.
19. Turkyilmaz, Y.; Senturk, A.; Bayrakdar, M.E. Employing Machine Learning Based Malicious Signal Detection for Cognitive Radio Networks. *Concurr. Comput.* **2023**, *35*, e7457. [CrossRef]
20. Li, K.; Hao, D. Cooperation Enforcement and Collusion Resistance in Repeated Public Goods Games. In Proceedings of the AAAI Conference on Artificial Intelligence, Palo Alto, CA, USA, 17 July 2019; pp. 2085–2092.
21. An, Q.; Cheng, S.; Li, C.; Li, L.; Peng, H. Game Theory-based Control Strategy for Trajectory Following of Four-wheel Independently Actuated Autonomous Vehicles. *IEEE Trans. Veh. Technol.* **2021**, *70*, 2196–2208. [CrossRef]
22. Guo, D.; Fu, M.; Li, H. Cooperation in Social Dilemmas: A Group Game Model with Double-layer Networks. *Future Internet* **2021**, *13*, 33. [CrossRef]
23. Hagen, A.; von Mouche, P.; Weikard, H.P. The Two-stage Game Approach to Coalition Formation: Where We Stand and Ways to Go. *Games* **2020**, *11*, 3. [CrossRef]
24. Rajendran, M.; Duraisamy, M. Distributed Coalition Formation Game for Enhancing Cooperative Spectrum Sensing in Cognitive Radio Ad Hoc Networks. *IET Netw.* **2020**, *9*, 12–22. [CrossRef]
25. Wang, J.; Guo, D.; Zhang, B.; Jia, L.; Tong, X. Spectrum Access and Power Control for Cognitive Satellite Communications: A Game-theoretical Learning Approach. *IEEE Access* **2019**, *7*, 164216–164228. [CrossRef]
26. Kishor, A.; Niyogi, R. A Game-theoretic Approach to Solve The Free-rider Problem. In Proceedings of the 2017 Tenth International Conference on Contemporary Computing (IC3), Noida, India, 10–12 August 2017; pp. 1–6.
27. Elnahas, O.; Elsabrouty, M.; Muta, O.; Furukawa, H. Game Theoretic Approaches for Cooperative Spectrum Sensing in Energy-harvesting Cognitive Radio Networks. *IEEE Access* **2018**, *6*, 11086–11100. [CrossRef]
28. Hao, X.; Cheung, M.H.; Wong, V.W.S.; Leung, V.C.M. Hedonic Coalition Formation Game for Cooperative Spectrum Sensing and Channel Access in Cognitive Radio Networks. *IEEE Trans. Wirel. Commun.* **1978**, *11*, 3968–3979.
29. Taylor, P.D.; Jonker, L.B. Evolutionary Stable Strategies and Game Dynamics. *Math. Biosci.* **2012**, *40*, 145–156. [CrossRef]

30. Cheng, D.; Xu T.; Qi, H. Evolutionarily Stable Strategy of Networked Evolutionary Games. *IEEE Trans. Neural. Netw. Learn. Syst.* **2014**, *25*, 1335–1345. [CrossRef]
31. Jiang, C.; Chen, Y.; Gao, Y.; Liu, K.R. Joint Spectrum Sensing and Access Evolutionary Game in Cognitive Radio Networks. *IEEE Trans. Wirel. Commun.* **2013**, *12*, 2470–2483. [CrossRef]
32. Sediq, A.B.; Gohary, R.H.; Schoenen, R.; Yanikomeroglu, H. Optimal Tradeoff between Sum-rate Efficiency and Jain's Fairness Index in Resource Allocation. *IEEE Trans. Wirel. Commun.* **2013**, *12*, 3496–3509. [CrossRef]

sensors

MDPI

Article

Cooperative Multiband Spectrum Sensing Using Radio Environment Maps and Neural Networks

Yanqueleth Molina-Tenorio [1], Alfonso Prieto-Guerrero [2], Rafael Aguilar-Gonzalez [3,4,*] and Miguel Lopez-Benitez [5,6]

1 Information Science and Technology Ph.D., Metropolitan Autonomous University, Mexico City 09360, Mexico; yanqueleth@xanum.uam.mx
2 Electrical Engineering Department, Metropolitan Autonomous University, Mexico City 09360, Mexico; apg@xanum.uam.mx
3 Faculty of Science, Autonomous University of San Luis Potosi, San Luis Potosi 78210, Mexico
4 Engineering Department, Arkansas State University Campus Queretaro, Queretaro 76270, Mexico
5 Department of Electrical Engineering and Electronics, University of Liverpool, Liverpool L69 3GJ, UK; mlopben@liverpool.ac.uk
6 ARIES Research Centre, Antonio de Nebrija University, 28040 Madrid, Spain
* Correspondence: rafael.aguilar@uaslp.mx

Abstract: Cogitive radio networks (CRNs) require high capacity and accuracy to detect the presence of licensed or primary users (PUs) in the sensed spectrum. In addition, they must correctly locate the spectral opportunities (holes) in order to be available to nonlicensed or secondary users (SUs). In this research, a centralized network of cognitive radios for monitoring a multiband spectrum in real time is proposed and implemented in a real wireless communication environment through generic communication devices such as software-defined radios (SDRs). Locally, each SU uses a monitoring technique based on sample entropy to determine spectrum occupancy. The determined features (power, bandwidth, and central frequency) of detected PUs are uploaded to a database. The uploaded data are then processed by a central entity. The objective of this work was to determine the number of PUs, their carrier frequency, bandwidth, and the spectral gaps in the sensed spectrum in a specific area through the construction of radioelectric environment maps (REMs). To this end, we compared the results of classical digital signal processing methods and neural networks performed by the central entity. Results show that both proposed cognitive networks (one working with a central entity using typical signal processing and one performing with neural networks) accurately locate PUs and give information to SUs to transmit, avoiding the hidden terminal problem. However, the best-performing cognitive radio network was the one working with neural networks to accurately detect PUs on both carrier frequency and bandwidth.

Keywords: multiband spectrum sensing; cognitive radios; radio environment maps; neural networks; cooperative sensor networks; real-time implementation

1. Introduction

Cognitive radio (CR) is a concept involving a communication device capable of knowing the spectral behavior in its environment and adapting to it. Taking advantage of spectral gaps (or holes) not utilized by licensed users, also known as primary users (PUs), CR technology allows nonlicensed or secondary users (SUs) to detect these available parts of the spectrum [1]. Specifically, the operation of CR involves four stages or functions: spectrum sensing, spectrum sharing, spectrum decision, and spectrum mobility. Spectrum sensing (SS) is a fundamental task of detecting one or more PUs; this stage shows whether the sensed spectrum is occupied or empty [2]. Usually, this task is carried out in single bands; however, the current paradigm of multiband spectrum sensing (MBSS) involves multiple bands that are not necessarily contiguous [3].

Many MBSS techniques issued from digital signal processing and machine learning (ML), such as wavelets, compressed sensing, energy detectors, and blind or semiblind methods, have been proposed, primarily in simulated scenarios [4–10]. Other works have been implemented in a real wireless communications environment [11–16], where software-defined radio (SDR) and universal software radio peripheral (USRP) technologies have been employed. Recently, SDR devices such as the HackRF One, the LimeSDR Mini, and the RTL-SDR have become extremely popular because of their affordability and good performances [17,18]. These generic communication devices offer radio equipment the flexibility of a programmable system, allowing the modifications to the communication system behavior simply by changing its software, permitting anyone, including hobbyists on a budget, access to the full radio spectrum.

In addition to knowing the PUs' behavior in a given frequency domain and avoiding the hidden terminal issue, knowledge of the behavior of the radio spectrum in its specific geographical area of influence is necessary. For this reason, the idea of including a geographic tool constructed with the radioelectric information provided by the SUs issued from a real environment is pertinent. Under this perspective, the ability to build a radio environment map (REM) has become very important in recent years. A REM is "a tool that combines information collected from the radio environment, such as received signal intensity, interference measurements, propagation conditions, etc., for specific locations and frequencies, with the aim of building a map that provides an overview of coverage of the network" [19]. Thus, REMs permitting the characterization of the position, directivity, power, and modulation type of PUs have become a challenging task in cognitive radio network (CRN) design [20]. Indeed, in [21], REMs were used to locate relevant PUs in a geographic region of interest, characterizing their positions, directivities, powers, and modulation types. Likewise, in [22], REMs were used to sense the spectrum based on an adaptive compressed spectrum-sensing algorithm, contributing spatial information to the network capable of adapting to the radio environment. REMs are very flexible tools, as shown in [23], where they are combined with ML to determine the effective coverage area perceived by a cognitive sensor network, correctly estimating it at around 92%.

Another important tool, owing to the computing power and the amount of available data, is neural networks (NNs), used widely from pattern recognition and image classification to financial market behavior prediction and autonomous vehicle driving [24]. For spectrum sensing, for example, in [25], an NN was implemented to obtain the local information on single-node spectrum detection (spatial and temporal features). The information (extracted features) from multiple nodes fed another NN, thus permitting cooperation in the CRN. On the basis of REM and NN paradigms, this work proposed a new CRN methodology, implementing an MBSS involving a network integrated with low-cost SDR devices in a controlled, realistic wireless communications environment.

This work developed a CRN for monitoring multiband spectrum and locating PUs with their characteristic bandwidth, carrier frequency, and power in specific areas. In addition, our CRN located the spectral opportunities where the SUs could be located and, unlike models of other studies, could be implemented in real time. In the following, Section 2 briefly lays out the theoretical basis of REMs and NNs, and Section 3 presents the authors' previous work in implementing the MBSS used as a basis for the new proposal presented in this work. Section 4 explains in detail the new proposed methodology, and Section 5 details the implemented real scenario. Finally, Section 6 offers the results, conclusions, and discussion [20].

2. Theoretical Basis of REMs and NNs

2.1. Construction of Radio Environment Maps

REMs are used in telecommunications research to represent the power distribution at a specific area emitted by different radio sources (one or more). A REM is constructed by collecting power measurements at different points in a specific area and interpolating them to produce a graphical representation (map) of the power distribution (or coverage) of the emitted signals, as shown in Figure 1. These maps are generally used to evaluate and predict the behavior of radio signals in a given environment, which is essential for capacity planning, communication system design, and troubleshooting related to interference. In addition, REMs are also used in CR applications, where devices can use the information provided by the REM to select the most suitable frequency to transmit data and avoid interference with other devices in the same location [20].

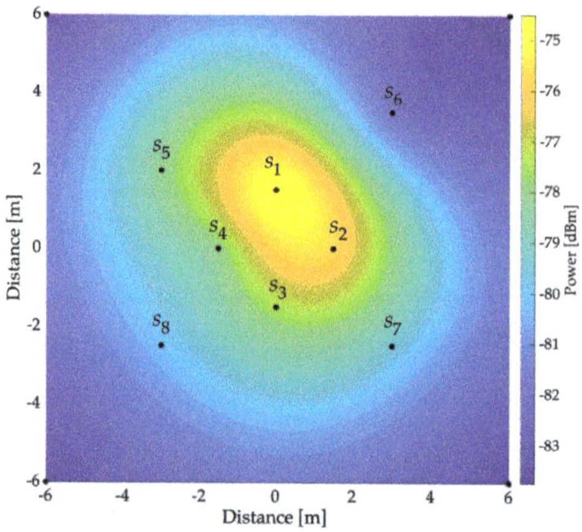

Figure 1. Example of a REM: Black dots indicate locations where the power measurements are taken using eight sensors. In this case, only a transmission source with −60 dBm is considered.

Collecting and interpolating power measurements are the two essential processes in constructing a *correct* map. In this case, interpolation methods [26] are critical for *correctly* estimating an unknown value between two or more known points (measurement points). In other words, a REM is simply a smoothed estimate of the power distribution based on a few known values of measured power. Two interpolation methods widely used to build REMs are inverse distance weighting (IDW) and Kriging, and both are explained below.

2.1.1. IDW Method

The inverse distance weighting interpolation method [27,28] is simple and easy to implement, and it is widely used in applications such as precipitation estimation, topographic data interpolation, and air pollution estimation. IDW is a classical interpolation method in spatial analysis and is most commonly used in geostatistical and mathematical interpolation [29].

IDW estimates the value of a variable at a specific point based on known values at nearby points. This method rests on the idea that nearby points have a greater impact on the estimate than more distant points. Thus, IDW uses a formula that assigns a weight to each known point based on the distance between them and the unknown point. The weights are determined by considering the inverse of the distance, so that the closest points have higher weights and the farther ones have lower weights. For the IDW method, we considered the data transposed vector $\mathbf{z} = (z(s_1), \ldots, z(s_n))^T$ as observations from a random process. In our case, $\mathbf{z}$ contained the average power at each point of the considered sampled region (see Figure 1)

$$\{z(s) : s \in D\}; D \subset \mathbb{R}^2, \tag{1}$$

at known geographic locations $s_1, \ldots, s_n$. The final estimate was obtained using the weighted sum of the known values in the nearby points, given by [28]:

$$Z(s) = \frac{\sum \left(\frac{s_i}{d(s_i,s)^p} \right) (w_i(s_i))}{\sum \frac{1}{d(s_i,s)^p}}, \tag{2}$$

where $Z(x)$ is the estimated value of the variable at the unknown point s, z_i is the known value of the variable at the i-th point s_i, $d(s_i, s)$ is the Euclidean distance between the unknown point s and the point s_i, $p \in \mathbb{R}$ is a smoothing parameter controlling the influence of the known points over the estimated values, and $w_i(s_i)$ is the weight assigned to each known point s_i in order to account for the uncertainty in the known data. In practice, a value of $p = 2$ or $p = 3$ is usually used. The estimated values of $Z(s)$ result in a mesh of unknown points. However, this method has limitations, such as a tendency to smooth out data variability and produce an inaccurate estimate in areas with few known points. In this work, $Z(s)$ represents the power spectral density (PSD) in the different points around the sensed area.

2.1.2. Kriging Method

The Kriging method [30] is an interpolation technique derived from regionalized variable theory. It depends on expressing the spatial variation of the property using the variogram, and it minimizes the estimated prediction errors [31]. The goal of Kriging is to find the most accurate and least uncertain estimate by considering not only the value of the variable at the known points but also the distribution of the variable in space and its correlation with the known points. Following [30], the Kriging method can be established by taking the points and the region mentioned in (1), assuming μ is known and

$$z(s) = \mu + \delta(s); s \in D, \tag{3}$$

are considered with known covariance function:

$$C(s,u) \equiv \mathrm{cov}(z(s), z(u)); s, u \in D, \tag{4}$$

where $\delta(\cdot)$ is a zero-mean stochastic process. Thus, the best linear unbiased predictor $z(s_0)$ is obtained by minimizing [32]

$$E\left(z(s_0) - \sum_{h=1}^{n} \lambda_h z(s_h) \right)^2, \tag{5}$$

with $\lambda_1, \ldots, \lambda_n$, subject to

$$\sum_{h=1}^{n} \lambda_h = 1. \tag{6}$$

The optimal values were computed using the method of Lagrange multipliers. For this, the mean-squared prediction error (Equation (5)) was obtained by

$$\begin{aligned} &E(z(s_0) - \hat{z}(s_0))^2 = \\ &= C(s_0, s_0) - c'C^{-1}c + (1 - c'C^{-1}1)^2 (1'C^{-1}1)^{-1}, \end{aligned} \tag{7}$$

where

$$c \equiv (C(s_0, s_1), \ldots, C(s_0, s_n))', \tag{8}$$

$$C \equiv (C(s_h, s_j)), \tag{9}$$

and

$$\hat{z}(s_0) = \gamma' \Gamma^{-1} Z + \left(1 - \gamma' \Gamma^{-1} 1\right) \left(1' \Gamma^{-1} 1\right)^{-1} \left(1' \Gamma^{-1} Z\right), \tag{10}$$

$$\begin{aligned} & E(z(s_0) - \hat{z}(s_0))^2 = \\ & = \gamma' \Gamma^{-1} \gamma - (1 - \gamma' \Gamma^{-1} 1)^2 (1' \Gamma^{-1} 1)^{-1}, \end{aligned} \tag{11}$$

where $\gamma \equiv (\gamma(s_0, s_1), \ldots, \gamma(s_0, s_n))'$, Γ is an $n \times n$ matrix whose $(i,j)^{th}$ element is $\gamma(s_i, s_j)$, and

$$2\gamma(s_i, s_j) \equiv C(s_i, s_i) + C(s_j, s_j) - 2C(s_i, s_j), \tag{12}$$

is the variogram [$\gamma(s_i, s_j)\gamma(s_i, s_j)$, named the semivariogram [33]]. The variogram, a valuable tool in modeling spatial variables, describes how the data are correlated with distance and, in doing so, allows us to accomplish spatial interpolation using the sampled data and the variogram information to estimate the variance of the values of the variable at the unsampled points [33,34]. In summary, Kriging was more precise than other interpolation methods, such as IDW or cubic interpolation [29], and suitable for applications where the data have a certain spatial dependence.

2.2. Neural Networks

Neural networks are artificial intelligence models inspired by the structure and functioning of the human brain. These networks comprise many nodes, known as „neurons", connected to each other through „synapses". Each neuron receives inputs from other neurons, processes them through an activation function, and sends its output to other neurons, as shown in Figure 2. The combination of inputs and connections between neurons allows a neural network to learn complex tasks from the selected training data. Neural networks are a powerful tool for machine learning because they can model complex relationships between inputs and outputs and make accurate inferences from real data. However, these tasks require a large amount of training data and are computationally intensive (sometimes impossible) to train and use in real-time implementations [35].

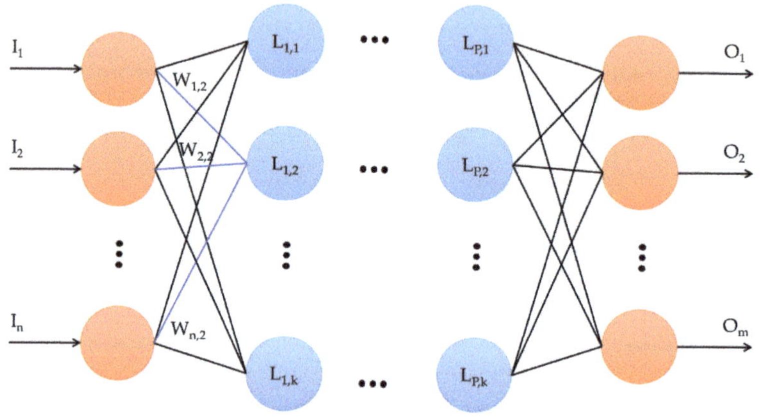

Figure 2. Neural network concept: the multilayer perceptron.

In this work, we used the multilayer perceptron (MLP), as shown in Figure 2. This type of artificial neural network consists of an input layer, one or more hidden layers, and one output layer. In the MLP, neurons in each layer are connected to neurons in the next layer, and they use an activation function to determine the neuron's output. Information flows only in one direction, from the input to the output layers, passing through the hidden intermediate layers between the input and the output [36]. The general formula for calculating the output of a multilayer perceptron is given by [37]

$$O = f(\sum(W_{i,j}\, I_i) + b), \quad (13)$$

where O is the output of the multilayer perceptron, f is the activation function, which can be a step function, sigmoid, rectified linear unit (ReLU), etc., $W_{i,j}$ are the synaptic weights connecting the input I_i to the current neuron, I_i is the input of the multilayer perceptron, and b is the bias, where $i = 1, \ldots, n$ and $j = 1, \ldots, k$, with n being the number of input features to the NN (corresponding to the number of neurons in the input layer), k the number of neurons in a given hidden layer, and m the number of provided outputs (i.e., the number of neurons in the output layer of the NN).

The formula is applied to every neuron in every network layer, including the input, hidden, and output layers. Each neuron in a hidden layer takes the output of all the neurons in the previous layer (i.e., assuming a fully connected network) as its input, and the output of one neuron in the output layer is the network's final output. During the training phase, the weights and the bias are adjusted to minimize the error between the network output and the desired output. This is achieved by a learning algorithm, such as the backpropagation algorithm [38], which updates the weights and bias in the direction of the downward gradient [39].

3. Preliminary Work

This section briefly describes the preliminary work on a novel MBSS technique based on the sample entropy (SampEn) developed and published by the authors in [39], constituting the basis of the new proposal. However, the MBSS technique presented in [39] was the result of the authors' previous efforts to refine and complement it with modules implemented in real time to, for instance, diminish the noise introduced by SDR devices. Hence, several comparative studies have been carried out in order to measure the performance of the proposed MBSS technique compared with other classical methods, such as energy detectors. The efficiency of locating the PUs by this MBSS technique stands out in SNR values close to 0 dB [11]. Table 1 highlights the principal contributions of these previous works.

Table 1. Main contributions of previous works.

Previous Work.	Contributions
"A Novel Multiband Spectrum Sensing Method Based on Wavelets and the Higuchi Fractal Dimension" [9].	- Two MBSS techniques: Wavelets and multiresolution analysis (MRA). Decision rule based on the Higuchi fractal dimension (HFD). - 95% effective in detecting PUs for SNR values higher than 0 dB. - Controlled simulated environment. - Blind technique.
"Multiband Spectrum Sensing Based on the Sample Entropy" [40].	- MBSS technique based on SampEn. - Technique implemented in a real environment of wireless communications. - Technique with real-time operation (updated every 100 ms). - Probability of 0.99 to correctly detect the PU for SNR values greater than 0 dB. - Cooperation of the different SUs to sense a wide range of frequency. - Blind technique.

Table 1. *Cont.*

Previous Work.	Contributions
"Real-Time Implementation of Multiband Spectrum Sensing Using SDR Technology" [11].	- MBSS technique based on MRA, ML, and HFD implemented in a real wireless communications environment. - Real-time operation (update every 100 ms). - Implemented with SDR devices. - A module is proposed for the elimination through the detailed coefficients obtained with the MRA. - Efficiency of 95% for SNR values greater than 0 dB. - Blind technique.
"Machine Learning Techniques Applied to Multiband Spectrum Sensing in Cognitive Radios" [41].	- Three machine learning methods are applied to the MBSS - Effectiveness of 98% in detecting PUs in the spectrum for SNR values greater than 0 dB. - Controlled simulated environment. - Blind technique.

The last MBSS technique from Table 1 was implemented in a real environment using low-cost SDR devices. Each connected SDR device was considered an SU, independent from others, processing its information locally in a determined single band to achieve ensemble wideband spectrum sensing. The MBSS technique used in this proposal, however, combines different SDR devices, namely the RTL-SDR [42], the HackRF One [43], and the LimeSDR mini [44], to conform each considered SU, forming a CR infrastructure, which permits having an MBSS in a specific location. Figure 3 below shows how this MBSS method works. The MBSS blocks are as follows:

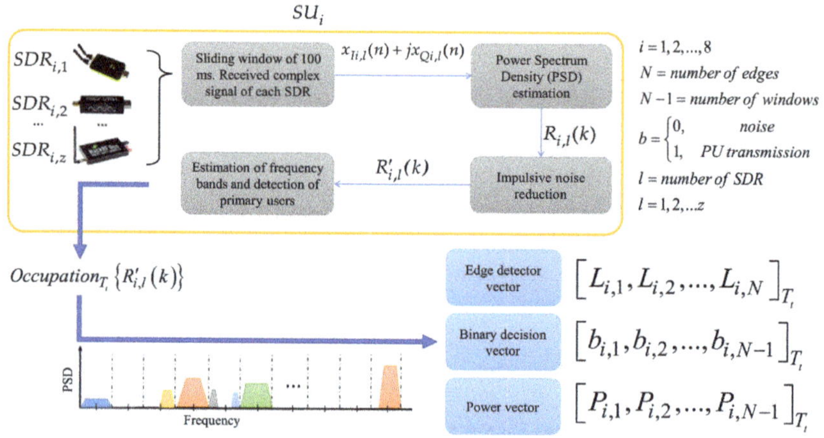

Figure 3. General scheme of implemented MBSS technique [40].

- **Sliding window of 100 ms.** In this block, the complex signal $x_{Ii,l}(n) + jx_{Qi,l}(n)$ from each SDR in the time domain was received and updated every 100 ms from the radio environment of the i-th SU integrated by z different SDR devices;
- **Power Spectrum Density (PSD) estimation.** In this module, the Welch method [45] was applied to each signal $x_{Ii,l}(n) + jx_{Qi,l}(n)$ in order to obtain, on a linear scale, the wideband PSD $R_{i,l}(k)$ from the SU ensemble;
- **Impulsive noise reduction.** In this block, impulsive noise, high-frequency noise, and abrupt changes (many of them generated by the SDR devices themselves) in the signal $R_{i,l}(k)$ were eliminated or diminished through discrete wavelets via the multiresolution analysis [46], resulting in the signal $R'_{i,l}(k)$;

- **Estimation of frequency bands and detection of primary users.** In this module, the MBSS technique based on SampEn, K-means algorithm [47] (permitting to optimize specific detection parameters), and discrete wavelets was applied to each SU's analyzed wideband spectrum in order to obtain the spectrum occupation given by $Occupation_{T_t}\{R'_{i,l}(k)\}$. This result included three vectors $[b_{i,1}, b_{i,2}, \ldots, b_{i,N-1}]_{T_t}$ which contained binary values indicating occupied ("1") and empty ("0") bands. The second vector was $[L_{i,1}, L_{i,2}, \ldots, L_{i,N}]_{T_t}$ and contained the corresponding computed boundaries for each detected band, and the third was $[P_{i,1}, P_{i,2}, \ldots, P_{i,N-1}]_{T_t}$, which contained the power for each detected band.

One of the primary motivations for this technique was to propose an MBSS method (i) adequate to correctly detect a primary user and the operating parameters, such as carrier frequency, bandwidth, and power, with (ii) computational complexity permitting a real-time implementation. Both aims were fulfilled, and the results were used to create a compact CRN capable of sharing spectral information with a central entity in order to estimate the area occupied by different detected PUs in the studied zone. A wireless communications environment was proposed considering these SDR devices.

4. New Proposal and Methodology

This section presents the new research proposal and the implemented methodology.

4.1. Proposal

Our previous work's novel idea was to consider a computer integrating different SDR devices detecting each PU in a single band to sense a wide frequency range, as shown in Figure 4a. This model was seen as the only entity containing different SUs. Our follow-up work proposed replicating this entity containing different connected SDR devices and creating a cooperative CR network for sensing a wide frequency range in a broader geographical region. Each entity was considered an SU containing different technologies for sensing a broad spectrum, as shown in Figure 4b. In addition, to sense the radio spectrum and uncover the PUs' behavior in a given geographical region, this cooperative CR network had the task of avoiding the hidden terminal problem. One of the main strengths of this new proposal (see Figure 4b) over previous work is the cooperation of the different SUs in geographically locating the PUs through the REMs and the ability to carry out this processing in real time.

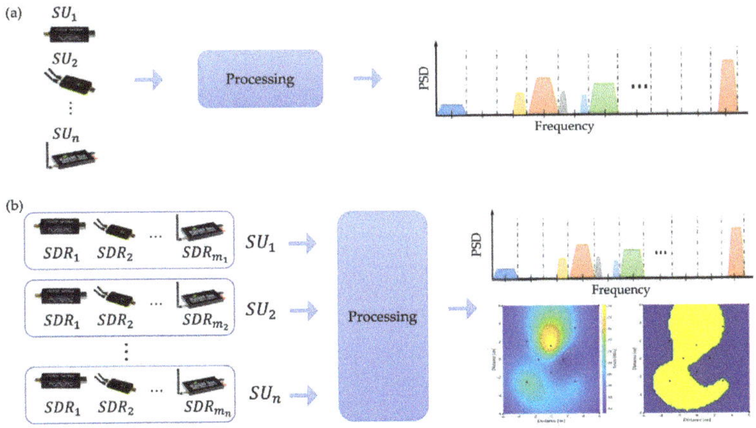

Figure 4. (**a**) Previous work and (**b**) new proposal.

4.2. Methodology

The proposed methodology can be outlined in three big blocks, shown in Figure 5:

- The collection of information obtained by SUs. For this, each SU locally processes the sensed data to be sent to a central entity, including the occupancy of the observed spectrum in its geographic location and the frequency band edges and estimated power vectors;
- The database for storing the information obtained by each SU at a specific time;
- The central entity overseeing the processing determines the geographic area occupied by the detected PUs in the radio spectrum.

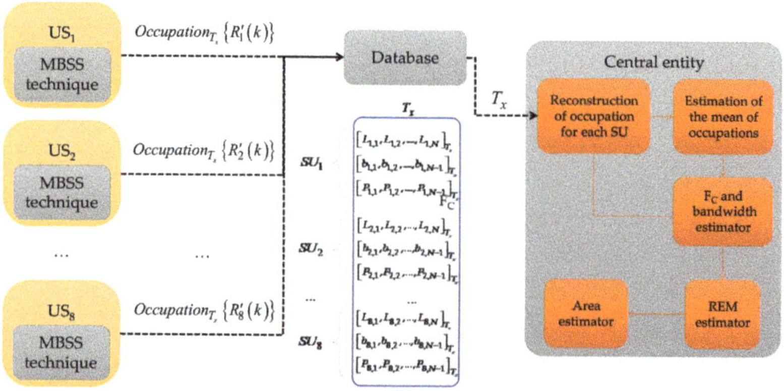

Figure 5. General scheme of the methodology.

Each block is described in detail below.

4.2.1. Collection of Information Locally by the SUs

The information collection was carried out every 100 ms for each SU, resulting in the $Occupation_{T_t}\{R'_{i,l}(k)\}$ of analyzed spectral signal, as mentioned in Section 3 and displayed in Figure 3. This information was conveyed by three vectors: (i) the edge detector vector $[L_{i,1}, L_{i,2}, \ldots, L_{i,N}]_{T_t}$, which stores the frequency edges where it is possible to find the presence of PUs, (ii) the binary decision vector $[b_{i,1}, b_{i,2}, \ldots, b_{i,N-1}]_{T_t}$, which stores a binary decision for corresponding delimited bands where noise or a possible PU transmission is detected, and (iii) the power vector $[P_{i,1}, P_{i,2}, \ldots, P_{i,N-1}]_{T_t}$, which stores the received average power corresponding to each classified window (i.e., each binary decision). The vectors in (ii) and (iii) are the same size.

4.2.2. Database

The edge detector, binary decision, and power vectors were stored in the database. It is important to note that the information stored in the database stemming from each SU was not necessarily similar; the lengths of the vectors were different for each SU. Indeed, each SU observed a different behavior of the radioelectric spectrum, because they were placed randomly in different geographic coordinates. To account for this discrepancy, all vectors were labeled with the exact sensing time T_x (see Table 2), i.e., synchronized. Information was uploaded to a server, saving only enough relevant data (three different vectors for each SU) in the database to permit fast storage and extraction, taking advantage of SUs' hardware performing the MBSS technique locally. Indeed, the three uploaded vectors for each SU take up very little memory space. In the best case (without even a PU in the sensed spectrum), the edge detection vector had a length of two, while the binary decision vector and the power vector had a length of one. In the worst case, when the SNR had a low value (close to 0 dB), multiple windowing appeared due to noise, thus increasing the vectors' lengths. However, regardless of the size of the three vectors, the information was compact and facilitated the reconstruction of the spectrum signal occupation, including the average power for each reconstructed window.

Table 2. Information that each SU shares in the database at each sensing time T_x.

	T_1	T_2	...	T_x
SU_1	$[L_{1,1}, L_{1,2}, \ldots, L_{1,N}]_{T_1}$ $[b_{1,1}, b_{1,2}, \ldots, b_{1,N-1}]_{T_1}$ $[P_{1,1}, P_{1,2}, \ldots, P_{1,N-1}]_{T_1}$	$[L_{1,1}, L_{1,2}, \ldots, L_{1,N}]_{T_2}$ $[b_{1,1}, b_{1,2}, \ldots, b_{1,N-1}]_{T_2}$ $[P_{1,1}, P_{1,2}, \ldots, P_{1,N-1}]_{T_2}$	...	$[L_{1,1}, L_{1,2}, \ldots, L_{1,N}]_{T_x}$ $[b_{1,1}, b_{1,2}, \ldots, b_{1,N-1}]_{T_x}$ $[P_{1,1}, P_{1,2}, \ldots, P_{1,N-1}]_{T_x}$
SU_2	$[L_{2,1}, L_{2,2}, \ldots, L_{2,N}]_{T_1}$ $[b_{2,1}, b_{2,2}, \ldots, b_{2,N-1}]_{T_1}$ $[P_{2,1}, P_{2,2}, \ldots, P_{2,N-1}]_{T_1}$	$[L_{2,1}, L_{2,2}, \ldots, L_{2,N}]_{T_2}$ $[b_{2,1}, b_{2,2}, \ldots, b_{2,N-1}]_{T_2}$ $[P_{2,1}, P_{2,2}, \ldots, P_{2,N-1}]_{T_2}$	...	$[L_{2,1}, L_{2,2}, \ldots, L_{2,N}]_{T_x}$ $[b_{2,1}, b_{2,2}, \ldots, b_{2,N-1}]_{T_x}$ $[P_{2,1}, P_{2,2}, \ldots, P_{2,N-1}]_{T_x}$
...	...	...	...	...
SU_8	$[L_{8,1}, L_{8,2}, \ldots, L_{8,N}]_{T_1}$ $[b_{8,1}, b_{8,2}, \ldots, b_{8,N-1}]_{T_1}$ $[P_{8,1}, P_{8,2}, \ldots, P_{8,N-1}]_{T_1}$	$[L_{8,1}, L_{8,2}, \ldots, L_{8,N}]_{T_2}$ $[b_{8,1}, b_{8,2}, \ldots, b_{8,N-1}]_{T_2}$ $[P_{8,1}, P_{8,2}, \ldots, P_{8,N-1}]_{T_2}$	...	$[L_{8,1}, L_{8,2}, \ldots, L_{8,N}]_{T_x}$ $[b_{8,1}, b_{8,2}, \ldots, b_{8,N-1}]_{T_x}$ $[P_{8,1}, P_{8,2}, \ldots, P_{8,N-1}]_{T_x}$

4.2.3. Central Entity

The main tasks of the central entity were (i) to indicate how many PUs appear in the spectrum compared with the perceived SUs, (ii) to build a REM for each detected PU using the location of each SU, the perceived power in each SU, the time of radio space monitoring as parameters, and finally (iii) to show the area covered by the PUs.

Two ways of implementing these tasks were developed, the first through *classical* digital processing and the second using machine learning techniques (specifically, the NN). Both approaches are detailed below.

4.2.3.1. Central Entity Implementing Classical Digital Signal Processing

Digital signal processing, which uses programmed algorithms to process and analyze data, is a well-established and stable technology that has been used for decades in a variety of applications. It is relatively easy to implement and can be very fast and efficient in situations where the datasets and tasks are specific. In this case, the central entity was implemented with classical techniques for reconstructing the spectrum signal occupation from the shared information (edge detector, binary decision, and power vectors) obtained by each SU in the database. The central entity processing the information is shown in Figure 6 and described by Algorithm 1.

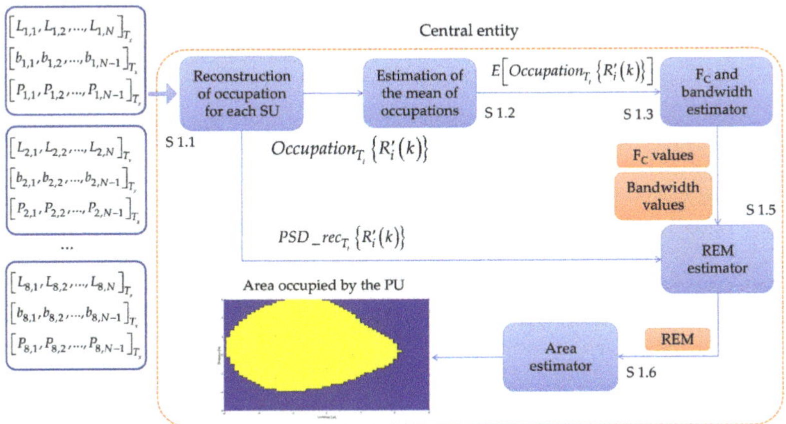

Figure 6. Implementation of the central entity using classical digital signal processing.

Algorithm 1 Central Entity Computed by Traditional Signal Processing

Step 1.1	The central entity collects the edge detector, binary decision, and power vectors of each PU at determined time instant t from the database. With this information, the central entity reconstructs the occupancy signal $Occupation_{T_t}\{R'_i(k)\}$ and constructs simultaneously, an approximation of the power spectral density $PSD\_rec_{T_t}\{R'_i(k)\}$ using only the average powers (power vector) associated to each SU. The length of these frames is set to 1024 samples.
Step 1.2	Once the reconstruction of the occupation of each SU is performed, the mean value $E[Occupation_{T_t}\{R'_i(k)\}]$ is computed.
Step 1.3	After computing the average value of the Occupation, it is possible to infer the presence of one or several PUs, giving two possible vectors. First, the vector $F_{C\_vector} = [F_{C1}, F_{C2}, \ldots]$ containing the central frequencies values of detected PUs. The length of this vector will be the number of PUs detected by the algorithm. A sec ond vector, containing the occupied bandwidths by each detected PU $B_{vector} = [B_{PU_1}, B_{PU_2}, \ldots]$, is also obtained. Through the number of sin gularities and their corresponding widths detected in $E[Occupation_{T_t}\{R_i'(k)\}]$, it is possible to estimate how many PUs are in the spectrum and their respective bandwidths $(B_{vector} = [B_{PU_1}, B_{PU_2}, \ldots])$. The F_{C_vector} is integrated by the central frequency values of estimated bandwidth for each detected PU.
Step 1.4	Knowing now the central frequency and the corresponding bandwidth of each detected PU, it is possible to locate them in each reconstructed PSD $PSD\_rec_{T_t}\{R_i'(k)\}$. The mean of the interval B_{PU_N} centered on F_{C_N} of the $PSD\_rec_{T_t}\{R_i'(k)\}$ is computed giving the scalar $aux\_psd_{i,N}$. In the strict sense, this scalar represents the average power in the carrier, sensed by each SU, where the PU_N is or should be.
Step 1.5	With the scalar $aux\_psd_{i,N}$ of each SU located in a specific coordinate, the REM will be built using a double interpolation. First interpolation is done through the IDW method and for the sec ond one, the Kriging method is applied. Due to the fact that in our case only eight geographical points are considered (i.e., only eight SUs are implemented), this double interpolation is carried out with the purpose of having a better precision to describe the behavior of the radio electric space in the environment described later. In this case, each REM is constructed with values $aux\_psd_{i,N}$ that specifically correspond to the average power of the bandwidth B_{PU_N}.
Step 1.6	Finally, the active area of each PU will be determined according to the information collected by each sec ondary entity SU_i. For this, a threshold of −80 dBm is used to classify the area estimated by the REMs that was chosen in [41] for a wireless environment. That is, regions in the REM that are above this threshold correspond to the active area of detected PUs.

4.2.3.2. Central Entity Implementing Classical Digital Signal Processing

The central entity was implemented using ML techniques, specifically NNs. NNs have the ability to learn from data and improve their performance as they absorb more information. This makes them particularly useful in applications where the data are complex, difficult to understand, and processed using traditional methods. Furthermore, NNs can perform tasks that traditional methods cannot, such as recognizing patterns in unstructured data or processing input signals that change over time. In this case, some modules in the central entity were replaced by NNs, as shown in Figure 7. The steps of this operation (Algorithm 2) are described below.

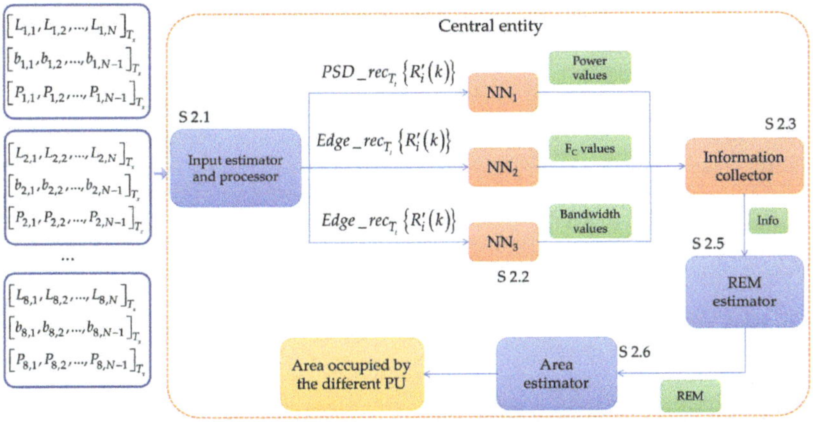

Figure 7. The central entity determining the REM construction parameters and area maps through NNs.

Algorithm 2 Central Entity Computed by NNs

Step 2.1 The central entity collects the data from each of the SUs in the database at a specific time T_x. In the module input estimator and processor, a processing is carried out in order to estimate the inputs driving a NNs ensemble. As mentioned above, the three vectors coming from each SU for time T_x do not have the same length as the vectors for time T_{x+t_1} or T_{x-t_2}. Even, the size of the vectors differs from one SU to another SU even though they are at the same time T_x. Based on this fact, this module oversees building the input vectors $PSD\_rec_{T_t}\{R_i'(k)\}$ and $Edge\_rec_{T_t}\{R_i'(k)\}$ driving the NNs. For this, these vectors must have always the same length (in our case this length is set to 13 samples). This block is detailed below.

Step 2.2 The vector $PSD\_rec_{T_t}\{R_i'(k)\}$ is evaluated by NN_1, at the same time the vector $Edge\_rec_{T_t}\{R_i'(k)\}$ is evaluated by NN_2 and by NN_3. This evaluation is performed sequentially, i.e., each vector of each SU will be evaluated by its corresponding NN one after the other until the result of the i − th connected SU be obtained. As a result of this step, NN_1 provides an approximate number of detected PUs and their powers. NN_2 gives an approximate number of detected PUs and their central frequencies. Finally, NN_3 returns an approximate number of detected PUs and their bandwidths. At the end of this section, it is discussed what would happen if the number of PUs detected by each NN is not the same.

Step 2.3 The information obtained from Step 2.2 is evaluated to determine the number of PUs in the spectrum and their corresponding power, bandwidth, and carrier frequency. This evaluation is detailed below.

Step 2.4 The information obtained from Step 2.3 will be shared with the REM estimator module. This block receives (i) the geographic coordinates of the SUs in the network, (ii) the data of the possible PUs in the spectrum (power, carrier, and bandwidth) and (iii) an identifier that corresponds to time T_x in which the spectrum was monitored.

The rest of the steps of this algorithm that include NNs corresponds to Steps 1.5 and 1.6 of the methodology described in Algorithm 1 (Section 4.2.3.1). The *input estimator and processor* (Step 2.1) and the *information collector* (Step 2.3) modules are described below.

Figure 8 shows the process of generating $PSD\_rec_{T_t}\{R'_i(k)\}$, the input vector of NN_1. First, power, edge detector, and binary decision vectors were used, with samples ranging in the intervals of [1–20], [1–20], and [2–21], respectively, according to the studies carried out in our preliminary work [39]. As mentioned above, using these vectors, we obtained (i) the frequency range that was sensed by each SU, (ii) the number of samples of the estimated power spectral density, and (iii) the number of windows in which the sensed spectrum was partitioned. An occupation value was assigned for each detected window, 1 for a possible PU transmission or 0 for the noise. Accompanying each occupation value was an average power value or the power vector element representing the power estimate in each detected window. The occupation and the power signal were reconstructed using these average values of each detected window, setting both vectors to a length of 1024 samples each.

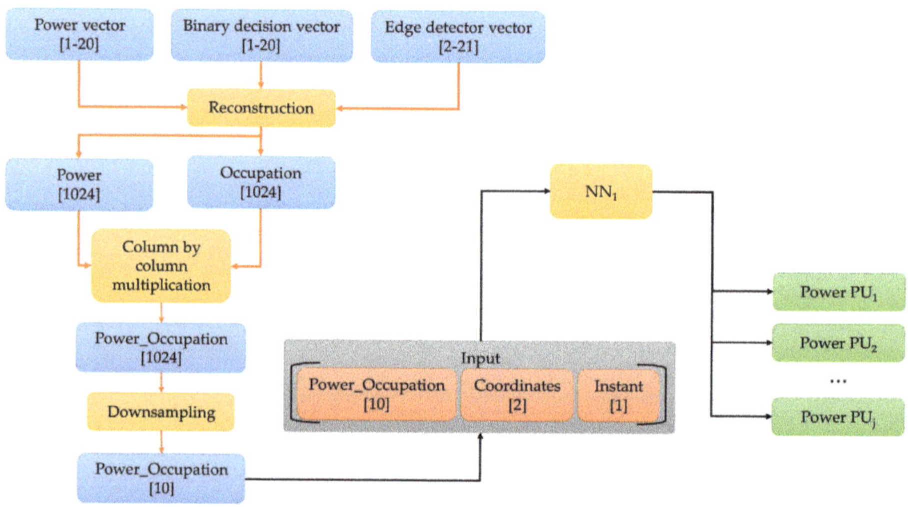

Figure 8. $Edge\_rec_{T_t}\{R'_i(k)\}$ vector construction.

After that, the power- and occupation-reconstructed vectors were multiplied column by column, resulting in the *Power_Occupation* vector of 1024 samples. In this last vector, locations with estimated noise had a value of 0, and those with the possible transmission of one or several PUs had a value corresponding to the average power of detected windows. A downsampling of the *Power_Occupation* vector was performed to reduce it to only 10 samples and to combine it with the coordinates of the SUs (to which the analyzed vectors correspond) and the time T_x at which the signal $Occupation_{T_t}\{R'_{i,l}(k)\}$ was sensed, thus finally integrating the input vector for NN_1. It should be noted that the time parameter is extremely important since the spectrum behaves differently over time.

Figure 9 shows the construction of the vector $Edge\_rec_{T_t}\{R'_i(k)\}$. Here, the occupancy and the frequency bands vectors corresponding to the frequency interval perceived by the corresponding SU were reconstructed. This vector was also multiplied column by column with the occupation vector, resulting in the *Freq_Occupation* vector being 1024 samples in length. This latter vector was assigned zeros where it corresponded to noise and nonzero values for the samples that corresponded to one or several PU transmissions. Again, downsampling was applied to reduce this vector to only 10 samples and join it with the coordinates of the SUs and the time T_x corresponding to the sensing period. This concatenation drives the input of NN_2 and NN_3.

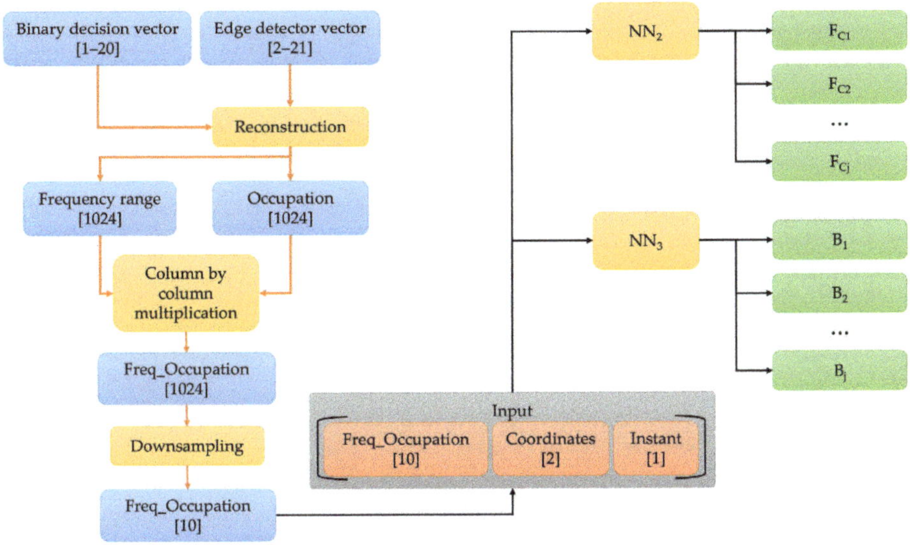

Figure 9. $Edge\_rec_{T_i}\{R'_i(k)\}$ vector construction.

The NN_1 output corresponds to the power of each PU detected by each SU. When the PU power value was less than -80 dBm, it was interpreted as no transmission, corresponding to noise. The output of NN_2 resulted in the carrier on which a possible PU was located. Finally, the NN_3 output delivered the transmission bandwidths corresponding to each detected PU in the analyzed spectrum.

The resulting outputs from NNs for each SU are shown in Figure 10. These outputs can be classed into three large blocks: powers, carrier frequencies, and transmission bandwidths of each PU detected by the corresponding SU. The *information collector* block analyzes the output of each SU making up the network and sends it to the module permitting the REM estimate. For example, Figure 10a shows that the power of PU_1 is less than -80 dBm; the *information collector* interprets that PU_1 does not correspond to a transmission and, by sharing the information delivered by SU_1 with the estimator of REM, assumes that SU_1 observes $(j-1)$ PUs (where j represents the number of possible PU transmissions). Figure 10b shows that the carrier frequencies of PU_1 and PU_j are outside the monitored space, so the *information collector* shares with the REM estimator that SU_2 only observes $(j-2)$ PUs. In Figure 10c, the B of PU_1 is a very small value (possibly corresponding to impulse noise); thus, the *information collector* shares with the REM estimator module that the SU_n observes $(j-1)$ PUs. It is important to point out that any of these combinations shown in Figure 10 may change the number of observed PUs, i.e., if the power of a first PU does not exceed the threshold of -80 dBm, the carrier of a second PU is not in a correct frequency range, and the B of a third PU tends to zero; then, these three PUs will be considered as noise.

The proposed methodology has been implemented in a real wireless communication environment. This controlled environment is explained in the next section.

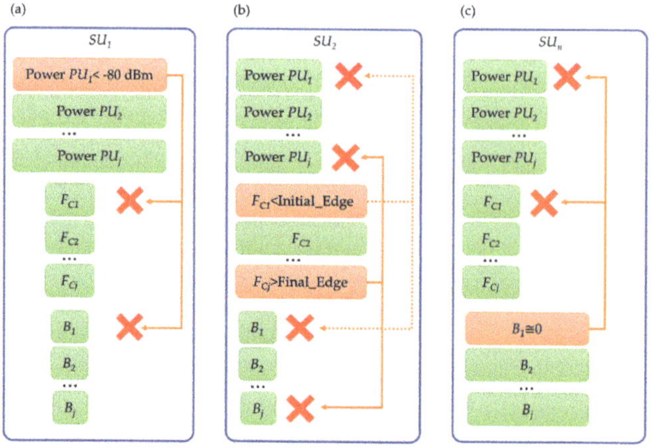

Figure 10. (**a**). The SU_1 observes j PUs, but PU_1 is discarded because it does not meet the threshold. (**b**) The SU_2 observes j PUs; nevertheless, PU_1 and PU_j are discarded because they are not in the correct frequency range. (**c**) The SU_n observes j PUs, but the PU_1 is discarded for not having a large enough bandwidth to be considered a transmission.

5. Real Wireless Communication Environment

Figure 11 shows the real, controlled environment implemented in our research. In this proposed scenario, we considered two PUs collocated at the center of the studied area while, at the same time, eight SUs were sensing their behavior in their geographical zone of influence. It is important to note that the SUs and PUs did not share the geographic coordinates among themselves, nor was that information used in spectrum sensing. PUs were located at the center, hoping most SUs could receive part of the Pus' signal. SUs were set randomly in the area of study.

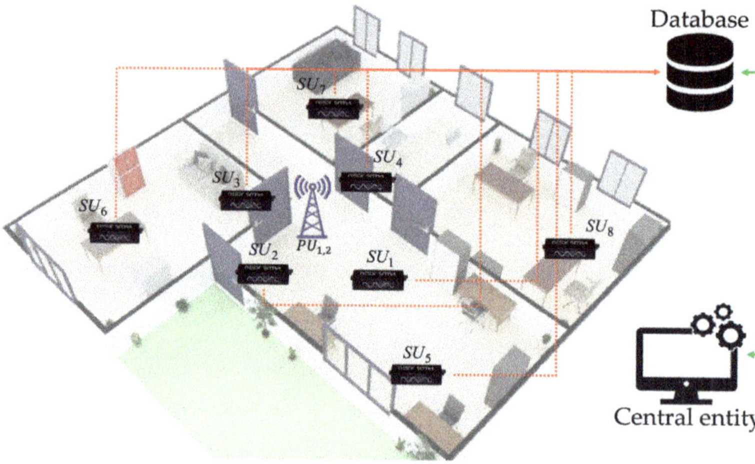

Figure 11. The real implemented scenario.

Table 3 specifies the involved parameters of both SUs and PUs. These SUs shared the channel occupancy with the database and the central entity in order to determine (i) how many PUs on average were observed in this environment, (ii) the B and the F_C in which they located the detected PUs, and (iii) finally the area occupied by the detected PUs.

Table 3. PU and SU main parameters.

Label	Device	F_C Tx (MHz)	F_C Rx (MHz)	Bandwidth (MHz)	Location Coordinate (X,Y) (m)
PU_1	Mini LimeSDR	699.5	-	0.5	(0, 0)
PU_2	HackRF ONE	700.5	-	1	(0, 0)
SU_1	RTL-SDR	-	700	2.4	(−1.5, 0)
SU_2	RTL-SDR	-	700	2.4	(0, 1.5)
SU_3	RTL-SDR	-	700	2.4	(1.5, 0)
SU_4	RTL-SDR	-	700	2.4	(0, −1.5)
SU_5	RTL-SDR	-	700	2.4	(−3, 2)
SU_6	RTL-SDR	-	700	2.4	(3, 3.5)
SU_7	RTL-SDR	-	700	2.4	(3, −2.5)
SU_8	RTL-SDR	-	700	2.4	(−3, −2.5)

PUs and SUs were deployed in an area 12 × 12 m^2, containing structures such as walls, doors, windows, columns, etc., as indicated in Figure 11.

6. Results

This section outlines the results of implementing the proposed methodology in a real wireless communication environment. They are divided into two subsections; the first corresponds to the central entity based on digital signal processing, and the second concerns the central entity based on NNs.

6.1. Results with a Central Entity Based on Digital Signal Processing

This section shows the results of the implemented scenario presented in Figure 11. Figure 12 shows the reconstruction of the occupation of each SU by the central entity. In this case, we can highlight the following important points:

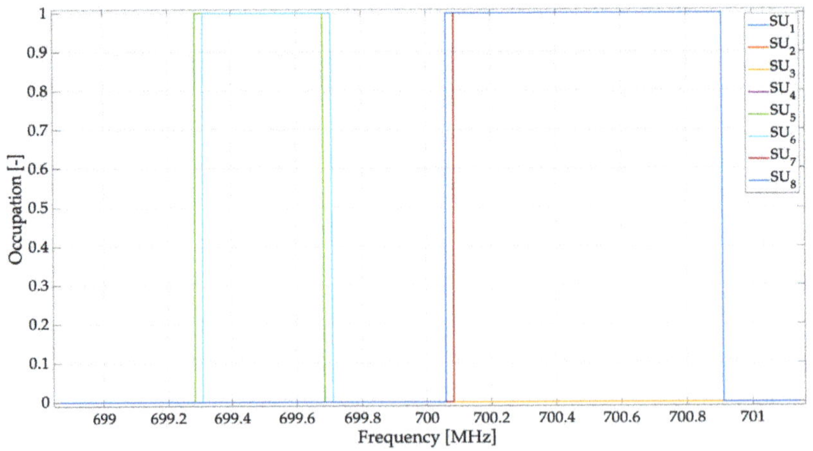

Figure 12. Spectrum occupancy for each SU.

- There were SUs who failed to perceive both PUs;
- The central frequency F_C of each PU was perceived at a different frequency by each SU;
- The bandwidth size B for each PU varied for each SU;

- SUs that were the farthest from the PUs failed to detect both PUs.

Figure 13 shows the result of constructing an approximation of the PSD for each SU, considering only mean power values for each spectrum section. These approximations were used for the estimation of the REM.

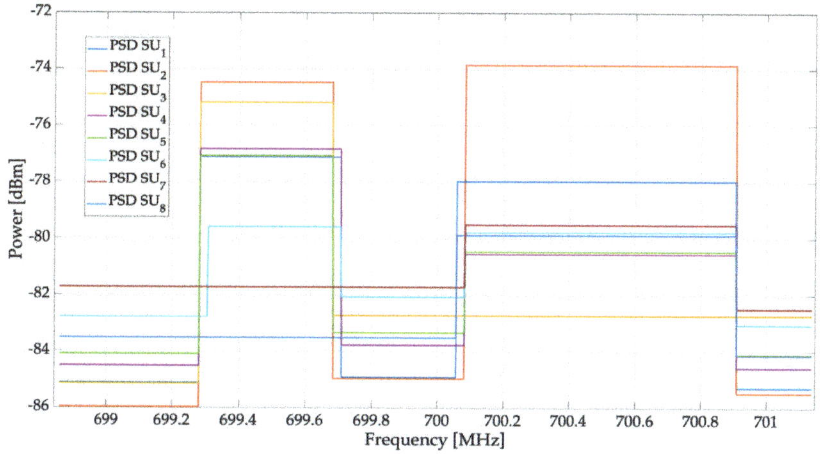

Figure 13. Approximated PSD for each SU.

Figure 14 shows the result of applying the module for estimating the average occupancy of each SU in addition to the result of the F_C and B estimation for each PU. The first value obtained was $F_{C1} = 699.48$ MHz with a bandwidth of $B_1 = 0.4$ MHz. $F_{C2} = 700.49$ MHz with a bandwidth of $B_2 = 0.825$ MHz was also obtained. Given that the exact values were $F_{C1} = 699.5$ MHz, $B_1 = 0.5$ MHz, $F_{C2} = 700.5$ MHz, and $B_2 = 1$ MHz, the values we obtained here show a good occupancy estimation of the MBSS method in conjunction with the central entity for a specific geographical zone covered by the PUs.

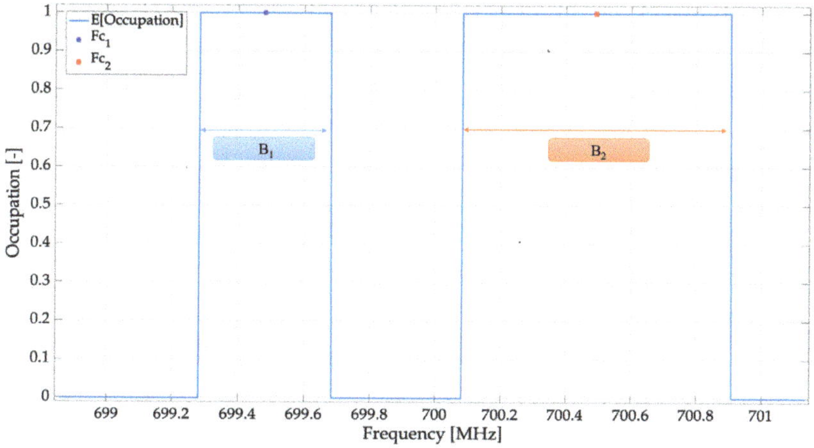

Figure 14. Average occupancy obtained by the central entity.

Figures 15a and 16a show the REM of PU_1 and PU_2, respectively. These maps were created based on the information collected by the SUs in their different locations. Figures 15b and 16b show the respective areas occupied by PU_1 and PU_2. Values of the occupied area for each PU are $area_{PU_1} = 60.76\,\text{m}^2$ and $area_{PU_2} = 56.39\,\text{m}^2$. These results

were obtained by placing a threshold $L = -80\,\text{dBm}$ in the obtained REM. In this way, the area with a power greater than this threshold was considered a space occupied by the PU.

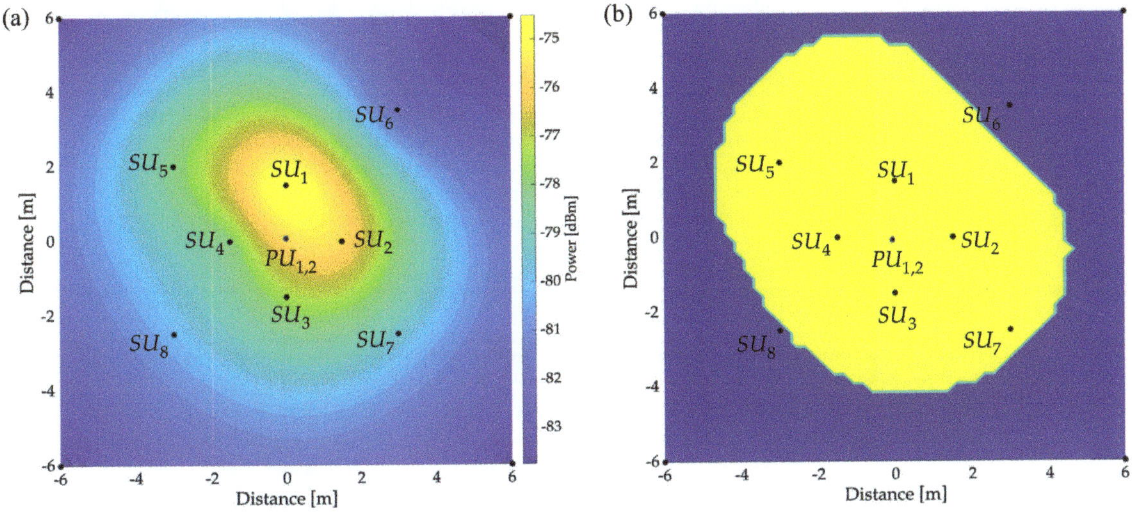

Figure 15. Primary transmission in the 699.48 MHz band with a B of 0.4 MHz: (**a**) REM and (**b**) occupied area.

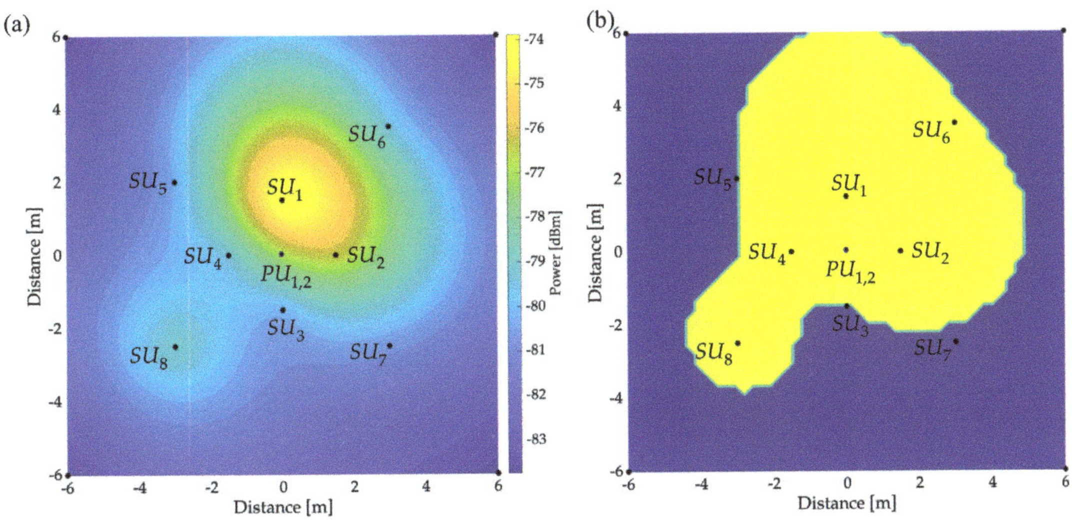

Figure 16. Primary transmission in the 700.49 MHz band with a B of 0.825 MHz: (**a**) REM and (**b**) occupied area.

Figure 15b shows that SU_6 and SU_8 did not observe the transmission of PU_1 (i.e., both SU_6 and SU_8 did not appear in this area of influence). This effect was a result of the structural distribution in which the implementation of the real scenario occurred. Nevertheless, an expected collaboration between SUs might improve this result.

6.2. Results with a Central Entity Based on NNs

This section details the results obtained using a central entity and NNs. First, the NN training is presented, and then the results obtained by applying ML algorithms to this stage are shown.

6.2.1. Training

For the NN_1 training, NN_2 and NN_3, the backpropagation algorithm, the Levenberg–Marquardt activation function, 1000 epochs, the mean square error as loss function, and a low learning rate were used. In addition, 9000 vectors of inputs and their corresponding outputs were used. To determine the best NN?, the number of layers that should be used, and how many neurons, 12 NNs were studied to carry out the work of NN_1, NN_2, and NN_3. Each NN was considered with the number of layers $la = 1, 2, 3, 4$ and with the number of neurons $ne = 16, 32, 64$. For each NN, there were 12 possible combinations between neurons and layers. They were performed to find the most convenient NN configuration for this work.

The results are presented for both the entity that uses digital signal processing and the one that uses the three NNs. As the first parameter, the training time used by each NN is shown in Table 4, indicating the time each NN spent in the training stage. Additionally, Figure 17 shows that the more neurons and layers an NN had, the longer its training time was. All the processing was carried out using the same computer (MacBook Pro with 8 GB RAM and a 1st-generation M1 processor), and the training time for each NN in its different versions tended to follow similar behavior.

Table 4. The training time of (a) NN_1, (b) NN_2, and (c) NN_3.

		Training Time NN_1 [min]		
(a)		Neurons		
		16	32	64
Layers	1	2	10	60
	2	10	25	180
	3	120	192	1500
	4	162	600	3984
		Training Time NN_2 [min]		
(b)		Neurons		
		16	32	64
Layers	1	2	6	16
	2	8	33	290
	3	15	80	850
	4	20	129	7610
		Training Time NN_3 [min]		
(c)		Neurons		
		16	32	64
Layers	1	2	5	16
	2	5	19	280
	3	11	75	875
	4	21	145	3628

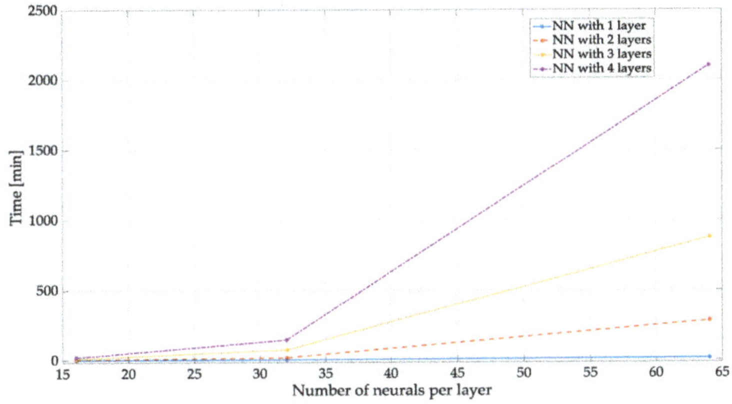

Figure 17. The training time for the different versions of the NN_3.

6.2.2. Statistics

This section analyzes the statistical results of the three NNs used in this proposal. To obtain these results, 27,000 entries were considered for each NN. Figure 18a presents the results of NN_1, responsible for granting the power value of the detected PUs, showing the power difference between the real values and those obtained with the different versions of NN_1. The difference between these values was, on average, -0.01 dBm for each version of NN_1, highlighting the accuracy of the predictive ability of the NN-based approach. Figure 18b only shows the average value, indicating that the margin of error between the expected value and those given by the different NNs was very small; the values were practically the same.

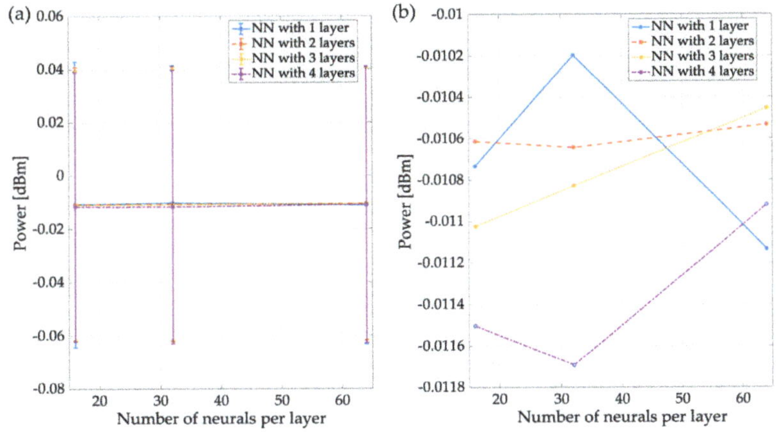

Figure 18. Difference between real values and those obtained by NN_1: (**a**) mean and standard deviation (STD) and (**b**) mean.

Figure 19 shows the result of NN_2, the network that gives the carrier value of each PU detected. In the proposed environment, the NN_2 detected two PUs, and Figure 19a shows its precision in detecting each PU carrier, with F_{C1}, on average, at 699.4 MHz, regardless of the configuration of NN_2. Figure 19b shows the precision of NN_2 in detecting the PU_2 carrier, indicating that, on average, the carrier value was 700.494 MHz. Even though the network with 2 layers and 64 neurons per layer deviated slightly from the ideal value (700.5 MHz) compared with the other NN configurations, it continued to provide reasonably accurate results. In both images, the value of applying digital signal processing can be seen in blue and is quite close to the ideal values of 699.4934 MHz and 700.4935 MHz for PU_1 and PU_2, respectively, thus resulting in a comparable level of accuracy of the NN-based approach.

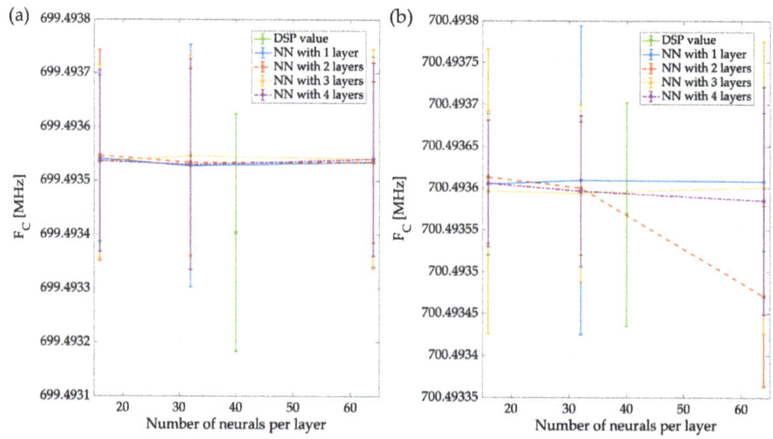

Figure 19. Mean and STD of the precision in detecting (**a**) the PU_1 carrier of the different NN_2 variants and (**b**) the PU_2 carrier of the different networks.

Figure 20 presents the results of NN_3, the network in charge of giving the B that each detected PU in the spectrum frames. In this figure, the average B of PU_1 is 0.48 MHz, regardless of the network configuration (see Figure 20a). Moreover, Figure 20b shows that notwithstanding the NN structure, the value of the B for PU_2 is practically 1 MHz, which perfectly coincides with the ideal value (see Table 3). The blue dot in the image indicates the result of digital signal processing, as $B_1 = 0.425$ MHz and $B_2 = 0.824$ MHz. For B_1, the margins of error/precision of DSP and NN were quite similar (only DSP overestimates while NN underestimates). For B_2, however, it seems that the NN was much more precise, suggesting that the method based on NNs tends to improve DSP or, in the worst case, provide a comparable absolute error.

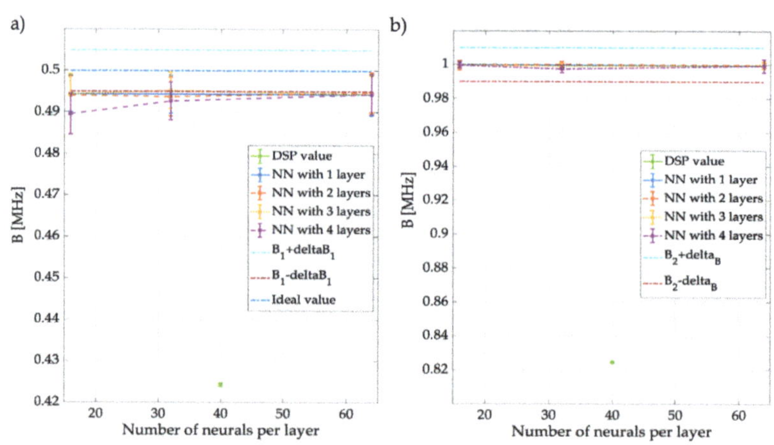

Figure 20. Mean and STD of the precision in detecting (**a**) the B_1 of different NN_3 variants and (**b**) the B_2 of the different networks.

To measure the precision of this methodology, the F1 score ($F1$) [48] was used as a common evaluation metric in the ML field for assessing the accuracy of binary classification models. This metric combines the accuracy and recall of the model into a single measure.

Precision refers to the ratio of true positives (TP) to the sum of true and false positives (FP), while recall, on the other hand, refers to the ratio of true positives to the sum of true positives and false negatives (FN). The $F1$ value expresses the harmonic mean of precision and completeness, giving more weight to low values. The $F1$ formula is

$$F1 = 2\frac{(precision * recall)}{(precision + recall)} \quad (14)$$

A value of $F1$ equal to 1 indicates that the accuracy and completeness are perfect, while an $F1$ of 0 indicates that the model is unable to correctly classify any of the samples. The $F1$ metric is valuable for comparing different binary classification models and selecting the best model for a given classification task. To determine this parameter (i.e., $F1$), the following four possible cases were considered (see Figure 21):

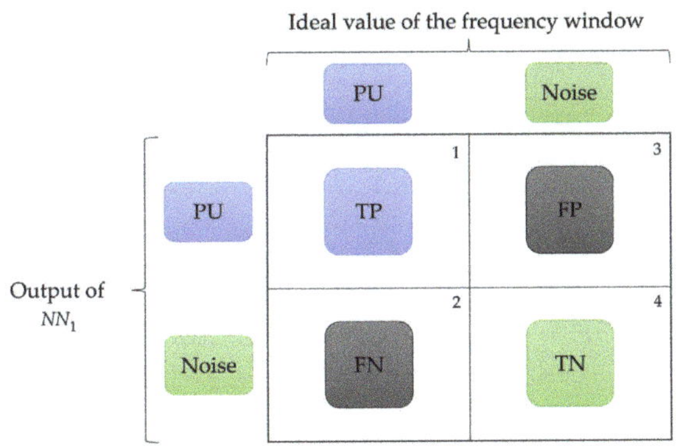

Figure 21. Evaluation outputs of detected windows.

1. A window corresponding to a PU transmission and classified as such by the SU is considered a true positive (TP) value;
2. A frequency window corresponding to a PU transmission classified as noise by the SU is a false negative (FN) value;
3. A window corresponding to noise classified by the SU as a PU transmission is a false positive (FP) value;
4. A frequency window corresponding to noise classified by the SU as such is a true negative (TN) value.

Figure 22 shows the $F1$ of the NN_1, which indicates whether the PUs were correctly located. In this image, all the NNs have a nearly perfect performance, around 0.98.

In the case of B, the $F1$ was estimated from the following cases (see Figure 23):

- The resulting NN_3 B matching the ideal bandwidth that corresponds to a transmission is a true positive (TP) case;
- The resulting NN_3 B corresponding to an ideal bandwidth close to zero is a true negative (TN) case;
- The resulting NN_3 B much greater than an ideal bandwidth that is close to zero is a false positive (FP) case;
- The resulting NN_3 B that should correspond to an ideal transmission B but is a value close to zero will be a false negative (FN) case.

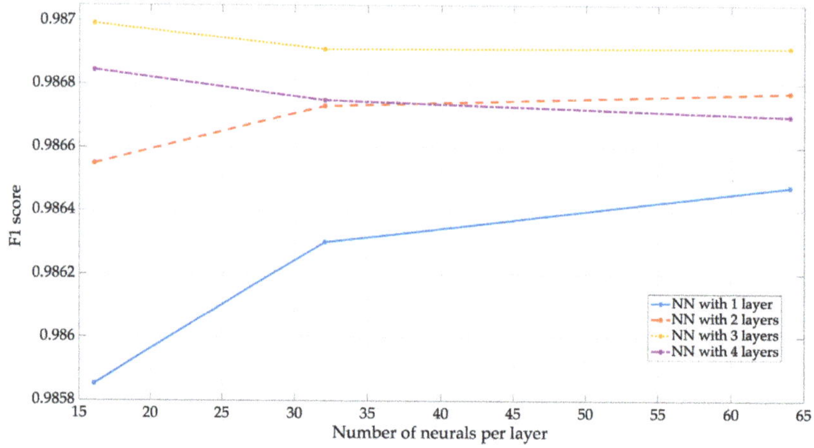

Figure 22. $F1$ score for NN_1.

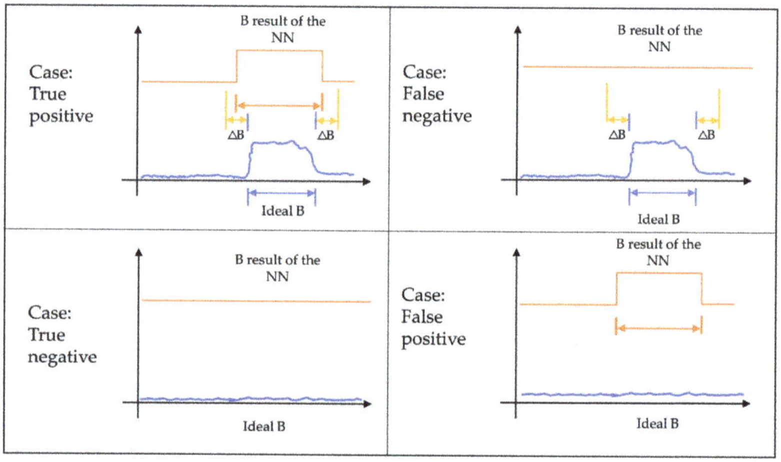

Figure 23. B assessment results.

Figure 23 shows a ΔB value, which is the difference between real and estimated bandwidths when there is a PU. When this parameter tends to grow, it provides flexibility so that the resulting NN_3 B matches an ideal B that corresponds to one transmission. However, when the value of ΔB is very small, the system becomes more inaccurate, since it only detects as TP those values that are close to the ideal value.

Figure 24 shows the $F1$ of NN_3. The NN with four layers and the NN with two layers both had an undesirable performance. This result could be explained by observing Figure 20, which shows the result of B_1 is very close to the minimum value $(ideal\_value - \Delta B)$. Nevertheless, this undesirable performance was observed only when a low number of neurons per layer (i.e., 16) were employed. By properly configuring the number of neurons per layer to a sufficiently high value (e.g., 32 or greater), accurate performance with $F1$ values around 0.95 and 0.96 was obtained, as well as with two and four layers in the NN. Figure 24 also suggests overfitting in the NNs with two and four layers. This is because the models for the 2-layer and 4-layer NNs, both with 16 neurons per layer, fit too well to the training data and, as a result, do not generalize well to the new data. Despite this, they have an $F1$ of 0.9 and 0.79, respectively.

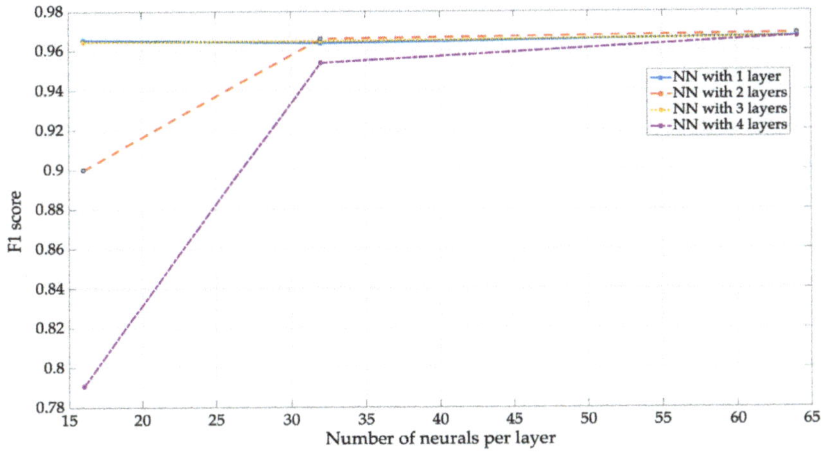

Figure 24. $F1$ of NN_3.

Figure 25 shows the ratio of $F1$ scores between training time for NN_1 and NN_3 (Figure 25a and 25b, respectively). The value obtained from this relationship indicates which NN provides the best trade-off based on the F1 accuracy result and the time invested in training the network to attain such accuracy. As can be noticed, using an NN with a single layer provides the best F1 performance for the training time required to obtain it. Further increasing the number of layers will increase the F1 score performance but will proportionately require a much longer training time, thus leading to a worse trade-off or relation between benefit (represented by F1 performance) and cost (represented by the required training time).

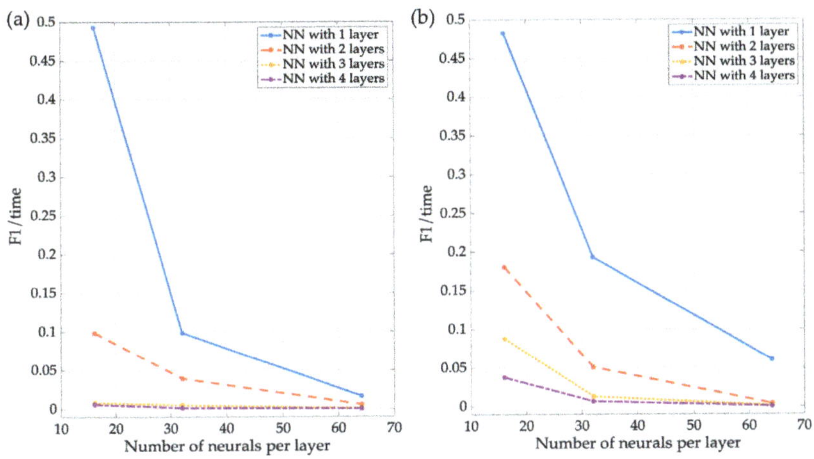

Figure 25. $F1$ and training time ratio of (**a**) NN_1 and (**b**) NN_3.

Finally, Figure 26 shows the behavior of the proposed methodology over time, during which the sensed spectrum was analyzed in a specific geographic area and where SUs collaborated to obtain the REM of the different detected PUs. As can be appreciated, the proposed methodology was able to characterize the dynamic temporal evolution of the REM in the geographical area of interest, thus providing a valuable tool for the study of CRNs.

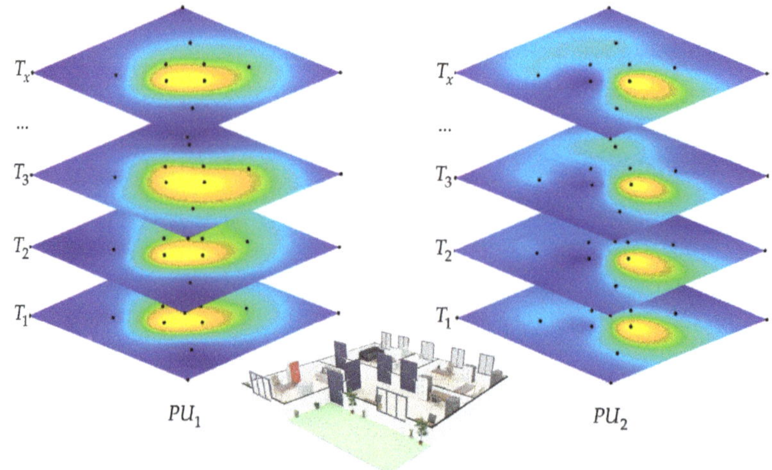

Figure 26. Result of the central entity.

7. Conclusions

In this work, a real wireless communication scenario was implemented to detect the occupancy of multiple PUs through several SUs via a central entity, permitting the determination of the area used by the PUs based on REM estimations. Apart from the REM constructions, other features, such as power, bandwidth, and central frequency of possible detected PUs from the multiband spectrum frames, were estimated by the considered SUs.

For this task, we proposed using neural networks to substitute the classical digital signal processing used in our preliminary work. It was expected that the performance and processing time would be faster. Clearly, the training stage of the NNs, as shown by the results, was a factor to be considered. In this work, higher precision was preferred to locate the PUs and less processing time for the central entity when using the NNs. However, it must be considered that NN training time was not negligible and could even be high, depending on the number of layers and neurons each NN contained. The difference between the real values and that of the NNs was minimal, and it could even be said they were practically the same. In some cases, the NN showed even better results than the digital signal processing, for instance, when detecting the PUs' carriers.

Neural networks are a powerful and useful tool in many applications but are not always the best option. In some cases, classical digital signal processing may be sufficient, while in others, neural networks can significantly improve performance and accuracy. Future work will aim to determine the optimal number of SUs needed to obtain the area occupied by the PUs, thus avoiding the excessive processing of the central entity.

Author Contributions: All authors conceptualized the original idea of how to use SDR devices and designed the proposed methodology. Y.M.-T. improved these proposals by adding some ideas on how to obtain the data and how to develop the necessary software. Y.M.-T. also performed all the experiments of this work. All authors have read and agreed to the published version of the manuscript.

Funding: This research received funding from the Mexican National Council of Humanities, Science, and Technology (CONAHCyT), Grant (no. 490180).

Institutional Review Board Statement: Not applicable.

Informed Consent Statement: Not applicable.

Data Availability Statement: Data sharing not applicable.

Acknowledgments: The authors are grateful to CONAHCYT, the Department of Electrical Engineering of UAM-Iztapalapa and Department of Electrical Engineering and Electronics of University of Liverpool, for providing the necessary support for this work.

Conflicts of Interest: The authors declare no conflict of interest.

Abbreviations

Cognitive radio	CR
Primary user	PU
Secondary user	SU
Spectrum sensing	SS
Multiband spectrum sensing	MBSS
Machine learning	ML
Software defined radio	SDR
Universal software radio peripheral	USRP
Radio environment map	REM
Cognitive radio network	CRN
Inverse Distance Weighting	IDW
Neural networks	NN
Multilayer perceptron	MLP
Multiresolution analysis	MRA
Rectified linear unit)	ReLU
Sample Entropy	SampEn
Power Spectrum Density	PSD
Step 1.1	S 1.1
Step 1.2	S 1.2
Step 1.3	S 1.3
Step 1.5	S 1.5
Step 1.6	S 1.6
Step 2.1	S 2.1
Step 2.2	S 2.2
Step 2.3	S 2.3
Step 2.5	S 2.5
Step 2.6	S 2.6
Bandwidth	B
Carrier frequency	Fc
Standard deviation	STD
F1 score	F1
True positive	TP
False positive	FP
False negative	FN
True negative	TN

References

1. Mitola, J.; Maguire, G.Q. Cognitive radio: Making software radios more personal. *IEEE Pers. Commun.* **1999**, *6*, 13–18. [CrossRef]
2. Akyildiz, I.F.; Lee, W.-Y.; Vuran, M.C.; Mohanty, S. NeXt generation/dynamic spectrum access/cognitive radio wireless networks: A survey. *Comput. Netw.* **2006**, *50*, 2127–2159. [CrossRef]
3. Hattab, G.; Ibnkahla, M. Multiband Spectrum Access: Great Promises for Future Cognitive Radio Networks. *Proc. IEEE* **2014**, *102*, 282–306. [CrossRef]
4. El-Khamy, S.E.; El-Mahallawy, M.S.; Youssef, E.S. Improved wideband spectrum sensing techniques using wavelet-based edge detection for cognitive radio. In Proceedings of the 2013 International Conference on Computing, Networking and Communications (ICNC), San Diego, CA, USA, 28–31 January 2013; pp. 418–423. [CrossRef]
5. Kumar, A.; Saha, S.; Bhattacharya, R. Wavelet transform based novel edge detection algorithms for wideband spectrum sensing in CRNs. *AEU-Int. J. Electron. Commun.* **2018**, *84*, 100–110. [CrossRef]
6. Diao, X.; Dong, Q.; Yang, Z.; Li, Y. Double-Threshold Cooperative Spectrum Sensing Algorithm Based on Sevcik Fractal Dimension. *Algorithms* **2017**, *10*, 96. [CrossRef]
7. Popoola, J.J.; van Olst, R. Application of neural network for sensing primary radio signals in a cognitive radio environment. In Proceedings of the IEEE Africon '11, Victoria Falls, Zambia, 13–15 September 2011; pp. 1–6. [CrossRef]

8. Shamsi, N.; Mousavinia, A.; Amirpour, H. A channel state prediction for multi-secondary users in a cognitive radio based on neural network. In Proceedings of the 2013 International Conference on Electronics, Computer and Computation (ICECCO), Ankara, Turkey, 7–9 November 2013; pp. 200–203. [CrossRef]
9. Molina-Tenorio, Y.; Prieto-Guerrero, A.; Aguilar-Gonzalez, R. A Novel Multiband Spectrum Sensing Method Based on Wavelets and the Higuchi Fractal Dimension. *Sensors* **2019**, *19*, 1322. [CrossRef] [PubMed]
10. Zaidawi, D.J.; Sadkhan, S.B. Blind Spectrum Sensing Algorithms in CRNs: A Brief Overview. In Proceedings of the 2021 7th International Engineering Conference "Research & Innovation amid Global Pandemic" (IEC), Erbil, Iraq, 24–25 February 2021; pp. 78–83. [CrossRef]
11. Molina-Tenorio, Y.; Prieto-Guerrero, A.; Aguilar-Gonzalez, R. Real-Time Implementation of Multiband Spectrum Sensing Using SDR Technology. *Sensors* **2021**, *21*, 3506. [CrossRef] [PubMed]
12. Politis, C.; Maleki, S.; Duncan, J.M.; Krivochiza, J.; Chatzinotas, S.; Ottesten, B. SDR Implementation of a Testbed for Real-Time Interference Detection with signal cancellation. *IEEE Access* **2018**, *6*, 20807–20821. [CrossRef]
13. Stewart, R.W.; Crockett, L.; Atkinson, D.; Barlee, K.; Crawford, D.; Chalmers, I.; Mclernon, M.; Sozer, E. A low-cost desktop software defined radio design environment using MATLAB, simulink, and the RTL-SDR. *IEEE Commun. Mag.* **2015**, *53*, 64–71. [CrossRef]
14. Hiari, O.; Mesleh, R. A Reconfigurable SDR Transmitter Platform Architecture for Space Modulation MIMO Techniques. *IEEE Access* **2017**, *5*, 24214–24228. [CrossRef]
15. Santos-Luna, E.; Prieto-Guerrero, A.; Aguilar-Gonzalez, R.; Ramos, V.; Lopez-Benitez, M.; Cardenas-Juarez, M. A Spectrum Analyzer Based on a Low-Cost Hardware-Software Integration. In Proceedings of the 2019 IEEE 10th Annual Information Technology, Electronics and Mobile Communication Conference (IEMCON), Vancouver, BC, Canada, 17–19 October 2019; pp. 0607–0612. [CrossRef]
16. Aghabeiki, S.; Hallet, C.; Noutehou, N.E.-R.; Rassem, N.; Adjali, I.; Mabrouk, M.B. Machine-learning-based spectrum sensing enhancement for software-defined radio applications. In Proceedings of the 2021 IEEE Cognitive Communications for Aerospace Applications Workshop (CCAAW), Cleveland, OH, USA, 21–23 June 2021; pp. 1–6. [CrossRef]
17. Selva, A.F.B.; Reis, A.L.G.; Lenzi, K.G.; Meloni, L.G.P.; Barbin, S.E. Introduction to the Software-defined Radio Approach. *IEEE Lat. Am. Trans.* **2012**, *10*, 1156–1161.
18. About RTL-SDR, rtl-sdr.com, 11 de Abril de 2013. Available online: https://www.rtl-sdr.com/about-rtl-sdr/ (accessed on 8 August 2022).
19. Koutlia, K.; Bojović, B.; Lagén, S.; Giupponi, L. Novel radio environment map for the ns-3 NR simulator. In Proceedings of the Workshop on ns-3, Virtual Event USA: ACM, New York, NY, USA, 23–24 June 2021; pp. 41–48. [CrossRef]
20. Li, J.; Ding, G.; Zhang, X.; Wu, Q. Recent Advances in Radio Environment Map: A Survey. In *Machine Learning and Intelligent Communications*; Gu, X., Liu, G., Li, B., Eds.; En Lecture Notes of the Institute for Computer Sciences, Social Informatics and Telecommunications Engineering; Springer International Publishing: Cham, Switzerland, 2018; Volume 226, pp. 247–257. [CrossRef]
21. Spooner, C.M.; Khambekar, N.V. Spectrum sensing for cognitive radio: A signal-processing perspective on signal-statistics exploitation. In Proceedings of the 2012 International Conference on Computing, Networking and Communications (ICNC), Maui, HI, USA, 30 January–2 February 2012; pp. 563–568. [CrossRef]
22. Zhang, X.; Zhao, Y.; Chen, H. Adaptive Compressed Wideband Spectrum Sensing Based on Radio Environment Map Dedicated for Space Information Networks. In *Wireless and Satellite Systems*; Jia, M., Guo, Q., Meng, W., Eds.; En Lecture Notes of the Institute for Computer Sciences, Social Informatics and Telecommunications Engineering; Springer International Publishing: Cham, Switzerland, 2019; pp. 128–138. [CrossRef]
23. Santana, Y.H.; Plets, D.; Alonso, R.M.; Nieto, G.G.; Martens, L.; Joseph, W. Radio Environment Map of an LTE Deployment Based on Machine Learning Estimation of Signal Levels. In Proceedings of the 2022 IEEE International Symposium on Broadband Multimedia Systems and Broadcasting (BMSB), Bilbao, Spain, 15–17 June 2022; pp. 1–6. [CrossRef]
24. Borisov, V.; Leemann, T.; Sessler, K.; Haug, J.; Pawelczyk, M.; Kasneci, G. Deep Neural Networks and Tabular Data: A Survey. *IEEE Trans. Neural Netw. Learn. Syst.* **2022**. *early access*. [CrossRef] [PubMed]
25. Xu, M.; Yin, Z.; Zhao, Y.; Wu, Z. Cooperative Spectrum Sensing Based on Multi-Features Combination Network in Cognitive Radio Network. *Entropy* **2022**, *24*, 129. [CrossRef] [PubMed]
26. Lam, N.S.-N. Spatial Interpolation Methods: A Review. *Am. Cartogr.* **1983**, *10*, 129–150. [CrossRef]
27. Burrough, P.A.; McDonnell, R.; Burrough, P.A. *Principles of Geographical Information System*; en Spatial Information Systems; Oxford University Press: Oxford, UK, 1998.
28. Harman, B.I.; Koseoglu, H.; Yigit, C.O. Performance evaluation of IDW, Kriging and multiquadric interpolation methods in producing noise mapping: A case study at the city of Isparta, Turkey. *Appl. Acoust.* **2016**, *112*, 147–157. [CrossRef]
29. Arseni, M.; Voiculescu, M.; Georgescu, L.P.; Iticescu, C.; Rosu, A. Testing Different Interpolation Methods Based on Single Beam Echosounder River Surveying. Case Study: Siret River. *ISPRS Int. J. Geo-Inf.* **2019**, *8*, 507. [CrossRef]
30. Matheron, G. Principles of geostatistics. *Econ. Geol.* **1963**, *58*, 1246–1266. [CrossRef]
31. Oliver, M.A.; Webster, R. Kriging: A method of interpolation for geographical information systems. *Int. J. Geogr. Inf. Syst.* **1990**, *4*, 313–332. [CrossRef]

32. Gundogdu, K.S.; Guney, I. Spatial analyses of groundwater levels using universal kriging. *J. Earth Syst. Sci.* **2007**, *116*, 49–55. [CrossRef]
33. Cressie, N.A.C. *Statistics for Spatial Data, Revised ed.*; John Wiley & Sons, Inc.: Hoboken, NJ, USA, 2015.
34. Isaaks, E.H.; Srivastava, R.M. *Applied Geostatistics*; Oxford University Press: New York, NY, USA, 1989.
35. Han, J.; Kamber, M. *Data Mining: Concepts and Techniques*, 3rd ed.; Elsevier: Burlington, MA, USA, 2012.
36. Wang, W.; Yang, Y. Development of convolutional neural network and its application in image classification: A survey. *Opt. Eng.* **2019**, *58*, 1. [CrossRef]
37. Haykin, S.S.; Haykin, S.S. *Neural Networks and Learning Machines*, 3rd ed.; Prentice Hall: New York, NY, USA, 2009.
38. Rumelhart, D.E.; Hinton, G.E.; Williams, R.J. Learning representations by back-propagating errors. *Nature* **1986**, *323*, 533–536. [CrossRef]
39. Schmidhuber, J. Deep learning in neural networks: An overview. *Neural Netw.* **2015**, *61*, 85–117. [CrossRef] [PubMed]
40. Molina-Tenorio, Y.; Prieto-Guerrero, A.; Aguilar-Gonzalez, R. Multiband Spectrum Sensing Based on the Sample Entropy. *Entropy* **2022**, *24*, 411. [CrossRef] [PubMed]
41. Molina-Tenorio, Y.; Prieto-Guerrero, A.; Aguilar-Gonzalez, R.; Ruiz-Boqué, S. Machine Learning Techniques Applied to Multiband Spectrum Sensing in Cognitive Radios. *Sensors* **2019**, *19*, 4715. [CrossRef]
42. Nooelec-Nooelec NESDR SMArt v4 SDR-Premium RTL-SDR w/Aluminum Enclosure, 0.5PPM TCXO, SMA Input. RTL2832U & R820T2-Based-Software Defined Radio. Available online: https://www.nooelec.com/store/sdr/nesdr-smart-sdr.html (accessed on 8 March 2021).
43. HackRF One-Great Scott Gadgets, 8 de Marzo de 2021. Available online: https://greatscottgadgets.com/hackrf/one/ (accessed on 8 March 2021).
44. LimeSDR Mini is a $135 Open Source Hardware, Full Duplex USB SDR Board (Crowdfunding)-CNX Software, CNX Software -Embedded Systems News, 18 de Septiembre de 2017. Available online: https://www.cnx-software.com/2017/09/18/limesdr-mini-is-a-135-open-source-hardware-full-duplex-usb-sdr-board-crowdfunding/ (accessed on 13 March 2022).
45. Welch, P. The use of fast Fourier transform for the estimation of power spectra: A method based on time averaging over short, modified periodograms. *IEEE Trans. Audio Electroacoust.* **1967**, *15*, 70–73. [CrossRef]
46. Mallat, S.G. A theory for multiresolution signal decomposition: The wavelet representation. *IEEE Trans. Pattern Anal. Mach. Intell.* **1989**, *11*, 674–693. [CrossRef]
47. Jain, A.K. Data clustering: 50 years beyond K-means. *Pattern Recognit. Lett.* **2010**, *31*, 651–666. [CrossRef]
48. Sasaki, Y. The Truth of the F-Measure, oct. 2007 [En Línea]. Available online: https://www.cs.odu.edu/~mukka/cs795sum11dm/Lecturenotes/Day3/F-measure-YS-26Oct07.pdf (accessed on 7 April 2007).

sensors

MDPI

Article

Deep Q-Learning-Based Buffer-Aided Relay Selection for Reliable and Secure Communications in Two-Hop Wireless Relay Networks

Cheng Zhang [1,2], Xuening Liao [1,2,3,*], Zhenqiang Wu [1,2], Guoyong Qiu [1,2], Zitong Chen [2] and Zhiliang Yu [4]

[1] Key Laboratory of Modern Teaching Technology, Ministry of Education, Xi'an 710062, China; zhang_cheng@snnu.edu.cn (C.Z.); zqiangwu@snnu.edu.cn (Z.W.); qgyqgy@snnu.edu.cn (G.Q.)
[2] School of Computer Science, Shaanxi Normal University, Xi'an 710119, China; zitong_@snnu.edu.cn
[3] Shaanxi Key Laboratory for Network Computing and Security Technology, Xi'an 710048, China
[4] School of Mathematics and Computer Science, Shaanxi University of Technology, Hanzhong 723001, China; yu@snut.edu.cn
* Correspondence: liaoxuening@snnu.edu.cn

Abstract: This paper investigates the problem of buffer-aided relay selection to achieve reliable and secure communications in a two-hop amplify-and-forward (AF) network with an eavesdropper. Due to the fading of wireless signals and the broadcast nature of wireless channels, transmitted signals over the network may be undecodable at the receiver end or have been eavesdropped by eavesdroppers. Most available buffer-aided relay selection schemes consider either reliability or security issues in wireless communications; rarely is work conducted on both reliability and security issues. This paper proposes a buffer-aided relay selection scheme based on deep Q-learning (DQL) that considers both reliability and security. By conducting Monte Carlo simulations, we then verify the reliability and security performances of the proposed scheme in terms of the connection outage probability (COP) and secrecy outage probability (SOP), respectively. The simulation results show that two-hop wireless relay network can achieve reliable and secure communications by using our proposed scheme. We also performed comparison experiments between our proposed scheme and two benchmark schemes. The comparison results indicate that our proposed scheme outperforms the max-ratio scheme in terms of the SOP.

Keywords: physical-layer security; buffer-aided relay selection; Markov decision process; deep Q-learning; secrecy outage probability; connection outage probability

1. Introduction

With the development of 5G and beyond, wireless networks are widely used in various fields, such as wireless sensor networks (WSNs) [1], cognitive radio networks (CRNs) [2], and the Internet of Things (IoTs) [3]. With the wide use of wireless networks, a large amount of confidential information is transmitted over each network every day. However, signals may be undecodable at the receiver end due to the fading of wireless signals, and may be intercepted by an eavesdropper due to the broadcast nature of wireless channels, leading to critical reliability and especially security issues in wireless networks. Any unauthorized attacker within the transmission range of a transmitter can receive the transmitted information, which can easily cause information leakage [4]. Therefore, the problem of reliable and secure communications in wireless networks urgently needs to be solved.

The traditional method to achieve secure communications in wireless networks is based on a cryptographic mechanism. The principle of cryptography is to encrypt confidential information with a secret key at the legitimate sender's end and then decrypt it with a secret key at the legitimate receiver's end [5]. As the secret key is deployed only on the legitimate transmitter and receiver, eavesdroppers cannot decrypt the encrypted information because of the lack of the secret key [6]. The disadvantage of cryptography is that its implementation requires the deployment of devices with a high level of computational performance due to the high level of computational complexity associated with encrypting and decrypting. It is impossible to require all devices connected to wireless networks to have high computational ability. In recent years, a new method called physical-layer security (PLS), which has low computational complexity, has been proposed and is used to aid cryptography in achieving secure communications in wireless networks with low computing capability. The principle of PLS is based on information theory, and uses the randomness of noise and wireless channels to achieve secure communications [7]. Compared with the cryptography method, PLS has a lower network resource overhead and computational complexity [8]. Therefore, PLS has become a promising technique that can help in enhancing security performance in wireless networks.

Common PLS techniques include beamforming [9], artificial noise [10], and relay selection [11]. The principle of beamforming is to achieve the directional transmission of signals by adjusting the transmission direction of the antennas to achieve PLS [12]. Artificial noise interferes with eavesdropping by sending noise [13]. Relay selection achieves PLS by selecting the appropriate relay nodes to which to transmit confidential information [14]. Compared with beamforming and artificial noise, the implementation complexity of relay selection is lower. Depending on whether the relay nodes are equipped with buffers or not, the relay selection technique is divided into conventional and buffer-aided relay selection [15]. In conventional relay selection, once relay nodes without buffers receive the signals, they have to immediately forward them to the next hop [16]. In contrast, buffer-aided relay selection can temporarily store the received signals in buffers instead of transmitting them immediately [17]. So, buffer-aided relay selection can achieve better security performance than that of the conventional relay selection without buffers, especially when the channel quality is poor [18]. Due to its low implementation complexity and good security performance, we used the buffer-aided relay selection technique to achieve reliable and secure communications in this paper.

Traditional buffer-aided relay selection selects the best relay by adopting a central node that collects the network information (e.g., the channel state information (CSI) of the legitimate link and the CSI of the eavesdropping links) and then selects the relays online on the basis of this information. However, it is difficult to achieve CSI of eavesdropping links, as the eavesdroppers always transmit no information, and too much energy, storage, and time are needed to conduct the relay selection, as there are several transmission patterns (i.e., source–relay, relay–destination, and source–destination transmissions) and many possible relay buffer states during the transmission. This is challenged when the central node is resource-limited. Unlike traditional methods, traditional Q-learning (TQL) and DQL define the Q-function, which can simplify the modeling of information transmission in buffer-aided relay selection by evaluating the gain of choosing a particular link to which to transmit signals in the current state in an integrated manner, especially DQL. By using neural networks to fit the Q-function, DQL can create the Q-function without storing it in a Q table, reducing both spatial and temporal complexity for buffer-aided relay selection. Therefore, we used the DQL method to propose a new buffer-aided relay scheme to achieve reliable and secure communications in two-hop wireless relay networks.

2. Related Work

For two-hop buffer-aided relay networks without eavesdroppers, the authors in [19] consider the reliability of wireless communications and proposed a novel buffer-aided relay selection scheme called the max-link scheme. In the max-link scheme, the signals at each hop are transmitted by the link with the maximum signal-to-noise ratio (SNR) to achieve reliable communications. The authors in [19] also established a theoretical analysis framework on the basis of a Markov chain (MC) for analyzing the outage performance of their proposed buffer-aided relay selection scheme. The authors in [20] combine social networks with two-hop wireless relay networks and investigate how to design a buffer-aided relay selection scheme to achieve reliable communications when there are untrusted relays in the network. Due to the introduction of buffers, the queuing delay of data packets at buffer-aided relay increases. To achieve reliable communication and reduce delay, the authors in [21] proposed a delay-sensitive buffer-aided relay selection based on channel-based greedy scheduling in vehicular networks. Because of the low implementation complexity of buffer-aided relay selection, the authors in [22] use buffer-aided relay selection to improve the reliability of bidirectional wireless sensor network communications. These buffer-aided relay selection schemes described above are all based on MP. The authors in [23] model buffer-aided relay selection as a Markov decision process (MDP) rather than the MP and exploit TQL to design a buffer-aided relay scheme. TQL evaluates all links by Q-function and selects the link with the maximum Q-function each time to transmit the signals and thus achieves reliable communications. Due to the excellent reliability performance of the proposed scheme based on TQL, the authors in [24,25] extend this work to vehicular networks, D2D communications and achieve reliable communications in vehicular networks and D2D communications. Although these TQL-based schemes can achieve reliable communications, too much storage space is needed to store the Q-table and a high time cost to look up the Q-function in the Q-table as all Q-functions have to be stored in the Q-table.

As research continued, researchers began to investigate how to achieve secure communications using buffer-aided relay selection when considering possible eavesdroppers in the network [26–32]. Based on the work in [19], the authors in [26] consider the case where a passive eavesdropper is present and proposed a new buffer-aided relay selection scheme to achieve secure communications by selecting the link with the maximum instantaneous secrecy capacity at each hop to transmit the signals. In real scenarios, not only illegal eavesdroppers eavesdrop signals, but also untrusted relay nodes can intercept the transmitted signals as well. These untrusted relay nodes are both cooperators and potential eavesdroppers of information transmission. In response to the presence of untrusted relay nodes, the authors in [27] propose a secure buffer-aided relay selection scheme that uses the AF mode to avoid the decoding of confidential information by untrusted relay nodes. The authors in [28] extend this work to a more general scenario where both potential eavesdropping nodes and passive eavesdroppers are present. The authors in [29] extend secure communications to bidirectional wireless relay network and design a buffer-aided relay selection scheme based on an achievable rate. In addition, to resist the eavesdroppers, buffer-aided relay selection is often combined with full duplex (FD) [30], cooperative jamming (CJ) [31] and energy harvesting (EH) [32] to achieve secure communications. Although these schemes can realize secure communications, they also increase the implementation complexity, which conflicts with the original intent of adopting buffer-aided relay selection.

All of the above related works only consider the reliability or the security performances of wireless communications, without considering both the security and reliability issues. In fact, it is very challenging to simultaneously achieve reliable and secure communications by using buffer-aided relay selection. It requires simultaneously taking into account the legitimate channel quality, eavesdropping channel quality, buffer queues, secrecy rate, etc. Therefore, buffer-aided relay selection based on traditional methods is difficult to achieve reliable and secure communications while maintaining low implementation complexity. With the development of deep learning (DL), DL has been applied to wireless relay networks [33–35]. A large number of researchers have started to use deep learning to study buffer-aided relay selection [36–43]. The authors in [36] model the buffer-aided relay selection as a multi-classification problem and uses a deep neural network (DNN) to predict the suitable link to transmit the signals. Inspired by [23], the authors in [37] utilize DQL to solve the buffer-aided relay selection problem, where a modified version of TQL is used. Different from TQL, DQL uses DNN to fit the Q-function instead of storing Q-function in the Q-table. Therefore, DQL has lower time complexity and space complexity compared to TQL [38]. The comparison experiments in [39] demonstrated that DQL has better learning results and lower complexity than those of TQL, and the implemented scheme via DQL is more suitable for practical scenarios, as the implemented scheme via DQL could work without prior information. On this basis, the authors in [40,41] realized reliable communications for IoTs [40] and CRNs [41] by using the DQL-based buffer-aided relay selection schemes. The authors in [40,41] extend their work further and use the proposed DQL-based buffer-aided relay selection scheme to realize reliable and secure communications in CRNs [42,43].

DQL makes it possible to achieve reliable and secure communications using buffer-aided relay selection. However, the works in [42,43] use DQL to address the issue of power allocation (PA) to achieve reliable and secure communications and they did not consider possible eavesdroppers in the network. Therefore, this paper explores how to achieve reliable and secure communications using only a DQL-based buffer-aided relay selection scheme in the more common two-hop wireless relay networks rather than CRNs. To highlight the contributions of this paper, we give a comparison of our work with related works in Table 1. The work of this paper is summarized as follows:

Table 1. The main features of our work and related works.

References and Our Work	Feature				
	System Model	**Eavesdropper**	**Method**	**Reliability**	**Security**
[19]	Two-hop DF relay network	✘	MC	✓	✘
[23]	Two-hop AF relay network	✘	TQL	✓	✘
[26]	Two-hop AF relay network	✘	MC	✘	✓
[40]	Delay-constrained DF relay IOT	✘	DQL	✓	✘
[43]	RF relay CRN	✓ (untrusted users)	DQL+PA	✓	✓
Our Work	Two-hop AF relay network	✓	DQL	✓	✓

✓ indicates that the factor is considered in the paper, and ✘ indicates that the factor is not considered in the paper.

- To propose a DQL-based buffer-aided relay selection scheme, we first analyze the communication model of a two-hop AF buffer-aided relay network with the presence of a passive eavesdropper and then model the information transmission process as an MDP.
- We then propose a DQL-based buffer-aided relay selection scheme to optimize the above MDP. In the proposed scheme, we consider both the legal channel states and eavesdropping channel states, buffer states, target rate and target secrecy rate and use DNNs to fit the Q-function and select the link with the maximum Q-function value each time.
- Finally, we verify the reliability and security performances of the proposed scheme by using Monte Carlo simulations. The reliability and security performances are measured by the COP and the SOP, respectively. Simulation results demonstrate that the proposed scheme can achieve reliable and secure communications. We also compare the COP and SOP of the proposed scheme with the max-link and max-

ratio schemes, respectively. The comparison results show that the proposed scheme outperforms max-ratio schemes in terms of security performance.

The remainder of this paper is organized as follows: Section 3 introduces the system model; Section 4 introduces the framework of information transmission based on MDP; Section 5 describes the proposed buffer-aided relay selection scheme; Section 6 shows the simulation results of proposed scheme; Section 7 concludes the contributions of this paper.

3. System Model

As depicted in Figure 1, this paper considers a two-hop AF buffer-aided relay network, which is composed of a source node S, a cluster of AF buffer-aided relay nodes R_k $(k \in \{1, 2, \cdots, K\})$, a destination node D and a passive eavesdropper node E. The source node S cannot communicate with the destination node D directly due to path loss and the long distance, so the signals from S must be forwarded by the buffer-aided relay node R_k. The number of AF buffer-aided relay nodes is K. Every relay node is in half-duplex (HD) mode and is equipped with a buffer queue Q_k of length L, so these relay nodes can store the received signals instead of forwarding them immediately to D. This paper assumes that the eavesdropping node E only eavesdrops the signals from R_k to D, and does not eavesdrop the signals from S to R_k.

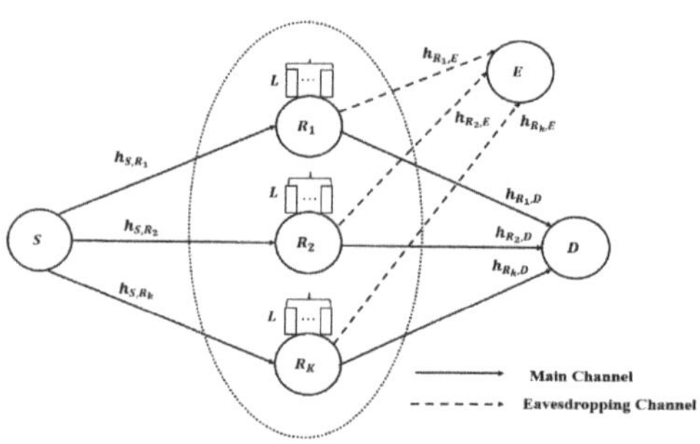

Figure 1. Illustration of the system model.

Without the decoding process at relays, AF relays can thus decrease the probability of being intercepted by potential eavesdroppers for transmitted signals [44]. Thus, we assume that all relays are AF relays in this paper to enhance the security of signals transmitted in the network.

We assume that all channels are independent and non-identically distributed quasi-static Rayleigh fading channels, including eavesdropping channels. In this paper, we use $h_{m,n}$ and $g_{m,n}$ to denote the channel coefficient and the channel gain between node m and node n, respectively, where $g_{m,n} = |h_{m,n}|^2$. Since all channels are Rayleigh channels [45], the channel gain follows the exponential distribution, which means that $E[|h_{m,n}|^2] = E[g_{m,n}] = \Omega_{m,n}$, where $E[.]$ is the expectation operator and $\Omega_{m,n}$ is the average channel gain. This paper assumes that the real-time CSI is completely known and sets the source node S as the central node, which receives the real-time CSI of all channels and buffers state information of all buffer-aided relay nodes then selects an appropriate link to transmit the signals according to relay selection schemes. Supposing at a time slot t, the central node selects an S to R_k link to transmit signals, the received signals $y_{R_k}(t)$ at R_k can be expressed as

$$y_{R_k}(t) = \sqrt{P_s} h_{S,R_k}(t) x_s(t) + n_{R_k}(t), \tag{1}$$

where P_s is the transmission power of the source node S, $x_s(t)$ is the signal sent by S at time t, and $n_{R_k}(t)$ is the additive white Gaussian noise (AWGN) noise with variance power σ^2 at R_k. According to (1), the instantaneous SNR of S to R_k link at time t is given by

$$\psi_{S,R_k}(t) = \frac{P_s|h_{S,R_k}(t)|^2}{\sigma^2}, k \in \{1,2,\cdots,K\}, \tag{2}$$

and the channel capacity of S to R_k link is $C_{S,R_k}(t) = \frac{1}{2}\log_2(1+\psi_{S,R_k}(t))$, $k \in \{1,2,\cdots,K\}$. The received signal $y_{R_k}(t)$ is stored in the corresponding buffer queue Q_k waiting for the transmission to the next hop. After waiting for t_1 time slots, the received signal $y_{R_k}(t)$ is amplified to resist path fading and then forwarded to the destination node D by the buffer-aided relay node R_k. Thus, at time slot $t' = t + t_1$, the signal $x_{R_k}(t')$ sent by the buffer-aided relay node R_k is represented as

$$x_{R_k}(t') = A_{R_k}(t')y_{R_k}(t), \tag{3}$$

where

$$A_{R_k}(t') = \frac{1}{\sqrt{P_s|h_{S,R_k}(t)|^2 + \sigma^2}} \tag{4}$$

is the amplification factor of the buffer-aided relay node R_k at time t', it is determined by the quality of the channel between source node S and the buffer-aided relay node R_k at time t. Due to the broadcast nature of wireless channel, eavesdropping nodes within the transmission range can also receive the transmitted signals. In this paper, we assume that the eavesdropping node only eavesdrops the signals sent by the buffer-aided relay nodes R_k to the destination node D. So the signals received by S and E can be expressed as

$$\begin{aligned} y_D(t') &= \sqrt{P_{R_k}}h_{R_k,D}(t')x_{R_k}(t') + n_D(t'), \\ y_E(t') &= \sqrt{P_{R_k}}h_{R_k,E}(t')x_{R_k}(t') + n_E(t'), \end{aligned} \tag{5}$$

respectively, where P_{R_k} is the transmission power of R_k, $n_D(t')$ and $n_E(t')$ are AWGN noises at D and E, respectively. According to (5), the instantaneous end-to-end SNR from S to D and from S to E can be derived as

$$\begin{aligned} \psi_{S,D}(t') &= \frac{P_sP_{R_k}|h_{S,R_k}(t)|^2|h_{R_k,D}(t')|^2}{(P_s|h_{S,R_k}(t)|^2 + P_{R_k}|h_{R_k,D}(t')|^2 + \sigma^2)\sigma^2}, \\ \psi_{S,E}(t') &= \frac{P_sP_{R_k}|h_{S,R_k}(t)|^2|h_{R_k,E}(t')|^2}{(P_s|h_{S,R_k}(t)|^2 + P_{R_k}|h_{R_k,E}(t')|^2 + \sigma^2)\sigma^2}, \end{aligned} \tag{6}$$

respectively. Thus, the end-to-end channel capacity from S to D and S to E can be given by

$$\begin{aligned} C_{S,D}(t') &= \frac{1}{2}\log_2(1+\psi_{S,D}(t')), \\ C_{S,E}(t') &= \frac{1}{2}\log_2(1+\psi_{S,E}(t')), \end{aligned} \tag{7}$$

respectively. The end-to-end secrecy rate from S to D is given by

$$C^{(s)}_{S,D}(t') = [\theta - C_{S,E}(t')]^+, \tag{8}$$

where $[z]^+ = \max(o,z)$, and θ is the target rate of the two-hop AF buffer-aided relay network.

4. The Framework of Information Transmission Based on MDP

To design a buffer-aided relay selection scheme that enables reliable and secure communications in two-hop wireless relay networks, we need to first analyze the information transmission process in two-hop wireless relay networks. Due to the Markovian property of the process of receiving and forwarding information in the buffers, the information transmission process in two-hop wireless relay networks can be modeled as an MDP to analyze. As shown in Figure 2, a complete MDP consists of a five-tuple (state s_t, action a_t, policy $\pi(a_t|s_t)$, reward $r(s_t, a_t)$), return U_t, environment and an agent. This section describes in detail how to model the process of information transmission in two-hop wireless relay networks as an MDP.

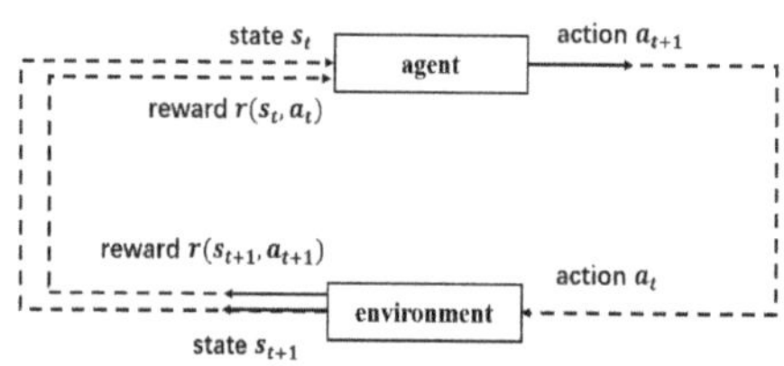

Figure 2. An MDP, which consists of state s_t, action a_t, policy $\pi(a_t|s_t)$, reward $r(s_t, a_t)$, return U_t, environment and an agent.

4.1. Agent and Environment

In the MDP, the agent can perceive the state of the environment, take actions according to the state and adjust the decisions based on the feedback of the environment. In the two-hop AF buffer-aided relay network, the central node is regarded as the agent in the MDP and the whole two-hop AF buffer-aided relay network is modeled as the environment in the MDP. The state of the environment will be changed by action of the agent, which can be perceived by the agent. In addition, the environment will give the agent feedback after each decision made by the agent.

4.2. State

For the two-hop AF buffer-aided relay network, this paper defines the state space $s(t)$ at time slot t as $s(t) = \{l(t), b(t)\}$, where $l(t)$ and $b(t)$ are the link states of all links and the buffer states of all buffer queues at time slot t, respectively. The link states $l(t)$ at time t are defined as

$$l(t) = \{l_{0,1}(t), l_{0,2}(t), \cdots, l_{0,K}(t), l_{1,1}(t), \cdots, l_{1,K}(t)\}, \tag{9}$$

where $j = 0, l_{0,k}(t)$ is the link state of S to the corresponding R_k link; $j = 1, l_{1,k}(t)$ is the link state of the corresponding R_k to D link. As we assume that the eavesdropping node E only intercept signals from the R_k to D link, only the reliability issue of the transmission link needs to be considered in the first hop. The value of $l_{0,k}(t)$ is taken as follows.

- $l_{0,k}(t) = 0$ denotes $C_{S,R_k}(t) \leq \theta$ and the corresponding link is unreliable. When $l_{0,k}(t) = 0$, the corresponding link can not transmit the signals at the target rate θ.
- $l_{0,k}(t) = 2$ denotes $C_{S,R_k}(t) \geq \theta$ and the corresponding link is reliable. When $l_{0,k}(t) = 2$, the corresponding link can transmit the signals at the target rate θ.

For an R_k to D link, the reliability and security of the link are both considered due to eavesdropping by E. The value of $l_{1,k}(t)$ is taken as follows.

- $l_{1,k}(t) = 0$ denotes $C_{S,D}(t) < \theta$ and the corresponding link is unreliable. When $l_{1,k}(t) = 0$, the corresponding link can not transmit the signals at the target rate θ.
- $l_{1,k}(t) = 1$ denotes $C_{S,D}(t) \geq \theta$, $C^{(s)}_{S,D}(t) < \zeta$ and the corresponding link is reliable but not secure, where ζ is the target secrecy rate. When $l_{1,k}(t) = 1$, the corresponding link

can transmit the signals with the target rate θ but cannot transmit the signals at the target secrecy rate ζ.

- $l_{1,K}(t) = 2$ denotes $C_{S,D}(t) \geq \theta$, $C^{(s)}_{S,D}(t) \geq \zeta$ and corresponding link is reliable and secure. When $l_{1,k}(t) = 2$, the corresponding link can transmit the signals at the target secrecy rate ζ.

Regarding one buffer-aided relay node R_k, there are two links, i.e., an S to R_k link and an R_k to D link, so the buffer state $b(t)$ at time t are defined as

$$b(t) = \{b_{0,1}(t), \cdots, b_{0,K}(t), b_{1,1}(t), \cdots, b_{1,K}(t)\}, \tag{10}$$

where $b_{j,k}(t) \in \{0, 1, \cdots, L\}$, $j \in \{0, 1\}$, $k \in \{1, 2, \cdots, K\}$, because the length of buffer queue is L. If the selected link is an S to R_k link and $b_{0,k}(t) = L$, the corresponding buffer-aided relay node R_k is unavailable at this time because its buffer queue Q_k is full, it can not receive the signals from S. If selected link is an R_k to D link and $b_{1,k}(t) = 0$, the corresponding buffer-aided relay node R_k is also unavailable at this time because its buffer queue Q_k is empty, it can not forward the signals to D.

According to the above analysis, we can conclude that the size of link state space $l(t)$ and buffer state space $b(t)$ are 6^K and $(L+1)^{2K}$, respectively. As $s(t) = \{l(t), b(t)\}$, the size of state space $s(t)$ is $(6(L+1)^2)^K$.

4.3. Action and Policy

In two-hop AF buffer-aided relay networks, the selection of a link for transmitting the signals is modeled as action in the MDP. The set of links that the agent can choose at time t is modeled by the action space $a(t)$.

At state s_t, if the agent selects an S to R_k link to transmit the signals, we denote $a_t = l_{(0,k)}$. If the agent selects an R_k to D link to transmit the signals, we denote $a_t = l_{(1,k)}$. It is worth noting that when the states of all S to R_k links are 0 and the states of all R_k to D links are not equal to 2, the agent will select no link to transmit the signals (i.e., a connection outage event occurs directly) and this case is denoted as $a_t = \emptyset$. Based on the analysis above, we also can deduce that the size of the action space $a(t)$ is $2K + 1$.

To guarantee the reliable and secure communications between legitimate users, if the link selected to transmit the signals is unreliable, then a connection outage event occurs. If the selected link is not secure, then a secrecy outage event occurs. Therefore, after the agent acts the action a_t, the environment may enter a new state s_{t+1}, or remain in the current state s_t due to the connection outage or secrecy outage. In addition, if the selected link is reliable and secure but the corresponding buffer is unavailable, a connection outage event also happens. Transmission is considered successful only if the selected link is reliable and secure (for an S to R_k link, the selected link is only required to be reliable) and the corresponding buffer is available. Table 2 shows the results of performing actions in different link states and buffer states.

In the MDP, the policy function $\pi(a_t|s_t)$ is the probability that the agent acts action a_t at state s_t and is denoted by

$$\pi(a_t|s_t) = P(a_t|s_t). \tag{11}$$

From (11), we can observe that $\pi(a_t|s_t)$ will affect the choice of an action and also the reward for the action.

Table 2. The results of performing actions in different link states and buffer states [1].

Action	Link State	Buffer State	Result
$l_{0,k}$	0	full	connection outage
$l_{0,k}$	0	not full	connection outage
$l_{0,k}$	2	not full	successful transmission
$l_{0,k}$	2	full	connection outage
$l_{1,k}$	2	empty	connection outage
$l_{1,k}$	2	not empty	successful transmission
$l_{1,k}$	1	empty	secrecy outage
$l_{1,k}$	1	not empty	secrecy outage
$l_{1,k}$	0	empty	connection outage
$l_{1,k}$	0	not empty	connection outage
$\varnothing$	$\forall k \in \{1,2,\cdots,K\}, l_{0,k}=0, l_{1,k} \neq 2$	any	connection outage

[1] Since TQL and DQL discussed in this paper are based on value iterations rather than policy iterations, the policy is not described in detail in this paper.

4.4. Reward and Return

The reward is the feedback given to the agent by the environment after the agent acts an action a_t in a state s_t, and is noted as $r(s_t, a_t)$. The reward can be divided into three categories: positive reward, negative reward and neutral reward.

- Positive reward: the selected link satisfies the transmission requirements, in which the target transmission rate θ and target secrecy transmission rate ζ are both considered, and the corresponding buffer-aided relay node is available.
- Negative reward: the selected link can not satisfy the transmission requirements or the corresponding buffer-aided relay node is unavailable.
- Neutral reward: no link is selected.

In the MDP, the accumulated reward from the beginning time t to the end time $t+n$ is called as the return, denoted by U_t. The expression of return U_t is given by

$$U_t = r(s_t, a_t) + \gamma * r(s_{t+1}, a_{t+1}) + \cdots + \gamma^n * r(s_{t+n}, a_{t+n}) = r(s_t, a_t) + \gamma * U_{t+1}, \tag{12}$$

where γ is the discount factor in the MDP. Moreover, the conditional expectation of the return U_t of acting action a_t in state s_t is defined as the action-value function $Q_\pi(s_t, a_t)$ and

$$Q_\pi(s_t, a_t) = E|U_t|s_t, a_t|, s_t \in s(t), a_t \in a(t), \tag{13}$$

which is used to evaluate the value of state s_t and action a_t. However, the action-value function is also influenced by the policy function π, and to eliminate the influence of the policy function π, we use the optimal action-value function $Q_*(s_t, a_t)$ (also known as the Q-function) to evaluate the value of state s_t and action a_t. The optimal action-value function is obtained by

$$Q_*(s_t, a_t) = \max_\pi Q_\pi(s_t, a_t), s_t \in s(t), a_t \in a(t). \tag{14}$$

In the MDP, the goal of the agent is to make the return U_t on each episode as high as possible, so the agent should select the link corresponding to the action with the maximum Q-function to transmit the signals each time.

With the above methods, we can model the process of information transmission in two-hop wireless relay networks as an MDP. Subsequently, we can use Q-learning algorithms to optimize the MDP for reliable and secure communications.

5. The Proposed Buffer-Aided Relay Selection Scheme

After modeling the process of information transmission as an MDP, we use Q-learning algorithms to optimize the transmission process and propose a new buffer-aided relay selection scheme based on it. Most of the existing schemes use TQL based on Q-table to optimize the transmission process and only consider reliability or security. Our proposed scheme utilizes DQL based on DNN to optimize the transmission process and considers both reliability and security. This section describes the principle of Q-learning algorithms and the steps of the proposed scheme based on DQL, respectively.

The principle of Q-learning algorithms including TQL and DQL is shown in Figure 3. The goal of the MDP is to make the return U_t of each episode as high as possible, so the agent should perform the action with the largest Q-function value each time. In TQL, the values of Q-function are stored in Q-table and updated by the Q-learning algorithms updates Q-function by

$$Q_*(s_t, a_t) = r(s_t, a_t) + \gamma * \max_{a \in a(t+1)} Q_*(s_{t+1}, a). \tag{15}$$

In the MDP, the size of state space $s(t)$ and action space $a(t)$ is $(6(L+1)^2)^K$ and $2K+1$, respectively. If we use the TQL based on Q-table to optimize the transmission process, it needs to occupy a lot of space to store a Q-table of $(6(L+1)^2)^K$ by $2K+1$ as in Table 3 and consume a lot of time to update Q-function and search the action with the maximum Q-function. In order to reduce the space occupation and lookup time, this paper uses DQL based on DNN rather than TQL based on Q-table to optimize the transmission process. DQL uses neural network to fit the Q-function without storing the Q-function in the Q-table, so DQL can save storage space.

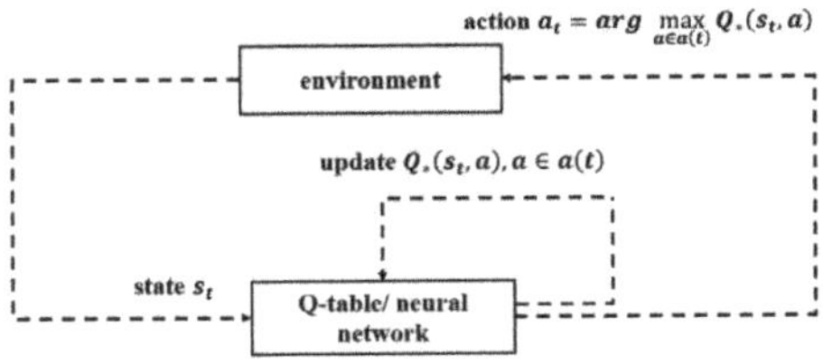

Figure 3. Framework of the Q-learning.

Table 3. The structure of the Q-table. The rows represent state space $s(t)$ and the columns represent action space $a(t)$.

Q-Table					
	s_1	s_2	s_3	$\cdots$	$s_{(6(L+1)^2)^K}$
a_1	$Q_*(s_1, a_1)$	$Q_*(s_2, a_1)$	$Q_*(s_3, a_1)$	$\cdots$	$Q_*(s_{(6(L+1)^2)^K}, a_1)$
a_2	$Q_*(s_1, a_2)$	$Q_*(s_2, a_2)$	$Q_*(s_3, a_2)$	$\cdots$	$Q_*(s_{(6(L+1)^2)^K}, a_2)$
a_3	$Q_*(s_1, a_3)$	$Q_*(s_2, a_3)$	$Q_*(s_3, a_3)$	$\cdots$	$Q_*(s_{(6(L+1)^2)^K}, a_3)$
$\vdots$	$\vdots$	$\vdots$	$\vdots$	$\vdots$	$\vdots$
a_{2K+1}	$Q_*(s_1, a_{2K+1})$	$Q_*(s_2, a_{2K+1})$	$Q_*(s_3, a_{2K+1})$	$\cdots$	$Q_*(s_{(6(L+1)^2)^K}, a_{2K+1})$

The proposed scheme based on DQL is divided into three phases, which are experience collection, training the network model and deploying it online. The steps of the proposed scheme are as follows.

5.1. Experience Collection

This phase focuses on collecting the experience needed to train the network model. Firstly, a DNN, which is called a prediction network, is initialized and used to fit the Q-function. The structure of the prediction network is shown in Figure 4.

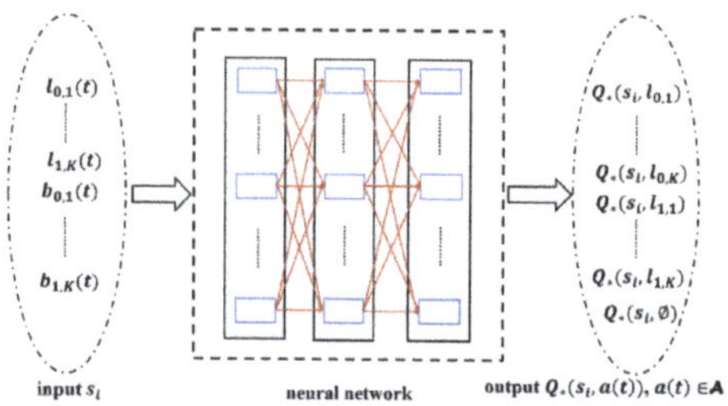

Figure 4. The structure of the prediction network. The input is state s_t, the output is Q-function for each action in action space $a(t)$, $Q_*(s_t, a), a \in a(t)$.

The input of the prediction network is state s_t, and the output is Q-function $Q_*(s_t, a)$, where $a \in a(t)$, which corresponds to a state action at time t. In this phase, the ϵ-greedy policy is used to select actions to balance the exploration–exploitation dilemma. The agent chooses the action with the Q-function with $1-\epsilon$ probability, and randomly chooses an action with ϵ probability, as shown in (16)

$$a_t = \begin{cases} \arg\max\limits_{a \in a(t)} Q_*(s_t, a), & prob.(1-\epsilon) \\ \text{random an action}, & prob.\epsilon \end{cases}, \tag{16}$$

where $0 < \epsilon \leq 1$. In this phase, the agent needs to explore the action space as much as possible, so ϵ is set as 1. In the training network model phase, the agent needs to train the network model by exploiting the collected experience, so as the number of training episodes increases, ϵ decreases to $\epsilon_{min} = 0.1$, and the attenuation factor $\varphi = 0.998$.

After the agent selects an action and enacts the selected action a_t, the state s_t moves to s_{t+1}, and the environment returns the reward $r(s_t, a_t)$, a sample $\{s_t, a_t, r(s_t, a_t), s_{t+1}\}$ is generated. The prediction network does not learn the sample immediately, but stores the sample in a buffer called as the replay buffer, which is depicted in Figure 5 and is used to store the generated experiences. The above steps of generating and collecting experience are repeated until the replay buffer is full and the prediction network starts to learn the experience.

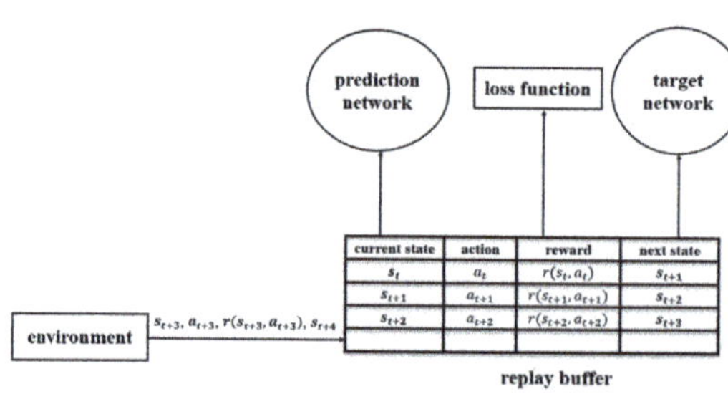

Figure 5. The structure of the replay buffer.

5.2. Training the Network Model

This phase uses the experience collected in the previous phase to train and update the network model. When the replay buffer is full, the agent starts to randomly select a batch of samples for training the network model. The trick is called experience replay, which can effectively reduce the correlation between samples and improve the convergence speed of the prediction network. To avoid bootstrapping of the prediction network, this paper introduces another neural network called the target network, which has the same structure as the prediction network. The input and the output of the prediction network are s_t and $Q_*(s_t, a), a \in a(t)$, respectively. Similarly, the input and output of the target network are s_{t+1} and $Q_*(s_{t+1}, a), a \in a(t+1)$, respectively. As the action a_t acts at state s_t and the reward $r(s_t, a_t)$ are available according to the sample $\{s_t, a_t, r(s_t, a_t), s_{t+1}\}$, we can obtain $Q_*(s_t, a_t)$ and $r(s_t, a_t) + \max_{a \in a(t+1)} Q_*(s_{t+1}, a)$. Next, we calculate the error between $Q_*(s_t, a_t)$ and $r(s_t, a_t) + \max_{a \in a(t+1)} Q_*(s_{t+1}, a)$ by using the loss function. This paper uses the mean square error (MSE) as the loss function of the prediction network and the target network. According to (16), the expression of the MSE loss function is obtained as

$$\varpi = \sum_{1}^{N} (r(s_t, a_t) + \gamma * \max_{a \in a(t+1)} Q_*(s_{t+1}, a) - Q_*(s_t, a_t))^2, \tag{17}$$

where N is the batch size. Then, we update the weights of the prediction network by using the MSE loss function and copy the weights of the prediction network to the target network periodically. Finally, we repeat the above steps of learning and updating for many episodes until the prediction network and target network converge. The framework of the experience collection phase and the training network model phase of the proposed scheme is shown in Figure 6.

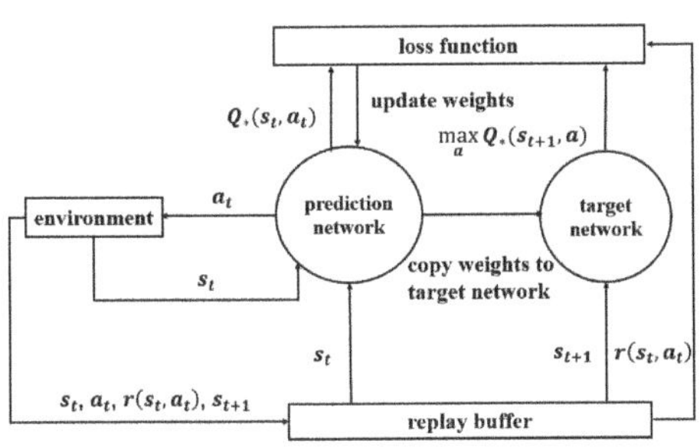

Figure 6. The framework of DQL with target network and experience replay.

5.3. Deployment Online

After the prediction network and the target network converge, we deploy the network model online. It is worth noting that both the experience collecting and training the network model phases are offline. In this phase, the prediction network directly estimates the $Q_*(s_t, a)$ corresponding to each action a, $a \in a(t)$ based on the current state s_t, and selects the action with the maximum $Q_*(s_t, a), a \in a(t)$ without training and updating the weights.

Finally, all the steps of the proposed buffer-aided relay selection scheme based on DQL are shown in Algorithm 1, where N_e is the number of training episodes and N_c is the capacity of the replay buffer.

Algorithm 1 The proposed buffer-aided relay scheme based on DQL

Initialize the environment for the two-hop AF buffer-aided relay network
Repeat:
for $i = 1, 2, \cdots, N_e$ **do**
  for $j = 1, 2, \cdots, N_c$ **do**
    (First phase: experience collection)
    At current state s_t, select action a_t according to ϵ-greedy policy.
    Act the selected action a_t, and return reward $r(s_t, a_t)$ and next state s_{t+1}.
    Generate a sample $s_t, a_t, r(s_t, a_t), s_{t+1}$, and store it in replay buffer.
  end for
  (Second phase: training the network model)
  Randomly select a batch of samples from replay buffer.
  According to s_t and a_t, get $Q_*(s_t, a_t)$ from the prediction network.
  According to $r(s_t, a_t)$ and s_{t+1}, get $r(s_t, a_t) + \max_{a \in a(t+1)} Q_*(s_{t+1}, a)$ from the target network.
  Calculate the loss between $Q_*(s_t, a_t)$ and $r(s_t, a_t) + \max_{a \in a(t+1)} Q_*(s_{t+1}, a)$ by the MSE loss function.
  Update the weights of prediction network.
  if i%100=0 **then**
    Copy the weights of the prediction network to the target network.
  end if
end for
(Third phase: deployment online)
Deploy the prediction network online.

6. Simulation Results and Discussion

This section verifies the reliability and security performances of the proposed scheme by using Monte Carlo simulations, and uses the COP and SOP to measure the reliability and security performances of the proposed scheme. In the two-hop AF buffer-aided relay network, the number of buffer-aided relay nodes is $K = 3$, the length of the buffer queue is $L = 3$, the average channel gain is set as $\Omega_{S,R_k} = \Omega_{R_k,D} = 30$ dB, and $\Omega_{R_k,E} = 5$ dB. Since the power of AWGN is normalized to unity, the ratio of transmitting power to noise is set as $P_s/\sigma^2 = R_{R_k}/\sigma^2 = 30$ dB. Furthermore, the target rate θ is set as 7 bps/Hz and the target secrecy rate ζ is set as 0.1 bps/Hz. In the proposed buffer-aided relay selection scheme based on DQL, the discount factor γ and learning rate v of the DQL are set as 0.9 and 0.1, respectively. The capacity N_c of replay buffer which stores samples and the batch size is set as 2000. In the phase of training the network model, the training episodes N_e is set as 20,000, the batch size in each episode is set as 128 and the target network updates its parameters every 100 episodes. After the training of the network model is completed, the reliability and security performances of the proposed scheme are verified by 1 million Monte Carlo simulations. The lower the COP and SOP, the higher the reliability and security performances. The expressions of COP and SOP are obtained by

$$COP = \frac{n_c^{'}}{1{,}000{,}000}, \quad SOP = \frac{n_s^{'}}{1{,}000{,}000}, \tag{18}$$

where $n_c^{'}$ and $n_s^{'}$ are the number of connection outage events and secrecy outage events that occurred in 1 million Monte Carlo simulations, respectively.

First, we verify the reliability and security performances of the proposed scheme. The simulation results are shown in Figure 7, where Figure 7a illustrates how the COP varies with the SNR P/σ^2 and Figure 7b shows how the SOP varies with the target secrecy rate ζ. From Figure 7a, we can observe that the COP decreases as the SNR increases. This is because an increase of SNR means a better transmission link and thus a lower COP. We can further see from Figure 7b that when the target secrecy rate ζ is set as 0.1 bps/Hz, the SOP can reach 10^{-4}, and the SOP increases as the target secrecy rate ζ increases. This is because as ζ increases, fewer legitimate channels can meet the requirements to secure transmissions, which will lead to more secrecy outage events. In conclusion, the simulation results in Figure 7 confirm that our proposed scheme can achieve reliable and secure communications in two-hop wireless relay networks.

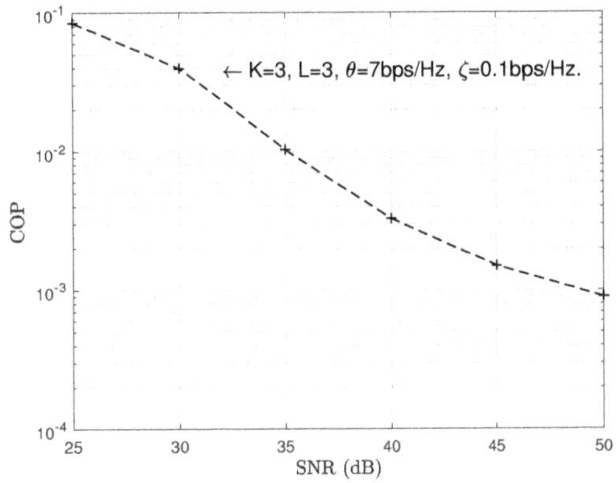

(**a**) COP vs. SNR.

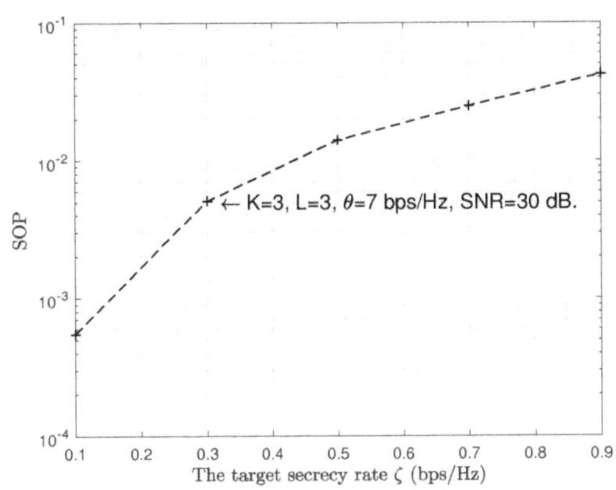

(**b**) SOP vs. the target secrecy rate ζ.

Figure 7. Analysis of the COP and SOP of the proposed scheme.

Then, we investigate the effect of the number of buffer-aided relay nodes K and the buffer length L on the reliability and security performances of the proposed scheme. Figures 8a and 9a investigate the effect of the number of buffer-aided relay nodes K and the buffer length L on the reliability performance of the proposed scheme, respectively. In Figures 8b and 9b, we, respectively, investigate the effect of the number of buffer-aided relay nodes k and the buffer length L on the security performance of the proposed scheme. We can observe from Figures 8a and 9a that the COP decreases gradually as the number of buffer-aided relay nodes K and the buffer length L increase. In Figures 8b and 9b, SOP also decreases as the number of buffer-aided relay nodes K and the buffer length L increase, respectively. The lower the COP and SOP, the higher the reliability and security. Resulting in Figures 8 and 9 indicate that the increase in the number of buffer-aided relay nodes

K and the buffer length L can improve the reliability and security performances of the proposed scheme. This is because the increase in the number of buffer-aided relay nodes K implies an increasing number of legitimate channels, and an increase in the buffer length L implies a lower probability of buffer-aided relay unavailability (the probability that a buffer is full).

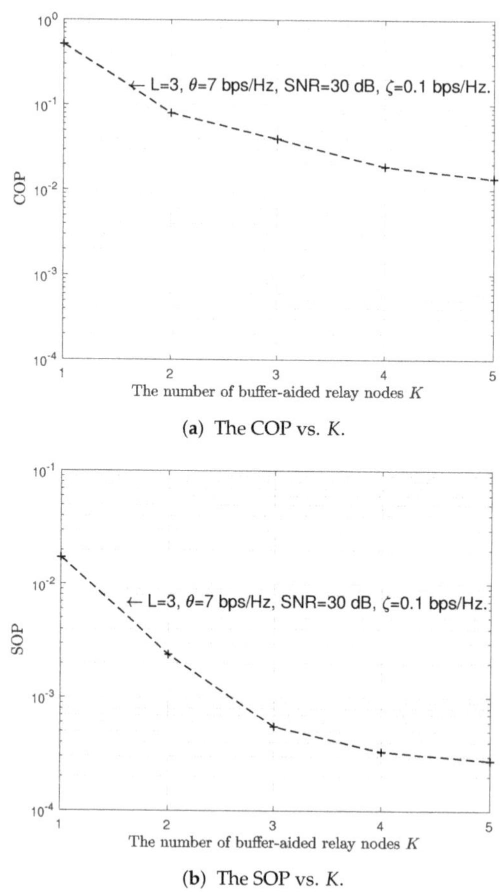

Figure 8. The impact of the number of buffer-aided relay nodes K on the COP and SOP of the proposed scheme.

Finally, we made a comparison between our proposed scheme and two benchmark schemes (i.e., the max-link scheme and the max-ratio scheme) regarding the COP and SOP, respectively. By setting $K = 3$, $L = 3$, $\theta = 7$ bps/Hz, and $\zeta = 0.1$ bps/Hz, we show in Figure 10a how the COP changes by varying the SNR from 25 dB to 50 dB. The results in Figure 10a show that the COP of our scheme is always lower than that of the max-link scheme. By setting $K = 3$, $L = 3$, $\theta = 7$ bps/Hz, and SNR = 30 dB, we then illustrate in Figure 10b how the SOP varies by varying the target secrecy rate ζ from 0.1 to 0.9. The results in Figure 10b show that the SOP of our scheme is always lower than that of the max-ratio scheme, indicating that the security performance of the two-hop buffer-aided wireless network can be improved by adopting the DQL. In addition, we investigate the differences between the proposed scheme implemented by DQL and TQL. The comparison results are also shown in Figure 10a and Figure 10b, respectively. The comparison results clearly show that, under the same conditions, the COP and SOP of the proposed scheme implemented by DQL are both lower than those of the proposed scheme implemented by TQL.

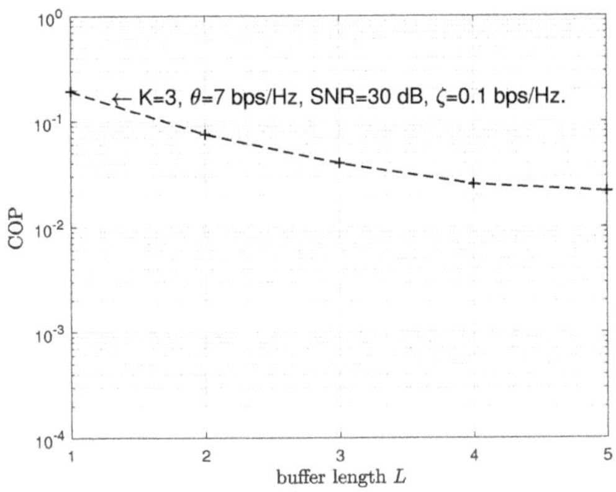

(**a**) The COP vs. *L*.

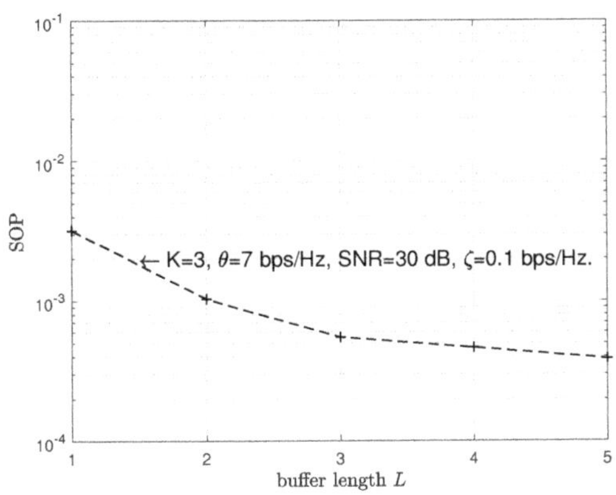

(**b**) The SOP vs. *L*.

Figure 9. The impact of the buffer length *L* on the COP and SOP of the proposed scheme.

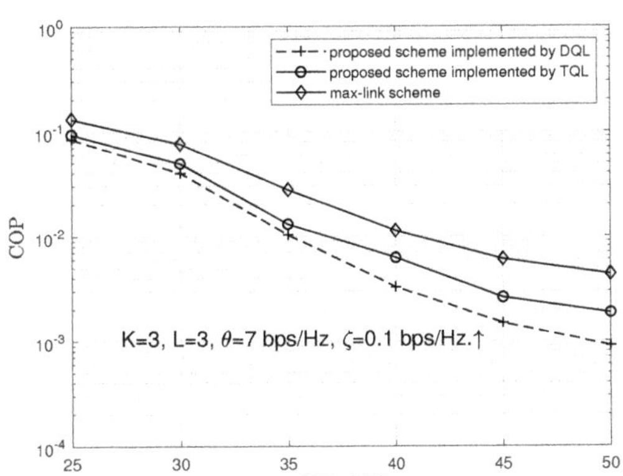

(**a**) Comparison results of the COP.

Figure 10. *Cont.*

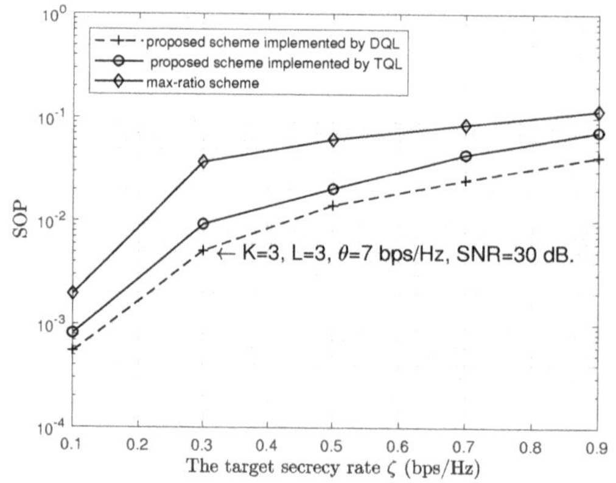

(**b**) Comparison results of the SOP.

Figure 10. Comparison results between the proposed scheme and other schemes.

7. Conclusions

This paper utilizes DQL to solve the problem of buffer-aided relay selection to achieve reliable and secure communications in a two-hop AF buffer-aided relay network with a passive eavesdropper. To propose the buffer-aided relay selection scheme, we first model the information transmission process in the network by applying an MDP. With the help of the MDP model, we then propose a novel buffer-aided relay selection scheme based on DQL to optimize the MDP. We finally verify the reliability and security performances of the proposed scheme by conducting Monte Carlo simulations and analyze how the network parameters affect the reliability and security performances of the concerned network in terms of the COP and the SOP. We also made a comparison between our proposed scheme and two benchmark buffer-aided relay selection schemes (i.e., the max-link scheme and the max-ratio scheme) regarding the COP and SOP, respectively. The results show that our proposed scheme can outperform the max-ratio scheme in terms of the SOP by 2.76 times.

Author Contributions: Conceptualization, C.Z. and X.L.; methodology, C.Z. and X.L.; software, C.Z. and Z.C.; writing—original draft preparation, C.Z.; writing—review and editing, X.L., Z.W., G.Q. and Z.Y.; supervision, Z.W.; funding acquisition, Z.W. and Z.Y. All authors have read and agreed to the published version of the manuscript.

Funding: This work was supported in part by the National Natural Science Foundation of China (62001273, 62002210), the Open Project Program of the Shaanxi Key Laboratory for Network Computing and Security Technology (NCST2021YB-02), the Fundamental Research Funds for the Central Universities (GK202103087) and the Scientific Research Plan of Shaanxi Provincial Department of Education (19JK0176).

Institutional Review Board Statement: Not applicable.

Informed Consent Statement: Not applicable.

Data Availability Statement: The data used to support the findings of this study is included within the article.

Conflicts of Interest: The authors declare no conflicts of interest.

Abbreviations

The following abbreviations are used in this manuscript:

AF	Amplify-and-forward
DQL	Deep Q-learning
COP	Connection outage probability
SOP	Secrecy outage probability
WSNs	Wireless sensor networks
CRNs	Cognitive radio networks
IoTs	Internet of Things
PLS	Physical-layer security
CSI	Channel-state information
TQL	Traditional Q-learning
MC	Markov chain
MDP	Markov decision process
FD	Full-duplex
CJ	Cooperative jamming
EH	Energy harvesting
DL	Deep learning
DNN	Deep neural network
PA	Power allocation
DF	Decode-and-forward
RF	Randomize-and-forward
HD	Half-duplex
AWGN	Additive white Gaussian noise
SNR	Signal-to-noise ratio
MSE	Mean square error

Symbols

The following symbols are used in this manuscript:

$h_{m,n}$	Channel coefficient between m and n
$g_{m,n}$	Channel gain between m and n
$\Omega m, n$	Average channel gain between m and n
$E[.]$	Expectation operator
$y_{R_k}(t)$	The received signal of R_k at time t
$x_S(t)$	The signal sent by S at time t
$n_{R_k}(t)$	AWGN noise of R_k at time t
$\psi_{S,R_k}(t)$	SNR of S to R_k link at t
$C_{S,R_K}(t)$	Channel capacity of S to R_k link
$A_{R_k}(t')$	Amplification factor of R_k at t'
$C^{(s)}_{S,D}(t')$	The end-to-end secrecy rate
θ	The target rate
ζ	The target secrecy rate
s_t	The state at time t
a_t	The action at time t
$s(t)$	State space
$a(t)$	Action space
γ	Discount factor
$Q_\pi(s_t, a_t)$	Action-value function
$Q_*(s_t, a_t)$	The optimal action-value function
ϵ	Exploration probability
ϖ	MSE loss function
N_e	Training episodes
N_c	Capacity of replay buffer

References

1. Ding, Y.; Yang, Y.; Jiang, W.; Liu, Y.; He, T.; Zhang, D. Nationwide Deployment and Operation of a Virtual Arrival Detection System in the Wild. *IEEE/ACM Trans. Netw.* **2023**, *31*, 574–589. [CrossRef]
2. Li, Z.; Wang, S.; Han, S.; Meng, W.; Li, C. Joint Design of Beam Hopping and Multiple Access Based on Cognitive Radio for Integrated Satellite-Terrestrial Network. *IEEE Netw.* **2023**, *37*, 36–43. [CrossRef]
3. Xie, H.; Xia, M.; Wu, P.; Wang, S.; Poor, H.V. Edge Learning for Large-Scale Internet of Things With Task-Oriented Efficient Communication. *IEEE Trans. Wirel. Commun.* **2023**, 1–16.
4. Bapatla, D.; Prakriya, S. Performance of Two-Hop Links With an Energy Buffer-Aided IoT Source and a Data Buffer-Aided Relay. *IEEE Internet Things J.* **2021**, *8*, 5045–5061. [CrossRef]
5. Yerrapragada, A.K.; Eisman, T.; Kelley, B. Physical Layer Security for Beyond 5G: Ultra Secure Low Latency Communications. *IEEE Open J. Commun. Soc.* **2021**, *2*, 2232–2242. [CrossRef]
6. Angueira, P.; Val, I.; Montalbán, J.; Seijo, O.; Iradier, E.; Sanz, P.; Fanari, L.; Arriola, A. A Survey of Physical Layer Techniques for Secure Wireless Communications in Industry. *IEEE Commun. Surv. Tutor.* **2022**, *24*, 810–838. [CrossRef]
7. Mitev, M.; Chorti, A.; Poor, H.V.; Fettweis, G.P. What Physical Layer Security Can Do for 6G Security. *IEEE Open J. Veh. Technol.* **2023**, *4*, 375–388. [CrossRef]
8. Huynh, P.; Phan, K.T.; Liu, B.; Ross, R. Throughput Analysis of Buffer-Aided Decode-and-Forward Wireless Relaying with RF Energy Harvesting. *Sensors* **2020**, *20*, 1222. [CrossRef]
9. Lu, X.; Xiao, L.; Li, P.; Ji, X.; Xu, C.; Yu, S.; Zhuang, W. Reinforcement Learning-Based Physical Cross-Layer Security and Privacy in 6G. *IEEE Commun. Surv. Tutor.* **2023**, *25*, 425–466. [CrossRef]
10. Arzykulov, S.; Celik, A.; Nauryzbayev, G.; Eltawil, A.M. Artificial Noise and RIS-Aided Physical Layer Security: Optimal RIS Partitioning and Power Control. *IEEE Wirel. Commun. Lett.* **2023**, 1–5.
11. Huang, H.; Hu, S. Generalized Relays Subsets Selection Algorithm in Cloud-Based 6G Large-Scale Relays Network. *IEEE Internet Things J.* **2022**, *9*, 24754–24766. [CrossRef]
12. Zhou, F.; Chu, Z.; Sun, H.; Hu, R.Q.; Hanzo, L. Artificial Noise Aided Secure Cognitive Beamforming for Cooperative MISO-NOMA Using SWIPT. *IEEE J. Sel. Areas Commun.* **2018**, *36*, 918–931. [CrossRef]
13. Vaishnavi, K.N.; Khorvi, S.D.; Kishore, R.; Gurugopinath, S. A Survey on Jamming Techniques in Physical Layer Security and Anti-Jamming Strategies for 6G. In Proceedings of the 2021 28th International Conference on Telecommunications (ICT), London, UK, 1–3 June 2021; pp. 174–179. [CrossRef]
14. Shukla, A.K.; Bhatnagar, M.R. Differential Modulation-Based Buffer-Aided Cooperative Relaying Network. *IEEE Syst. J.* **2022**, 1–12.
15. Xu, P.; Wang, Y.; Chen, G.; Krikidis, I.; Wong, K.K. Novel Mode Selection Schemes for Buffer-Aided Cooperative NOMA System. *IEEE Trans. Veh. Technol.* **2022**, *72*, 866–880.
16. Ikhlef, A.; Michalopoulos, D.S.; Schober, R. Max-Max Relay Selection for Relays with Buffers. *IEEE Trans. Wirel. Commun.* **2012**, *11*, 1124–1135. [CrossRef]
17. Hamamreh, J.M.; Furqan, H.M.; Arslan, H. Classifications and Applications of Physical Layer Security Techniques for Confidentiality: A Comprehensive Survey. *IEEE Commun. Surv. Tutor.* **2019**, *21*, 1773–1828. [CrossRef]
18. Guo, W.; Qureshi, N.M.F.; Siddiqui, I.F.; Shin, D.R. Cooperative Communication Resource Allocation Strategies for 5G and Beyond Networks: A Review of Architecture, Challenges and Opportunities. *J. King Saud Univ.-Comput. Inf. Sci.* **2022**, *34*, 8054–8078. [CrossRef]
19. Krikidis, I.; Charalambous, T.; Thompson, J.S. Buffer-Aided Relay Selection for Cooperative Diversity Systems without Delay Constraints. *IEEE Trans. Wirel. Commun.* **2012**, *11*, 1957–1967. [CrossRef]
20. Gong, Y.; Chen, G.; Xie, T. Using Buffers in Trust-Aware Relay Selection Networks With Spatially Random Relays. *IEEE Trans. Wirel. Commun.* **2018**, *17*, 5818–5826. [CrossRef]
21. Alam, M.Z.; Abkenar, F.S.; Adhicandra, I.; Murali, S.; Jamalipour, A. Low-Delay Path Selection for Cluster-Based Buffer-Aided Vehicular Communications. *IEEE Trans. Veh. Technol.* **2020**, *69*, 9356–9363. [CrossRef]
22. Adanvo, V.F.; Mafra, S.; Montejo-Sánchez, S.; Fernández, E.M.G.; Souza, R.D. Buffer-Aided Relaying Strategies for Two-Way Wireless Networks. *Sustainability* **2022**, *14*, 13829. [CrossRef]
23. Jadoon, M.A.; Kim, S. Relay selection algorithm for wireless cooperative networks: A learning-based approach. *IET Commun.* **2017**, *11*, 1061–1066. [CrossRef]
24. Wang, X.; Jin, T.; Hu, L.; Qian, Z. Energy-Efficient Power Allocation and Q-Learning-Based Relay Selection for Relay-Aided D2D Communication. *IEEE Trans. Veh. Technol.* **2020**, *69*, 6452–6462. [CrossRef]
25. Dong, Y.; Zhang, F.; Joe, I.; Lin, H.; Jiao, W.; Zhang, Y. Learning for Multiple-Relay Selection in a Vehicular Delay Tolerant Network. *IEEE Access* **2020**, *8*, 175602–175611. [CrossRef]
26. Chen, G.; Tian, Z.; Gong, Y.; Chen, Z.; Chambers, J.A. Max-Ratio Relay Selection in Secure Buffer-Aided Cooperative Wireless Networks. *IEEE Trans. Inf. Forensics Secur.* **2014**, *9*, 719–729. [CrossRef]
27. Mekkawy, T.; Yao, R.; Qi, N.; Lu, Y. Secure Relay Selection for Two Way Amplify-and-Forward Untrusted Relaying Networks. *IEEE Trans. Veh. Technol.* **2018**, *67*, 11979–11987. [CrossRef]

28. Zhang, C.; Liao, X.; Wu, Z.; Qiu, G. Buffer-Aided Relay Selection for Wireless Cooperative Relay Networks with Untrusted Relays. In Proceedings of the 2021 International Conference on Networking and Network Applications (NaNA), Lijiang, China, 29 October–1 November 2021; pp. 69–74. [CrossRef]
29. Nie, Y.; Lan, X.; Liu, Y.; Chen, Q.; Chen, G.; Fan, L.; Tang, D. Achievable Rate Region of Energy-Harvesting Based Secure Two-Way Buffer-Aided Relay Networks. *IEEE Trans. Inf. Forensics Secur.* **2021**, *16*, 1610–1625. [CrossRef]
30. Srirutchataboon, G.; Sugiura, S. Physical Layer Security of Buffer-Aided Hybrid Virtual Full-Duplex and Half-Duplex Relay Selection. In Proceedings of the 2022 IEEE 95th Vehicular Technology Conference: (VTC2022-Spring), Helsinki, Finland, 19–22 June 2022; pp. 1–5. [CrossRef]
31. He, J.; Liu, J.; Su, W.; Shen, Y.; Jiang, X.; Shiratori, N. Jamming and Link Selection for Joint Secrecy/Delay Guarantees in Buffer-Aided Relay System. *IEEE Trans. Commun.* **2022**, *70*, 5451–5468. [CrossRef]
32. Wang, Y.; Yin, H.; Zhang, T.; Yang, W.; Shang, X.; Shen, Y. Secure Transmission for Energy Harvesting Sensor Networks with a Buffer-Aided Sink Node. *IEEE Internet Things J.* **2021**, *9*, 6703–6718.
33. Mao, Q.; Hu, F.; Hao, Q. Deep Learning for Intelligent Wireless Networks: A Comprehensive Survey. *IEEE Commun. Surv. Tutor.* **2018**, *20*, 2595–2621. [CrossRef]
34. Zhang, C.; Patras, P.; Haddadi, H. Deep Learning in Mobile and Wireless Networking: A Survey. *IEEE Commun. Surv. Tutor.* **2019**, *21*, 2224–2287. [CrossRef]
35. Buenrostro-Mariscal, R.; Santana-Mancilla, P.C.; Montesinos-López, O.A.; Nieto Hipólito, J.I.; Anido-Rifón, L.E. A Review of Deep Learning Applications for the Next Generation of Cognitive Networks. *Appl. Sci.* **2022**, *12*, 6262. [CrossRef]
36. Zhang, Z.; Lu, Y.; Huang, Y.; Zhang, P. Neural Network-Based Relay Selection in Two-Way SWIPT-Enabled Cognitive Radio Networks. *IEEE Trans. Veh. Technol.* **2020**, *69*, 6264–6274. [CrossRef]
37. Zhou, D.; Yan, B.; Li, C.; Wang, A.; Wei, H. Relay selection scheme based on deep reinforcement learning in wireless sensor networks. *Phys. Commun.* **2022**, *54*, 101799. [CrossRef]
38. Rezwan, S.; Choi, W. A Survey on Applications of Reinforcement Learning in Flying Ad-Hoc Networks. *Electronics* **2021**, *10*, 449. [CrossRef]
39. Zhu, J.; Song, Y.; Jiang, D.; Song, H. A New Deep-Q-Learning-Based Transmission Scheduling Mechanism for the Cognitive Internet of Things. *IEEE Internet Things J.* **2018**, *5*, 2375–2385. [CrossRef]
40. Huang, C.; Chen, G.; Gong, Y. Delay-Constrained Buffer-Aided Relay Selection in the Internet of Things With Decision-Assisted Reinforcement Learning. *IEEE Internet Things J.* **2021**, *8*, 10198–10208. [CrossRef]
41. Huang, C.; Zhong, J.; Gong, Y.; Abdullah, Z.; Chen, G. Novel deep reinforcement learning-based delay-constrained buffer-aided relay selection in cognitive cooperative networks. *Electron. Lett.* **2020**, *56*, 1148–1150. [CrossRef]
42. Huang, C.; Chen, G.; Gong, Y.; Xu, P. Deep Reinforcement Learning Based Relay Selection in Delay-Constrained Secure Buffer-Aided CRNs. In Proceedings of the GLOBECOM 2020—2020 IEEE Global Communications Conference, Taipei, Taiwan, 7–11 December 2020; pp. 1–6.
43. Huang, C.; Chen, G.; Gong, Y.; Han, Z. Joint Buffer-Aided Hybrid-Duplex Relay Selection and Power Allocation for Secure Cognitive Networks With Double Deep Q-Network. *IEEE Trans. Cogn. Commun. Netw.* **2021**, *7*, 834–844. [CrossRef]
44. El-Zahr, S.; Abou-Rjeily, C. Threshold Based Relay Selection for Buffer-Aided Cooperative Relaying Systems. *IEEE Trans. Wirel. Commun.* **2021**, *20*, 6210–6223. [CrossRef]
45. Xu, P.; Quan, J.; Chen, G.; Yang, Z.; Li, Y.; Krikidis, I. A Novel Link Selection in Coordinated Direct and Buffer-Aided Relay Transmission. *IEEE Trans. Wirel. Commun.* **2022**, *22*, 3296–3309.

sensors

MDPI

Article

A High-Precision Vehicle Detection and Tracking Method Based on the Attention Mechanism

Jiandong Wang, Yahui Dong, Shuangrui Zhao * and Zhiwei Zhang

School of Computer Science and Technology, Xidian University, Xi'an 710071, China
* Correspondence: zhaoshuangrui@xidian.edu.cn

Abstract: Vehicle detection and tracking technology plays an important role in intelligent transportation management and control systems. This paper proposes a novel vehicle detection and tracking method for small target vehicles to achieve high detection and tracking accuracy based on the attention mechanism. We first develop a new vehicle detection model (YOLOv5-NAM) by adding the normalization-based attention module (NAM) to the classical YOLOv5s model. By exploiting the YOLOv5-NAM model as the vehicle detector, we then propose a real-time small target vehicle tracking method (JDE-YN), where the feature extraction process is embedded in the prediction head for joint training. Finally, we present extensive experimental results to verify our method on the UA-DETRAC dataset and to demonstrate that the method can effectively detect small target vehicles in real time. It is shown that compared with the original YOLOv5s model, the mAP value of the YOLOv5-NAM vehicle detection model is improved by 1.6%, while the MOTA value of the JDE-YN method improved by 0.9% compared with the original JDE method.

Keywords: vehicle detection; vehicle tracking; attention mechanism; data association

1. Introduction

With the rapid development of intelligent transportation systems, the explosive growth of vehicular communication services has led to the shortage of radio spectrum resources for vehicular network communications [1,2]. Cognitive radio-enabled vehicular networks can provide additional spectrum resources for vehicular communications; thus, they have the potential to be widely adopted in many scenarios, such as road transportation, railway transportation, aerospace, as well as military fields [3–5]. Vehicle detection and tracking is a critical technology that collects road traffic videos and exploits image processing approaches to guide traffic operations in cognitive radio-enabled vehicular networks [6]. Different from traditional techniques such as radar, lidar, RFID, or LASAR, vehicle detection and tracking has its own advantages in terms of real-time, low-cost, and high performance, and has shown effectiveness with a carefully designed detection and tracking model for intelligent traffic management [7]. However, the complex environment and changeable weather conditions pose an open problem for high-precision vehicle detection and tracking [8–11].

1.1. Literature Review

1.1.1. Vehicle Detection

Many works have been devoted to investigating high-precision vehicle detection methods by now [12–15]. These works mainly focus on the two categories of vehicle detection technology based on traditional image processing technology or convolutional neural networks (CNN). The former category usually requires manual participation and guidance with poor robustness, while the latter one uses the CNN network to extract vehicle features, which is more robust and is suitable for many complex scenarios. In particular, the vehicle detection based on CNN can be further divided into two sub-categories: one-stage or two-stage vehicle detection methods [16]. One-stage vehicle detection methods, including RetinaNet [17], CFENet [18], CornerNet [19], YOLO [20], SSD [21], etc., can directly send the original image into the CNN model for vehicle feature extraction and obtain the position and category of the vehicle through regression and classification. Two-stage vehicle detection methods such as Cascade R-CNN [22], Faster R-CNN [23], Mask R-CNN [24], R-FCN [25], and SPP-Net [26] first extract the candidate regions of the vehicle in the image and then send them into the CNN model for feature extraction and classification.

Compared with the two-stage vehicle detection methods, the one-stage vehicle detection methods have the advantages of saving time and low cost. However, the vehicle classification accuracy of these methods is relatively low since they perform the object localization and classification tasks on an input image with a single neural network only one time. Moreover, it is notable that in real-time video analysis tools such as traffic monitoring, only Yolo can be used because of its high-speed detection. Based on the initial version of YOLO in [20], several algorithms including YOLOv2 [27], YOLOv3 [28], YOLOv4 [29], and YOLOv5 [30] were continuously developed to further improve the classification accuracy. Particularly, the open-source version of YOLOv5 can fully extract the features of the vehicle with high detection accuracy, and thus, has achieved a double harvest of speed and precision.

1.1.2. Vehicle Tracking

With the help of vehicle models, the multi-target vehicle tracking technology based on CNN has also attracted much attention [31]. In [32], Bewley et al. selected the Faster R-CNN algorithm as the target detector and proposed a real-time tracking algorithm (SORT) to simultaneously track multiple targets based on the Kalman filter and the Hungarian matching algorithm. Unfortunately, the SORT algorithm only considers the motion characteristics of the target in the data association stage since the Kalman filter will have problems such as probability dispersion and data association failure when the targets are occluded. To address this problem, the authors in [33] considered both motion characteristics and appearance characteristics of the targets and proposed the improved DeepSORT algorithm, where the appearance features of the target through a CNN model are extracted after the detector detects the target. To further improve the complexity of the DeepSORT algorithm, Wang et al. [34] developed the JDE algorithm, which takes the target feature extraction network into the target detection network and directly outputs the target location and appearance features in the detection network. By using RetinaNet as the vehicle detector and embedding the instance-level vehicle feature extraction network in the detector model, Lu et al. [35] established the multi-target vehicle tracking model RetinaTrack to integrate vehicle motion characteristics and appearance characteristics for data association.

1.2. Contributions

It is worth noting that all methods in the aforementioned literature mainly focus on how to design a more efficient CNN network model to improve the detection speed of the target detection algorithm as much as possible. Although there exist several real-time vehicle detection and tracking methods, none of these approaches are sufficient to solve the vehicle detection and tracking problem of small target vehicles accurately. How to improve the detection and tracking accuracy of occluded small vehicles and reduce the number of vehicle identity switching instances has become an urgent problem [36]. As a step further towards the solution of the problem, this paper proposes a novel vehicle detection and tracking method for small target vehicles to achieve high detection and tracking accuracy based on the attention mechanism. Our main contributions are as follows:

- We develop a vehicle detection model YOLOv5-NAM based on the YOLOv5 model. By adding the spatial attention mechanism and channel attention mechanism module into the model, the detection accuracy of small vehicles is improved. An SD-NMS method based on the idea of the penalty function is further proposed to solve the problem of missed detection of vehicles in dense scenes.

- We propose a real-time multi-vehicle tracking method JDE-YN based on the JDE algorithm, which improves the tracking accuracy of the vehicles and reduces the number of vehicle identity switching instances. Based on direction correction, we also develop a cascade matching algorithm to solve the problem of vehicle identity switching caused by occlusion.

The rest of this paper is organized as follows. We introduce the system model of detection-based vehicle tracking in Section 2 and provide a detailed description of our vehicle detection and tracking method in Section 3. Section 4 presents the experiments and analysis of results, and Section 5 concludes the paper.

2. System Model

As shown in Figure 1, we consider a vehicle detection and tracking model, which consists of a detector and a tracker. The detector is responsible for detecting vehicles in each frame, while the tracker aims to correlate vehicles in adjacent frames to form a complete vehicle trajectory.

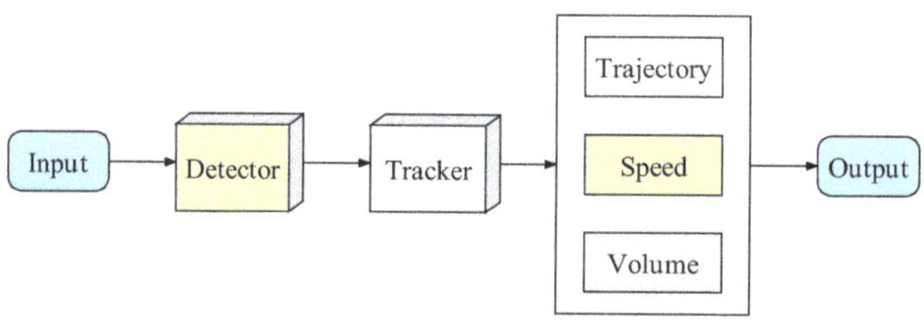

Figure 1. The detection-based vehicle tracking model.

2.1. Detection Model

Recently, the YOLO detection model in [20] was iterated from YOLOv1 to YOLOv5, which integrates the existing advanced research results and improves the detection accuracy and speed of the model. For a better understanding of our vehicle detection model, we first introduce the 6.0 version of the YOLOv5s network model, as shown in Figure 2. Here, based on different functions, the whole model can be divided into four parts: Input, Backbone, Neck, and Prediction Head.

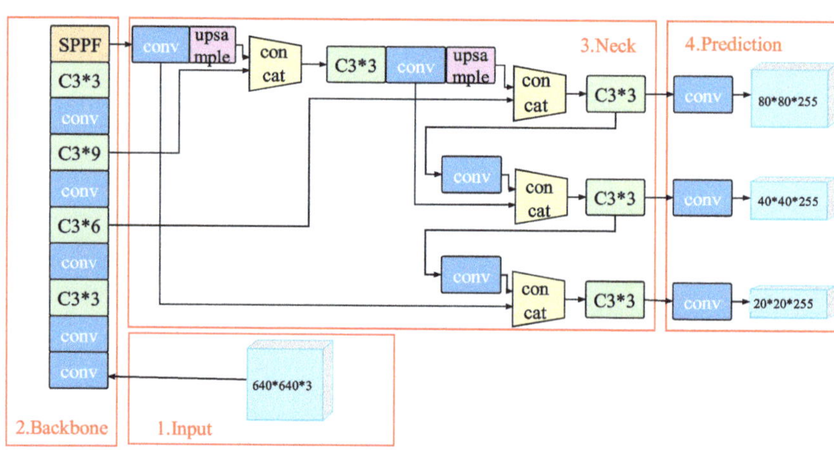

Figure 2. The YOLOv5 network model.

2.1.1. Input

The Input part provides data for model training and performing preprocessing operations on the experimental dataset. The YOLOv5 algorithm mainly uses Mosaic data enhancement and adaptive anchor box strategy on the Input to preprocess the original dataset. Mosaic data enhancement is borrowed from CutMix data enhancement [37]. The adaptive anchor box strategy solves the problem of differences in the aspect ratio of different targets. For example, the aspect ratio of the anchor box of the pedestrian is often greater than 1, while the aspect ratio of the anchor box of vehicles is just the opposite. The YOLOv5 algorithm introduces an adaptive anchor box strategy at the Input model that can automatically modify the preset anchor box during training.

2.1.2. Backbone

The Backbone is an indispensable part of the vehicle detection network model, which extracts the features of the original image through different CNN operations for subsequent classification and regression operations. After several years of development, researchers have designed some excellent Backbones, such as VGG, Darknet53, ResNet, and MobileNet. The Backbone of YOLO series algorithms is based on Darknet53 for extended research. The latest YOLOv5 algorithm Backbone consists of Conv, C3, and SPPF modules. The structures of C3 and SPPF are shown in Figures 3 and 4, respectively.

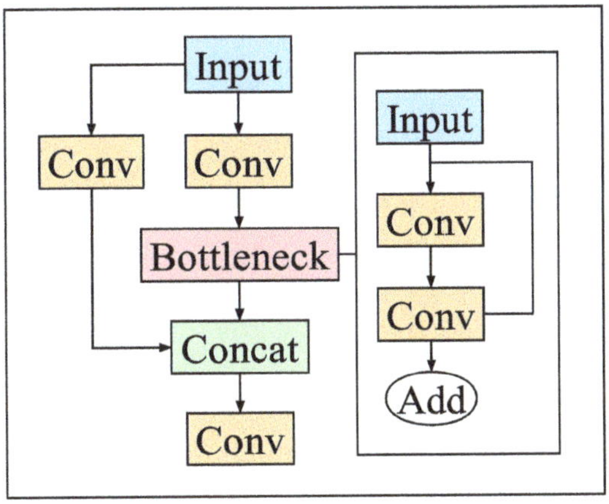

Figure 3. C3 structure.

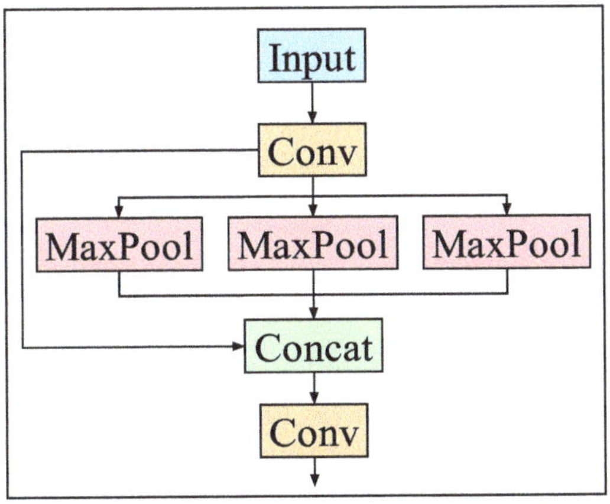

Figure 4. SPPF structure.

2.1.3. Neck

The Neck is between the Backbone and the Prediction Head, mainly to fuse the image features extracted by the Backbone. The YOLOv5 algorithm selects the FPN+PAN network structure to fuse image features in the Neck, as shown in Figure 5. The FPN network is responsible for passing the high-level semantic information down and merging with the Backbone. The PAN structure is responsible for the low-level feature information upward and merging with the FPN network. The final output feature map is the input feature map of the Prediction Head for target classification and prediction.

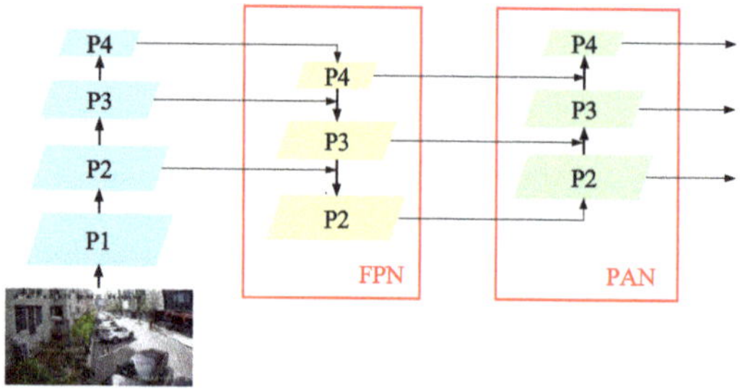

Figure 5. The FPN+PAN structure.

2.1.4. Prediction Head

The Prediction Head is the most important part of the vehicle detection network model. Its function is to expand the feature map into a one-dimensional vector and perform vehicle detection and classification operations on this one-dimensional vector. The YOLOv5 algorithm has three Prediction Heads to detect large targets, medium targets, and small targets. The Prediction Head of $20 \times 20 \times 255$ has the largest receptive field and is used to predict large targets. The Prediction Head of $40 \times 40 \times 255$ is used to predict medium targets. The Prediction Head of $80 \times 80 \times 255$ has the smallest receptive field and is used to predict small targets. The loss function will continually correct the accuracy of classification and regression, and the latest YOLOv5 algorithm chooses GIoU_Loss for training. There are many redundant target boxes in the output result of the Prediction Head, and the non-maximum suppression operation is used to screen out the detection box with high confidence as the final detection result of the target.

2.2. Tracking Model

The Kalman filter (KF) is an optimization method to solve the optimal state estimation of the system [38]. It usually serves as the tracker model for vehicle tracking tasks. By constructing the state transition equation and observation equation of the system, the KF algorithm continuously observes and estimates the state of the system, so that the observed value and the estimated value are as close to the real value as possible, to obtain the optimal estimation of the target motion. The general expressions of the state transition equation and observation equation of the KF algorithm are defined as follows:

$$x_t = Ax_{t-1} + Eu_{t-1} + q_t, \quad (1)$$

$$y_t = Hx_t + r_t, \quad (2)$$

where x_t and y_t are the system state real value and observed value at time t, respectively; A is the system state transition matrix; H is the system state observation matrix; E and u_{t-1} are the system model parameters; and q_t and r_t are the system state transition noise and observation noise, respectively, which are subject to Gaussian distribution. The realization process of the algorithm mainly includes three steps: prediction, filter estimation, and parameter update. The specific process is as follows:

(1) Calculate the system state estimated value $x_t{}'$ through the system state transition matrix A, and calculate its covariance matrix $\Omega_t{}'$ with the system state real value:

$$x_t{}' = Ax_{t-1} + Eu_{t-1}, \quad (3)$$

$$\Omega_t{}' = A\Omega_{t-1}A^T + Q, \quad (4)$$

where Q is the covariance matrix of the system state transition noise q_t.

(2) Calculate the KF gain $K_t{}'$ of the system at time t through the system state observe matrix H and the covariance matrix $\Omega_t{}'$ obtained in (4), and use $K_t{}'$ to estimate the system state at this time:

$$K_t{}' = \Omega_t{}'H^T\left(H\Omega_t{}'H^T + R\right)^{-1}, \quad (5)$$

$$x_t = x_t{}' + K_t{}'(y_t - Hx_t{}'), \quad (6)$$

where R is the covariance matrix of the system state observation noise r_t.

(3) Update the covariance matrix Ω_t of the system state estimated value and the system state real value to prepare for the next recursion:

$$\Omega_t = (1 - K_t{}')\Omega_t{}'. \quad (7)$$

By setting the initial system state estimation value x_t and the system state observation value y_t, and repeating the above algorithm process, we can obtain the latest values of x_t and y_t, and obtain the optimal estimation of the system state continuously. In the vehicle tracking tasks, the motion state estimation model of the vehicle is established based on the KF algorithm to correct the vehicle position, which makes up for the deficiency of the vehicle detector and improves the tracking accuracy of the vehicle tracking algorithms.

3. Vehicle Detection and Tracking Method

By integrating the NAM attention module into the YOLOv5s network model, we build a high-precision, real-time vehicle detection model YOLOv5-NAM and optimize its loss function and NMS method to improve the detection accuracy. Then, we propose a high-precision, real-time vehicle tracking method JDE-YN, where the appearance feature extraction network of the vehicle is embedded in the detection head of the YOLOv5-NAM model.

3.1. The YOLOv5-NAM Vehicle Detection Model

As an extension of the CBAM attention module, NAM is a lightweight and efficient new attention module achieved by redesigning the channel and spatial attention submodule based on normalization technology [39]. While the past research work in this field has focused on improving the image of the significant characteristics and ignores the problem of the striking feature of the image, the NAM attention module chooses the weight of the image channel and spatial information as a measure of image characteristics of significance, through a normalized scaling factor to represent the importance of the weight, and suppress the unimportant channel information and pixel information in the image.

Figure 6 shows the structure of the NAM channel attention submodule. The channel attention module is mainly for the model to acquire the ability of "what to see", corresponding to the classification problem of the vehicle target detection task. Given the input feature map F_1, the module first calculates the scale factor $\gamma_0, \gamma_1, \ldots, \gamma_n$ of each channel of F_1 through the regular normalization operation; then, it calculates the weight value of the scale factor of each channel of F_1 as follows:

$$\omega_i = \frac{\gamma_i}{\sum_{j=0}\gamma_j}. \quad (8)$$

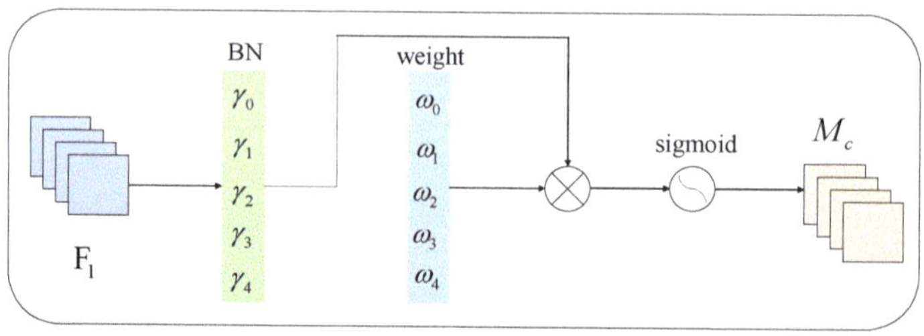

Figure 6. Channel attention submodule of NAM.

The weight value of each channel is applied to the original feature map F_1 as a penalty term, and the final image channel weight coefficient M_c is obtained through the sigmoid activation function.

Figure 7 shows the structure of the NAM spatial attention submodule. The spatial attention module is mainly for the model to acquire the ability of "where to look", corresponding to the positioning problem in the vehicle detection task. Given the input feature map F_2, the module first calculates the scale factor $\lambda_0, \lambda_1, \ldots, \lambda_n$ of each pixel in F_2 through normalization operation; then, it calculates the weight value of the scale factor of each pixel in F_2 as follows:

$$\omega_i = \frac{\lambda_i}{\sum_{j=0} \lambda_j}. \tag{9}$$

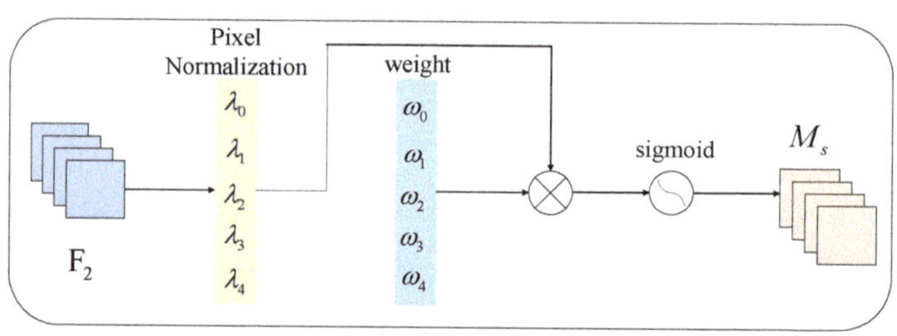

Figure 7. Spatial attention submodule of NAM.

The weight value of each pixel is applied to the original feature map F_2 as a penalty term, and the final image channel weight coefficient M_s is obtained through the sigmoid activation function.

As shown in Figure 8, we add the NAM attention module to the Neck of the YOLOv5s network model to build a YOLOV5-NAM vehicle detection model, where we aim to improve the detection ability for small vehicles, increase the vehicle detection accuracy, and reduce the rate of missed detection and error detection. Note that we adopt the mosaic data enhancement strategy at the Input for image preprocessing. Here, the size of the input image is 640*640*3 and the Backbone network is consistent with the original YOLOv5s model. The Neck still selects the FPN+PAN structure but the difference is that the convolution module before the last three Concat connection modules is replaced with the conv_NAM module to make the model pay attention to more features of the vehicles so as to improve the overall detection accuracy of the model.

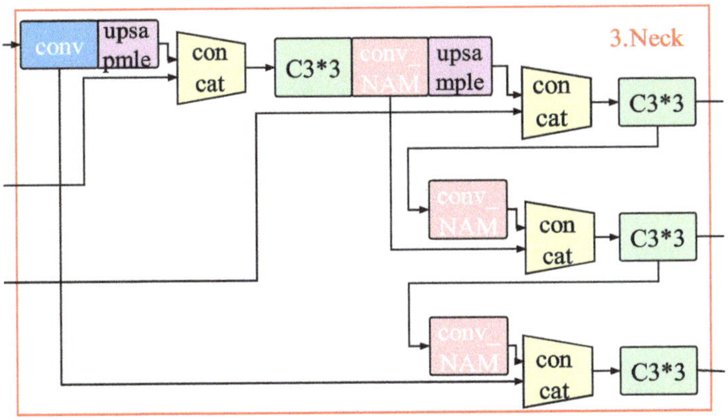

Figure 8. The Neck of the YOLOv5-NAM network model.

3.2. The Loss Function

The loss function of the YOLOv5 model training mainly includes three parts: the target classification loss, the target bounding box loss, and the target confidence loss. Formally, the YOLOv5 loss function is given by

$$L_{YOLOv5} = L_{lcls} + L_{lobj} + L_{lbox}, \tag{10}$$

where the target classification loss and the target confidence loss used the cross-entropy loss function for training, and the target bounding box loss used the Generalized Intersection over Union (GIoU) loss function for training.

The calculation process of cross-entropy loss of target classification loss in the YOLOv5s model is as follows:

$$\begin{aligned} L_{lcls} = &\sum_{i=0}^{S^2}\sum_{j=0}^{B} 1_{i,j}^{obj} \sum_{c \in classes} [y_i(c) log(p_i(c)) \\ &+ (1 - y_i(c)) log(1 - p_i(c))]. \end{aligned} \tag{11}$$

The calculation process of cross-entropy loss of target confidence loss in the YOLOv5s model is as follows:

$$\begin{aligned} L_{lobj} = &\sum_{i=0}^{S^2}\sum_{j=0}^{B} 1_{i,j}^{noobj} [(y_i - 1) log(1 - p_i) - y_i log(p_i)] \\ &+ \sum_{i=0}^{S^2}\sum_{j=0}^{B} 1_{i,j}^{obj} [y_i log(p_i) + (1 - y_i) log(1 - p_i)]. \end{aligned} \tag{12}$$

Compared with the original loss function of the YOLOv5s model, we choose the CIoU_Loss as the target bounding box's loss function for model training. The GIoU_Loss only considers the overlap area between the target bounding boxes and ignores the influence of other influencing factors on model training. Based on the GIoU_Loss, the CIoU_Loss [40] adds the center point distance and the aspect ratio between the target bounding boxes. The calculation formula is as follows:

$$L_{CIoU} = 1 - \frac{d^2(P, G)}{c^2} - \alpha v. \tag{13}$$

The complete calculation process of the loss function for the YOLOv5-NAM model training is as follows:

$$
\begin{aligned}
L = {} & L_{CIoU} \\
& + \sum_{i=0}^{S^2}\sum_{j=0}^{B} 1_{i,j}^{noobj}[(y_i - 1)log(1 - p_i) - y_i log(p_i)] \\
& + \sum_{i=0}^{S^2}\sum_{j=0}^{B} 1_{i,j}^{obj}[y_i log(p_i) + (1 - y_i)log(1 - p_i)] \\
& + \sum_{i=0}^{S^2}\sum_{j=0}^{B} 1_{i,j}^{obj} \sum_{c \in classes} [y_i(c)log(p_i(c)) \\
& + (1 - y_i(c))log(1 - p_i(c))].
\end{aligned} \tag{14}
$$

3.3. The SD-NMS Method

The Non-Maximum Suppression (NMS) algorithm is a post-processing method for vehicle detection tasks. To increase the detection accuracy and reduce the missed detection rate in the anchor-based vehicle detection model, researchers usually generate many candidate boxes in the initial stage and then match different scores for different candidate boxes in the analysis stage, which results in a large number of redundant boxes. How to select the correct prediction boxes from the candidate boxes as the output boxes is a key step to determine the detection accuracy of the vehicle detection model.

Algorithm 1 is the pseudocode of the NMS algorithm, where **B** is a candidate set of the boxes position, **S** is a candidate set of the boxes socre, N_t is a screening threshold, and **D** is a set of the prediction boxes. When set **B** is not null, select the maximum score m from set **S**; add the corresponding candidate box b_m in **D** into temporary set **M**; and move elements from set **B** to set **D**, which are in set **M**. Exclude the candidate box whose IoU value with b_m is greater than N_t in set **B** and set **S**. Loop iteration until set **B** is empty; the final output **D** and **S** is the income forecast box position and the corresponding score.

Algorithm 1 The Non-Maximum Suppression algorithm

```
While B ≠ empty
m ← argmax S
M ← b_m
D ← D ∪ M; B ← B − M
For b_i ∈ B
If iou(M, b_i)
B ← B − b_i; S ← S − s_i
```

The NMS algorithm can obtain better suppression effects in single-target and sparse multi-target scenes. However, the distance between the two targets is close in the dense multi-target case, and there is often a big overlap area between the real targets, as shown in Figure 9. If the NMS algorithm is directly applied, the detection boxes of the vehicle may be deleted by mistake, resulting in missed detection.

Figure 9. Vehicle overlapping.

To solve this problem, we further propose a Soft-DIoU NMS (SD-NMS) method based on DIoU-NMS [40]. The overall process of this method is consistent with Algorithm 1; we choose the DIoU distance as the new distance metric. The first step is to select the maximum score of the prediction boxes from the candidate boxes set **S**. The difference is that the IoU-NMS method directly deleted the candidate boxes with an IoU value over N_t. We will give a penalty item based on the initial score for the candidate boxes whose DIoU value is greater than N_t, and the penalty intensity is determined by the DIoU value. The calculation process is as follows:

$$\hat{S_i} = \begin{cases} S_i & DIoU(B, b_i) < N_t, \\ S_i(1 - |DIoU(B, b_i)|) & DIoU(B, b_i) \geq N_t. \end{cases} \tag{15}$$

We can obtain a group of new candidate boxes by continuously punishing candidate boxes and then deleting the low confidence prediction boxes by a uniform confidence threshold, which increases the detection accuracy and reduces the missed detections.

3.4. The Vehicle State Estimation Model

We reconstruct the vehicle state estimation model by introducing process noise into the system. The motion state of the vehicle is expressed as an 8-dimensional vector:

$$S = \left[x, y, h, r, v_x, v_y, v_h, v_r\right]^T, \tag{16}$$

where x and y are the horizontal and vertical coordinates of the vehicle center point in the image; h is the height of the vehicle boxes; r is the aspect ratio of the vehicle boxes; and v_x, v_y, v_h, and v_r are the speed components in four directions, respectively.

Suppose S_t is the position of the target vehicle at time t, v_t is the speed of the target vehicle at time t, and the acceleration of the target vehicle is a; then, the position of the target vehicle and the speed of the target vehicle at time $t+1$ is given by

$$S_{t+1} = S_t + v_t \Delta t + \frac{1}{2} a \Delta t^2, \tag{17}$$

$$v_{t+1} = v_t + a \Delta t, \tag{18}$$

where Δt is the interval from t to $t+1$.

Therefore, we can build the vehicle state prediction model as follows:

$$S_{t+1} = AS_t + Ea, \tag{19}$$

where A is the vehicle state transition matrix and E is the noise control matrix in the process of vehicle driving. Combining with the 8-dimensional vehicle state features we established in (16), it is easy to obtain the following:

$$A_t = \begin{bmatrix} 1 & 0 & 0 & 0 & \Delta t & 0 & 0 & 0 \\ 0 & 1 & 0 & 0 & 0 & \Delta t & 0 & 0 \\ 0 & 0 & 1 & 0 & 0 & 0 & \Delta t & 0 \\ 0 & 0 & 0 & 1 & 0 & 0 & 0 & \Delta t \\ 0 & 0 & 0 & 0 & 1 & 0 & 0 & 0 \\ 0 & 0 & 0 & 0 & 0 & 1 & 0 & 0 \\ 0 & 0 & 0 & 0 & 0 & 0 & 1 & 0 \\ 0 & 0 & 0 & 0 & 0 & 0 & 0 & 1 \end{bmatrix}, \tag{20}$$

$$E_t = \left[\frac{1}{2}\Delta t^2 \; \frac{1}{2}\Delta t^2 \; \frac{1}{2}\Delta t^2 \; \frac{1}{2}\Delta t^2 \; \Delta t \; \Delta t \; \Delta t \; \Delta t\right]^T. \tag{21}$$

Here, we can build the vehicle state observation model as follows:

$$P_t = HS_t, \tag{22}$$

where H is the system state observation matrix. Consider that we can only observe the vehicle position in the image through the video and the vehicle speed cannot be directly obtained. The observation matrix H is as follows:

$$H = \begin{bmatrix} 1 & 0 & 0 & 0 & 0 & 0 & 0 & 0 \\ 0 & 1 & 0 & 0 & 0 & 0 & 0 & 0 \\ 0 & 0 & 1 & 0 & 0 & 0 & 0 & 0 \\ 0 & 0 & 0 & 1 & 0 & 0 & 0 & 0 \end{bmatrix}, \tag{23}$$

In view that vehicle tracking in road traffic scenarios is susceptible to adverse weather and other external factors, we introduce system noise and observation noise to modify the model, and the complete vehicle state estimation model expression is as follows:

$$S_{t+1} = A_t S_t + C_t a + Q, \tag{24}$$

$$P_t = HS_t + R. \tag{25}$$

Based on the above analysis, we can use the YOLOv5-NAM vehicle detector to achieve multi-target vehicle tracking.

3.5. The Cascade Matching Algorithm Based on Direction Correction

The traditional multi-target tracking algorithm usually chooses the Hungarian matching algorithm as the data association strategy of the target between adjacent frames. The algorithm cost matrix is built based on the target motion characteristics and appearance characteristics. The distance measurement method of the target motion feature is the Mahalanobis distance between the output result of the state estimation model and the output result of the detector. Mahalanobis distance weakens the correlation between variables and makes the distance measurement of the target more accurate. The distance measurement method of target appearance characteristics is the cosine distance between the output result of the state estimation model and the output of the feature extraction network. Cosine distance depicts the appearance of the similarity between the target and makes up for the deficiency of the distance measurement method of target motion characteristics when the target motion is uncertain.

The inter-class similarity and the intra-class difference between the vehicles will lead to the failure of vehicle data association and also vehicle identity switching when the vehicles occlude each other during the driving process. In Figure 10, track 2 and track 17, respectively, belong to the black car and the white car in frame 115. The black car occluded the white car in frame 132, resulting in the white car miss detection. At the time, a white vehicle appeared and was mistakenly associated with track 17. The occlusion ended in frame 137 but the algorithm had assigned track 17 to the white vehicle; so, the white car had to be reassigned to another, causing the vehicle identity to switch. In panels (b), (d), and (f), track 32 and track 35 belong to the car and taxi in frame 187. The car and taxi occluded each other in frame 188. The occlusion ended in frame 190 but the data association algorithm assigned track 32 to the taxi and track 35 to the car, causing the vehicle identity to switch between the taxi and the car.

It is not difficult to find that when the vehicle identity switches due to occlusion, there is a "jump" phenomenon in the vehicle trajectory. In other words, the vehicle undergoes an abnormal angular rotation, which is almost impossible to occur in the driving environment. Therefore, we propose a cascade matching algorithm based on direction correction. Whether the result of data association is reasonable or not is judged by integrating the directional features of vehicles. If reasonably performed, the data association result is confirmed;

otherwise, we predict that maybe occlusion occurred, cancel the data association result, and maintain Kalman filter prediction results, as shown in Figure 11.

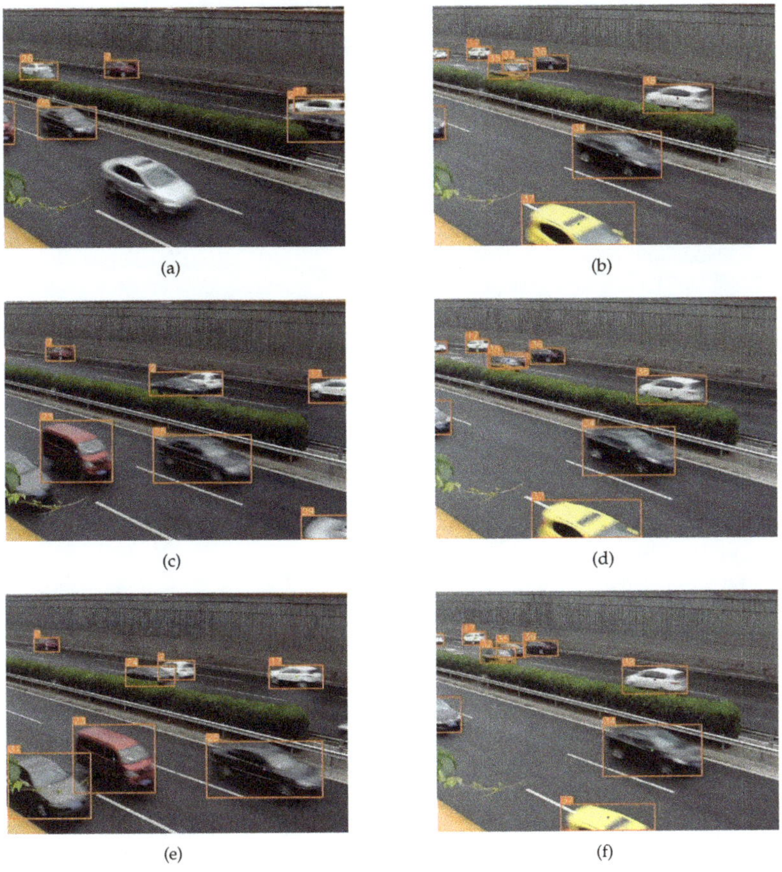

Figure 10. Data association failure. (a) frame 115; (b) frame 187; (c) frame 132; (d) frame 188; (e) frame 137; (f) frame 190.

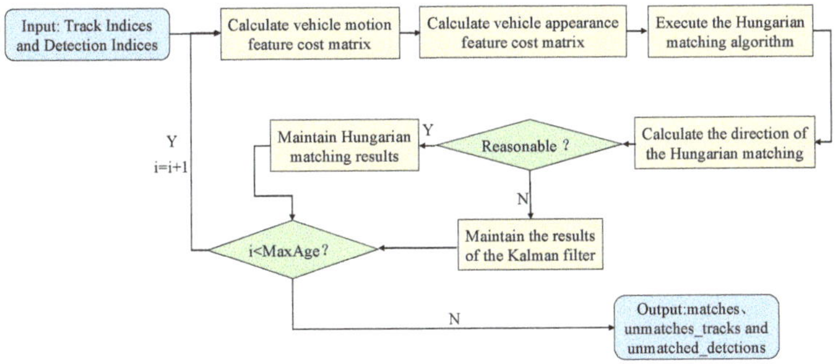

Figure 11. The cascade matching algorithm based on direction correction.

4. Detection and Tracking Experience Results

4.1. Detection Results on UA-DETRAC

The UA-DETRAC dataset is a large, open-source dataset for multi-vehicle detection and tracking [41]. The dataset was taken from the video surveillance of 24 real roads in Beijing and Tianjin, covering four different weather conditions: sunny, rainy, cloudy, and night. The researcher divided the vehicles into four types: cars, buses, vans, and other vehicles. On this basis, they marked 8250 vehicles and 1.21 million labels in total.

Since the UA-DETRAC dataset comes from multiple videos, the vehicle difference in most adjacent video frames is not evident. Direct application in the training of the YOLOv5-NAM model will lead to data redundancy. Therefore, we extract the original video every five frames, which can reduce the sample size of the dataset and reduce the training time of the model. At the same time, this method can also avoid the over-fitting phenomenon caused by repeated learning of the same feature. We select 16,542 images and 102,562 vehicle labels from the original UA-DETRAC train set as the new train set and select 11,235 images and 86,559 vehicle labels from the original UA-DETRAC test set as the new test set. Based on the above vehicle dataset, we set the initial learning rate as 0.01, the batch number as 32, and the epoch as 100. Here, we set a relatively high confidence level to compare the visual experiment results.

Table 1 shows the influence of network structure improves on model performance. As can be seen from the table, after adding the NAM attention module, the number of network layers in this model increases from 213 to 219 layers. The model size increases from 13.7 M to 13.9 M. The FPS of the model slightly decreases by 0.4. However, the overall mAP value of the model increases by 0.9%, indicating that the YOLOv5-NAM vehicle detection model is effective and the detection accuracy of the model increases with the sacrifice of a small amount of space and speed.

Table 1. The influence of network structure.

Model	Layer	Size	FPS	mAP
YOLOv5s	213	13.7M	55	0.503
YOLOv5-NAM	219	13.9M	54.6	0.512

Table 2 shows the influence of the loss function on improving the model performance. We can see from the table that the loss function does not directly influence the model layers, the model size, and the FPS indicators. The reason is that the loss function acts on the model training stage, while the number of model layers and model size are determined before the model training. The FPS indicator is the product of the model testing stage; so, it is not affected by the loss function. The direct influencing factor of the loss function is the value of mAP. When we use the CIoU_Loss to replace the GIoU_Loss, the YOLOv5s and YOlOv5-NAM models' mAP values increase by 0.2% and 0.3%, respectively. This shows that replacing the loss function training model can improve the detection accuracy of the model.

Table 2. The influence of loss function.

Model	Loss	Layer	FPS	mAP
YOLOv5s	GIoU	213	55	0.503
YOLOv5s	CIoU	213	55	0.505
YOLOv5-NAM	GIoU	219	54.6	0.512
YOLOv5-NAM	CIoU	219	54.6	0.515

Table 3 shows the influence of the NMS method on model performance. As can be seen from the table, the model layers and model size are not affected by the NMS method, which verifies that the non-maxima processing is a post-processing method and acts on the model testing stage. We can also find that the FPS index of the model decreases slightly after replacing the NMS method of the YOLOv5s and YOLOv5-NAM models with the SD-NMS method; the FPS indexes of the YOLOv5s and YOLOv5-NAM models decrease by 3 and 2.8, respectively. It is mainly because the SD-NMS method adopts the idea of cyclic suppression, which requires a longer processing time than the NMS method. However, after replacing the NMS method with the SD-NMS method, the mAP of the YOLOv5s and YOLOv5-NAM models increases by 0.2%. It is worth mentioning that when the targets are denser, the improvement effect will be even higher.

Table 3. The influence of NMS method.

Model	NMS	FPS	mAP
YOLOv5s	IoU-NMS	55	0.503
YOLOv5s	SD-NMS	52	0.505
YOLOv5-NAM	IoU-NMS	54.6	0.512
YOLOv5-NAM	SD-NMS	51.8	0.514

To further analyze the influence of the method on the detection accuracy of the four vehicle categories—cars, buses, vans, and other types of vehicles—we also analyze the detection accuracy of the YOLOv5s and YOLOv5-NAM model, and the SD-NMS method and the NMS method, for single-vehicle targets. The experimental results are shown in Table 4, where the evaluation indicators are the AP value of a single category and the mAP of all categories. According to the Table, after adding the NAM attention mechanism into the original YOLOv5s model, the overall mAP value of the model increases by 1.4%. The AP value of cars increases by 0.3%, the AP value of buses increases by 2.7%, the AP value of vans slightly decreases, and the AP value of other vehicles increases by 3.6%. When the detection results of the YOLOv5s algorithm are post-processed by the SD-NMS method, the overall mAP value of the model does not change significantly, only increasing by 0.1%, while the AP value of different types of vehicles varies. The AP value of cars increases by 1.5%, the AP value of others increases by 0.4%, and the AP value of buses and vans slightly decrease. When the NAM attention mechanism and the SD-NMS method are added to the YOLOv5s algorithm, the overall mAP value of the model increases by 1.6%, among which the AP of cars increases by 2.3%, the AP of buses increases by 1.3%, the AP of other vehicles increases by 4.7%, and the AP of vans has a slight decrease.

Table 4. The detection results on UA-DETRAC dataset.

Model	NMS	Cars	Buses	Vans	Other	mAP
YOLOv5s	IoU-NMS	0.697	0.433	0.743	0.139	0.503
YOLOv5-NAM	IoU-NMS	0.7	0.46	0.731	0.175	0.517
YOLOv5s	SD-NMS	0.712	0.424	0.735	0.143	0.504
YOLOv5-NAM	SD-NMS	0.72	0.446	0.726	0.186	0.519

The experimental results show that the NAM attentional mechanism module has a significant effect on improving the overall mAP value of the model. No matter the NAM attention mechanism or the SD-NMS method, the AP values of other vehicles of a single class are improved to varying degrees, mainly because other vehicles in the UA-DETRAC dataset are mostly marked instances of vehicles with a small volume or serious occlusion, which is consistent with the fact that our proposed method can improve the detection accuracy of vehicles with small targets; the van is mostly a large target, so the accuracy of our proposed method is slightly lost. This suggests that our proposed method has a certain effect on improving the overall detection accuracy of vehicles and a significant effect on improving the detection accuracy of small target vehicles.

The visualization results of the YOLOv5s model and the YOLOv5-NAM vehicle detection model are shown in Figure 12, where panels (a) and (c) are the detection results of the YOLOv5s model training on the COCO dataset and panels (b) and (d) are the detection results of the YOLOv5-NAM vehicle detection model training on the UA-DETRAC dataset. It is not difficult to find that after retraining on the UA-DETRAC dataset, the YOLOv5-NAM vehicle target detection model greatly improves the detection accuracy of vehicles. The detection accuracy of a single-vehicle target is above 0.9 on average, which greatly improves the detection accuracy of vehicles with small targets. The new model also improves the error detection of the YOLOv5s model to some extent. For example, the YOLOv5s algorithm mistakenly detects the van as a bus in subgraph (c) but the YOLOv5-NAM algorithm correctly detects the van in subgraph (d).

To eliminate the influence of the dataset, we retrain the YOLOv5s model on the UA-DETRAC dataset. Panels (e) and (f) in Figure 12 are the detection results of the YOLOv5s and YOLOV5-NAM model after retraining on the UA-DETRAC dataset, respectively. As can be seen from the figure, the vehicle detection accuracy of the YOLOv5-NAM model is improved to varying degrees compared with the YOLOv5s model, especially for small target vehicles, which confirms that the detection effect of our proposed method is indeed improved for small target vehicles.

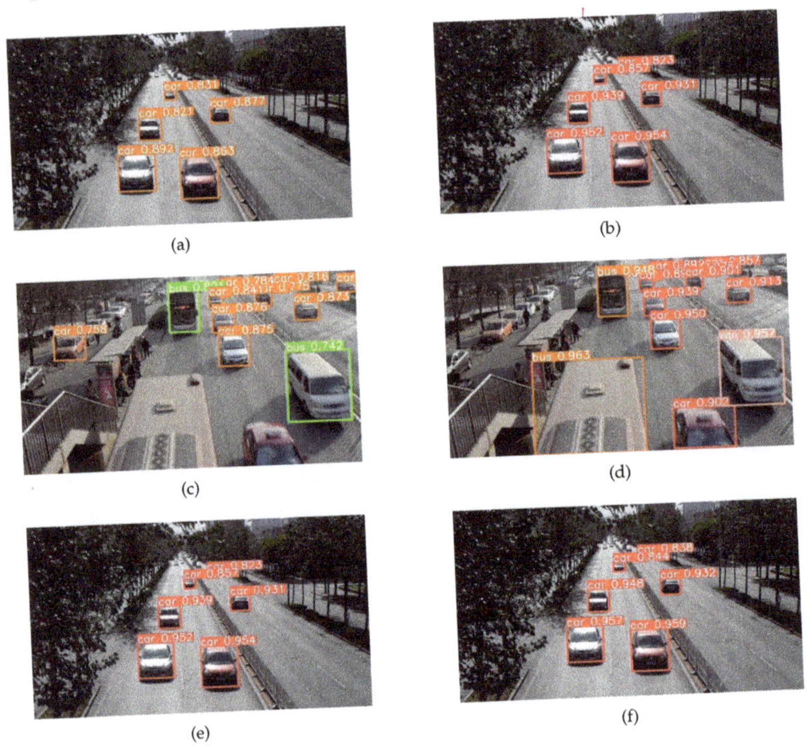

Figure 12. The visual detection results of the YOLOv5s and the YOLOv5-NAM models.

4.2. Detection Results on COCO

To verify that the YOLOv5-NAM model also has a certain reference significance for general object detection tasks, we retrain the YOLOv5-NAM model on the COCO2017 dataset. The training results are compared with the training results of mainstream target detection algorithms on this dataset. The evaluation index is the AP value officially provided by COCO. The results are shown in Table 5, where AP_{50} and AP_{75} are the AP values when the IoU values are 0.5 and 0.75, respectively; AP_S, AP_M, and AP_L are the AP value of small targets, medium targets, and big targets, respectively.

Table 5. Detection results on the COCO dataset.

Model	Backbone	AP	AP_{50}	AP_{75}	AP_S	AP_M	AP_L
Faster R-CNN [23]	VGG-16	0.219	0.427	–	–	–	–
Mask R-CNN [24]	R-101-FPN	0.382	0.603	0.417	0.201	0.411	0.502
R-FCN [25]	R-101	0.299	0.519	–	0.108	0.328	0.450
SSD [21]	VGG-16	0.288	0.485	0.303	0.109	0.318	0.435
YOLOv3 [28]	Darknet-53	0.33	–	–	–	–	–
RetinaNet [17]	X-101-FPN	0.390	0.594	0.417	0.226	0.434	0.509
FCOS [42]	X-101-FPN	0.421	0.621	0.452	0.256	0.449	0.520
CenterNet [43]	HG-104	0.421	0.611	0.459	0.241	0.455	0.528
YOLOv4 [29]	CSP	0.435	0.657	0.473	0.267	0.467	0.533
YOLOv5s [30]	CSP	0.355	0.55	–	–	–	–
YOLOv5x [30]	CSP	0.472	0.666	–	–	–	–
Ours	CSP-NAM	0.367	0.561	0.378	0.231	0.424	0.492

From Table 5, the AP value of our proposed method is higher than some classical object detection algorithms such as Faster R-CNN, SSD, and YOLOv3, and slightly lower than the RetainNet and CenterNet algorithms. Our research is mainly oriented to road traffic scenarios and has high requirements for the real-time performance of the algorithm; therefore, we choose the lightest YOLOv5s model for improvement. However, the YOLOv5x model with a higher AP value is not chosen for improvement, which is also the main reason why the AP value of our proposed method is slightly lower. If the improved idea of the paper is applied to the YOLOv5x model, the AP value will be greatly improved but the reasoning speed of the model will be greatly reduced, which does not apply to our scenario. Although the AP value of our proposed method is lower than the CenterNet and RetinaNet algorithms, the detection accuracy of the small target AP_S is higher than the CenterNet and the gap between our proposed method and the RetinaNet is also significantly reduced, which indicates that our proposed method has a good effect on improving the detection accuracy of small targets.

4.3. Tracking Results on UA-DETRAC

We choose the UA-DETRAC dataset for model training, choose 60 video sequences as the training set for model training, and choose 40 video sequences as the test set to measure the training results of the model. Every video sequence is split into a single frame and encapsulated in the same folder, and all image annotation information under each folder is encapsulated in an xml file. The xml file is listed in the unit of video frames and describes in detail the vehicle ID, vehicle position coordinates, vehicle speed, vehicle rotation angle, and vehicle type in each frame. In order to facilitate the training of the YOLOv5-NAM vehicle target detection model, we transform the xml file annotation format into VOC annotation format through the custom tools, retain the vehicle ID and location coordinate information, and set the vehicle type to car uniformly.

We use kmeans algorithm to obtain a set of optimal anchor boxes on UA-DETRAC dataset in advance, which is $[(25,20),(33,28),(44,34),(53,47),(68,38),(178,213),(96,49),(131,62),(97,89),(159,104),(255,134),(72,61)]$. During the experiment, the weight of the YOLOv5s pre-training model was used for training, and the initial learning rate was set to 0.01, the training loss of the model was optimized by Adam function, the batch processing size was set to 32, and the image size was set to 960*540. The weight of the trained model is tested on the test set for vehicle target tracking, and some of the target vehicle tracking results are shown in Figure 13.

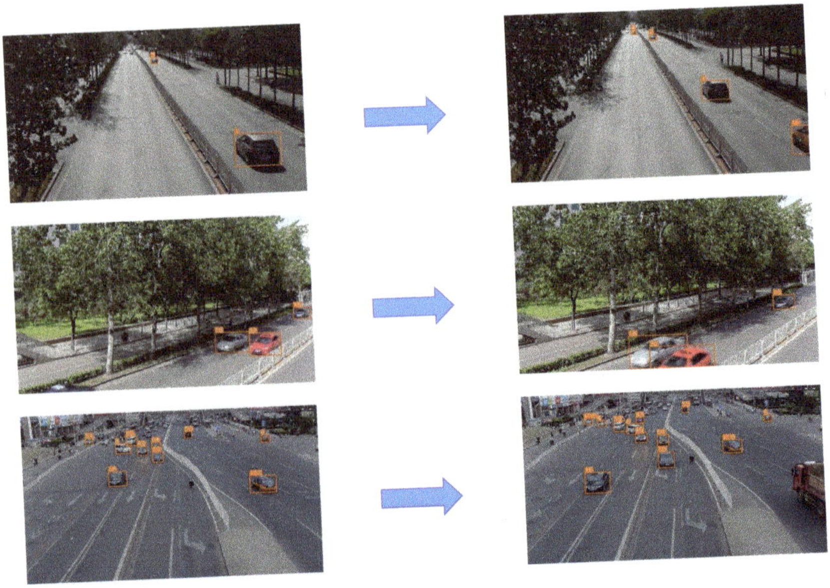

Figure 13. The vehicle target tracking results.

To verify the effectiveness of the proposed multi-target vehicle tracking method, we also carry out comparative experiments with benchmark methods such as SORT, DeepSORT, and JDE. The experimental dataset was the UA-DETRAC vehicle tracking dataset, and the experimental parameters remained unchanged. The evaluation indexes of experimental results are MOTA, MOTP, MT, ML, IDs, and the real-time processing speed FPS of the algorithm; the results are shown in Table 6.

Table 6. The tracking results on UA-DETRAC dataset.

Method	MOTA	MOTP	MT	ML	IDs	FPS
SORT [32]	0.302	0.371	0.178	0.245	868.8	287.6
DeepSORT [33]	0.302	0.369	0.298	0.202	512.7	25.6
JDE [34]	0.334	0.387	0.345	0.156	489.9	66.7
Ours	0.343	0.386	0.367	0.143	414.8	64.6

From Table 6, the MOTA and MOTP values of the four methods in the table are low on the UA-DETRAC dataset, maintaining between 0.3 and 0.4, which is mainly because the UA-DETRAC dataset covers a variety of weather environments and vehicle targets of different scales, which challenge the performance of the vehicle target detector, resulting in low MOTA and MOTP values. However, our proposed method still achieves the highest MOTA value of 0.343, and the MOTP value is only 0.1 percentage points lower than the highest JDE algorithm. Observation of the IDs column is not hard to find, and our proposed method obtained the identity of the switching times at least; this is mainly because our method integrates the directional features of vehicles in the data association stage, which reduces the error rate of data correlation, thereby reducing the switching times, and proving the effectiveness of our data correlation strategy. By observing the FPS column, it can be found that the SORT algorithm has the highest FPS but its IDs are too high to be of practical application value. The DeepSORT algorithm can significantly reduce IDs by integrating vehicle appearance features for data association; however, its FPS value also drops sharply, which cannot meet the requirements of real-time applications. The JDE method embedded the vehicle appearance feature extraction network into the detection network, which greatly improved the FPS value of the algorithm. Our proposed method is further improved based on the JDE algorithm, under the premise of losing a small part of FPS; the IDs of vehicle targets are further reduced by 15% compared with the JDE algorithm, which can support the application requirements of vehicle target tracking with high accuracy and high real-time performance.

5. Conclusions

This paper studied the small target vehicle detection and occlusion vehicle tracking. Based on the classical YOLOv5s model, we developed a new vehicle detection model (YOLOv5-NAM). By exploiting the YOLOv5-NAM model as the vehicle detector, we also proposed a real-time small target vehicle tracking method (JDE-YN). Our experiment results on the UA-DETRAC and COCO datasets demonstrated that the mAP value increased by 1.6% compared with the YOLOv5s model, the MOTA value of vehicle tracking increased by 0.9% compared with the JDE algorithm, and the identity switching times of vehicles decreased by 15%. It indicates that our proposed method can effectively detect small target vehicles and track multi-vehicles in real-time and efficiently, which has a certain promotion effect on promoting in-depth research in the vehicle detection and tracking field.

Author Contributions: Conceptualization, J.W.; methodology, J.W. and Y.D.; investigation, Y.D.; writing—original draft preparation, S.Z.; writing—review and editing, S.Z.; supervision, Z.Z.; funding acquisition, Z.Z. All authors have read and agreed to the published version of the manuscript.

Funding: This work was supported in part by the National Key R/D Program of China (Grant No. 2018YFE0207600), in part by the Natural Science Foundation of China (NSFC) under Grant 61972308, and in part by Natural Science Basic Research Program of Shaanxi (Program No. 2019JC-17).

Institutional Review Board Statement: Not applicable.

Informed Consent Statement: Not applicable.

Data Availability Statement: The data used to support the findings of this study are included within the article.

Conflicts of Interest: The authors declare no conflict of interest.

References

1. Zheng, K.; Jia, X.; Chi, K.; Liu, X. DDPG-based joint time and energy management in ambient backscatter-assisted hybrid underlay CRNs. *IEEE Trans. Commun.* **2022**. [CrossRef]
2. Liu, X.; Xu, B.; Wang, X.; Zheng, K.; Chi, K.; Tian, X. Impacts of sensing energy and data availability on throughput of energy harvesting cognitive radio networks. *IEEE Trans. Veh. Technol.* **2022**. [CrossRef]
3. Abdel-Basset, M.; Moustafa, N.; Hawash, H.; Razzak, I.; Sallam, K.M.; Elkomy, O.M. Federated Intrusion Detection in Blockchain-Based Smart Transportation Systems. *IEEE Trans. Intell. Transp. Syst.* **2022**, *23*, 2523–2537. [CrossRef]
4. Yan, P.; Liu, X.; Wang, F.; Yue, C.; Wang, X. LOVD: Land Vehicle Detection in Complex Scenes of Optical Remote Sensing Image. *IEEE Trans. Geosci. Remote. Sens.* **2022**, *60*, 1–13. [CrossRef]
5. Mao, S.; Liu, L.; Zhang, N.; Dong, M.; Zhao, J.; Wu, J.; Leung, V.C. Reconfigurable intelligent surface-assisted secure mobile edge computing networks. *IEEE Trans. Veh. Technol.* **2022**, *71*, 6647–6660. [CrossRef]
6. Messoussi, O.; de Magalhães, F.G.; Lamarre, F.; Perreault, F.; Sogoba, I.; Bilodeau, G.A.; Nicolescu, G. Vehicle Detection and Tracking from Surveillance Cameras in Urban Scenes. In *International Symposium on Visual Computing*; Springer: Cham, Switzerland, 2021; pp. 191–202.
7. Zhang, Y.; Zhou, Y.; Lu, H.; Fujita, H. Traffic Network Flow Prediction Using Parallel Training for Deep Convolutional Neural Networks on Spark Cloud. *IEEE Trans. Ind. Inf.* **2020**, *16*, 7369–7380. [CrossRef]
8. Zou, B.; Qin, J.; Zhang, L. Vehicle Detection Based on Semantic-Context Enhancement for High-Resolution SAR Images in Complex Background. *IEEE Geosci. Remote Sens. Lett.* **2022**, *19*, 4503905. [CrossRef]
9. Li, K.; Wan, G.; Cheng, G.; Meng, L.; Han, J. Object detection in optical remote sensing images: A survey and a new benchmark. *ISPRS J. Photogramm. Remote Sens.* **2020**, *159*, 296–307. [CrossRef]
10. Shamsolmoali, P.; Chanussot, J.; Zareapoor, M.; Zhou, H.; Yang, J. Multipatch Feature Pyramid Network for Weakly Supervised Object Detection in Optical Remote Sensing Images. *IEEE Trans. Geosci. Remote Sens.* **2022**, *60*, 5610113. [CrossRef]
11. Liu, L.; Zhao, M.; Yu, M.; Jan, M.A.; Lan, D.; Taherkordi, A. Mobility-aware multi-hop task offloading for autonomous driving in vehicular edge computing and networks. *IEEE Trans. Intell. Transp. Syst.* **2022**. [CrossRef]
12. Dai, Y.; Yu, J.; Zhang, D.; Hu, T.; Zheng, X. RODFormer: High-Precision Design for Rotating Object Detection with Transformers. *Sensors* **2022**, *22*, 2633. [CrossRef] [PubMed]

13. Xu, W.; Huang, H.; Pan, S. Using feature alignment can improve clean average precision and adversarial robustness in object detection. In Proceedings of the 2021 IEEE International Conference on Image Processing (ICIP), Anchorage, AK, USA, 19–22 September 2021; pp. 2184–2188.
14. Chai, W.; Lu, Y.; Velipasalar, S. Weighted average precision: adversarial example detection for visual perception of autonomous vehicles. In Proceedings of the 2021 IEEE International Conference on Image Processing (ICIP), Anchorage, AK, USA, 19–22 September 2021; Volume 2021, pp. 804–808.
15. Hong, H.; He, K.; Gan, Z.; Jiang, G. Stability Precision Error Correction of Photoelectric Detection by Unmanned Aerial Vehicle. *J. Sens.* **2021**, *2021*, 5564448. [CrossRef]
16. Hoffmann, J.E.; Tosso, H.G.; Santos, M.M.D.; Justo, J.F.; Malik, A.W.; Rahman, A.U. Real-Time Adaptive Object Detection and Tracking for Autonomous Vehicles. *IEEE Trans. Intell. Veh.* **2021**, *6*, 450–459. [CrossRef]
17. Lin, T.Y.; Goyal, P.; Girshick, R.; He, K.; Dollár, P. Focal Loss for Dense Object Detection. *IEEE Trans. Pattern Anal. Mach. Intell.* **2020**, *42*, 318–327. [CrossRef] [PubMed]
18. Zhao, Q.; Wang, Y.; Sheng, T.; Tang, Z. Comprehensive Feature Enhancement Module for Single-Shot Object Detector. In Proceedings of the 14th Asian Conference on Computer Vision, Perth, Australia, 2–6 December 2018; pp. 325 –340.
19. Law, H.; Deng, J. CornerNet: Detecting Objects as Paired Keypoints. *Int. J. Comput. Vis.* **2020**, *128*, 642–656. [CrossRef]
20. Redmon, J.; Divvala, S.; Girshick, R.; Farhadi, A. You Only Look Once: Unified, Real-Time Object Detection. In Proceedings of the 2016 IEEE Conference on Computer Vision and Pattern Recognition (CVPR), Las Vegas, NV, USA, 27–30 June 2016; pp. 779–788. . [CrossRef]
21. Liu, W.; Anguelov, D.; Erhan, D.; Szegedy, C.; Reed, S.; Fu, C.Y.; Berg, A.C. SSD: Single Shot MultiBox Detector. In Proceedings of the 14th European Conference, Amsterdam, The Netherlands, 11–14 October 2016; pp. 21–37.
22. Cai, Z.; Vasconcelos, N. Cascade R-CNN: Delving Into High Quality Object Detection. In Proceedings of the 2018 IEEE/CVF Conference on Computer Vision and Pattern Recognition, Salt Lake City, UT, USA, 18–23 June 2018; pp. 6154–6162.
23. Ren, S.; He, K.; Girshick, R.; Sun, J. Faster R-CNN: Towards Real-Time Object Detection with Region Proposal Networks. *IEEE Trans. Pattern Anal. Mach. Intell.* **2017**, *39*, 1137–1149. [CrossRef] [PubMed]
24. He, K.; Gkioxari, G.; Dollár, P.; Girshick, R. Mask R-CNN. In Proceedings of the 2017 IEEE International Conference on Computer Vision (ICCV), Venice, Italy, 22–29 October 2017; pp. 2980–2988. [CrossRef]
25. Dai, J.; Li, Y.; He, K.; Sun, J. R-FCN: Object Detection via Region-based Fully Convolutional Networks. In Proceedings of the Advances in Neural Information Processing Systems 29 (NIPS 2016), Barcelona, Spain, 5–10 December 2016; pp. 379–387.
26. He, K.; Zhang, X.; Ren, S.; Sun, J. Spatial Pyramid Pooling in Deep Convolutional Networks for Visual Recognition. *IEEE Trans. Pattern Anal. Mach. Intell.* **2015**, *37*, 1904–1916. [CrossRef]
27. Redmon, J.; Farhadi, A. YOLO9000: Better, Faster, Stronger. In Proceedings of the 2017 IEEE Conference on Computer Vision and Pattern Recognition (CVPR), Honolulu, HI, USA, 21–26 July 2017; pp. 6517–6525. [CrossRef]
28. Redmon, J.; Farhadi, A. YOLOv3: An incremental improvement. *arXiv* **2018**, arXiv:1804.02767.
29. Bochkovskiy, A.; Wang, C.; Liao, H.M. YOLOv4: Optimal Speed and Accuracy of Object Detection. *arXiv* **2020**, arXiv:2004.10934.
30. Jocher, G. YOLOv5-Master. Available online: https://github.com/ultralytics/yolov5 (accessed on 4 March 2022).
31. Isaac-Medina, B.K.S.; Poyser, M.; Organisciak, D.; Willcocks, C.G.; Breckon, T.P.; Shum, H.P.H. Unmanned Aerial Vehicle Visual Detection and Tracking using Deep Neural Networks: A Performance Benchmark. In Proceedings of the 2021 IEEE/CVF International Conference on Computer Vision Workshops (ICCVW), Montreal, BC, Canada, 11–17 October 2021; pp. 1223–1232. [CrossRef]
32. Bewley, A.; Ge, Z.; Ott, L.; Ramos, F.; Upcroft, B. Simple online and realtime tracking. In Proceedings of the 2016 IEEE International Conference on Image Processing (ICIP), Phoenix, AZ, USA, 25–28 September 2016; pp. 3464–3468. [CrossRef]
33. Wojke, N.; Bewley, A.; Paulus, D. Simple online and realtime tracking with a deep association metric. In Proceedings of the 2017 IEEE International Conference on Image Processing (ICIP), Beijing, China, 17–20 September 2017; pp. 3645–3649. [CrossRef]
34. Wang, Z.; Zheng, L.; Liu, Y.; Li, Y.; Wang, S. Towards Real-Time Multi-Object Tracking. In Proceedings of the 16th European Conference, Glasgow, UK, 23–28 August 2020; pp. 107–122.
35. Lu, Z.; Rathod, V.; Votel, R.; Huang, J. RetinaTrack: Online Single Stage Joint Detection and Tracking. In Proceedings of the 2020 IEEE/CVF Conference on Computer Vision and Pattern Recognition (CVPR), Seattle, WA, USA, 13–19 June 2020; pp. 14656–14666. [CrossRef]
36. Avar, E.; Avar, Y.Z. Moving vehicle detection and tracking at roundabouts using deep learning with trajectory union. *Multimed. Tools. Appl.* **2022**, *81*, 6653–6680. [CrossRef]
37. Yun, S.; Han, D.; Chun, S.; Oh, S.J.; Yoo, Y.; Choe, J. CutMix: Regularization Strategy to Train Strong Classifiers with Localizable Features. In Proceedings of the 2019 IEEE/CVF International Conference on Computer Vision (ICCV), Seoul, Republic of Korea, 27 October–2 November 2019; pp. 6022–6031. [CrossRef]
38. Kalman, R.E. A New Approach to Linear Filtering and Prediction Problems. *J. Basic Eng. (ASME)* **1960**, *82*, 35–45. [CrossRef]
39. Liu, Y.; Shao, Z.; Teng, Y.; Hoffmann, N. NAM: Normalization-based Attention Module. *arXiv* **2021**, arXiv:2111.12419.
40. Zheng, Z.; Wang, P.; Liu, W.; Li, J.; Ye, R.; Ren, D. Distance-IoU Loss: Faster and Better Learning for Bounding Box Regression. In Proceedings of the AAAI Conference on Artificial Intelligence, Honolulu, HI, USA, 27 January–1 February 1 2019; pp. 12993–13000. [CrossRef]

41. Wen, L.; Du, D.; Cai, Z.; Lei, Z.; Chang, M.C.; Qi, H.; Lim, J.; Yang, M.H.; Lyu, S. UA-DETRAC: A new benchmark and protocol for multi-object detection and tracking. *Comput. Vis. Image Underst.* **2015**, *193*, 102907. [CrossRef]
42. Tian, Z.; Shen, C.; Chen, H.; He, T. FCOS: Fully Convolutional One-Stage Object Detection. In Proceedings of the 2019 IEEE/CVF International Conference on Computer Vision (ICCV), Seoul, Republic of Korea, 27 October–2 November 2019; pp. 9626–9635. [CrossRef]
43. Zhou, X.; Wang, D.; Krähenbühl, P. Objects as Points. *arXiv* **2019**, arXiv:1904.07850.

MDPI AG
Grosspeteranlage 5
4052 Basel
Switzerland
Tel.: +41 61 683 77 34

Sensors Editorial Office
E-mail: sensors@mdpi.com
www.mdpi.com/journal/sensors